DICTIONARY OF
MEDICINAL PLANTS

DICTIONARY OF
MEDICINAL PLANTS

A.V.S.S. Sammbamurty

Reader, Department of Botany,
Sri Venkateswara College, New Delhi - 110 021

CBS

CBS Publishers & Distributors Pvt. Ltd.

New Delhi • Bengaluru • Chennai • Kochi • Kolkata • Mumbai

Bhubaneswar • Hyderabad • Jharkhand • Nagpur • Patna • Pune • Uttarakhand • Dhaka

ISBN: 81-239-1289-7

First Edition: 2006
Reprint: 2009, 2020

Published by **Satish Kumar Jain** and produced by **Varun Jain** for
CBS Publishers & Distributors Pvt. Ltd.,
4819/XI Prahlad Street, 24 Ansari Road, Daryaganj, New Delhi - 110002
delhi@cbspd.com, cbspubs@airtelmail.in • www.cbspd.com
Ph.: 23289259, 23266861, 23266867 • Fax: 011-23243014

Corporate Office: 204 FIE, Industrial Area, Patparganj, Delhi - 110 092
Ph: 49344934 • Fax: 011-49344935
E-mail: publishing@cbspd.com • publicity@cbspd.com

Branches:
• *Bengaluru:* 2915, 17th Cross, K.R. Road, Bansankari 2nd Stage,
  Bengaluru - 70 • Ph: +91-80-26771678/79 • Fax: +91-80-26771680
  E-mail: cbsbng@gmail.com, bangalore@cbspd.com
• *Chennai:* No. 7, Subbaraya Street, Shenoy Nagar, Chennai - 600030
  Ph: +91-44-26681266, 26680620 • Fax: +91-44-42032115
  E-mail: chennai@cbspd.com
• *Kochi:* Ashana House, 39/1904, A.M. Thomas Road, Valanjambalam,
  Ernakulum, Kochi • Ph: +91-484-4059061-65
  Fax: +91-484-4059065 • E-mail: cochin@cbspd.com
• *Kolkata:* 6-B, Ground Floor, Rameshwar Shaw Road, Kolkata - 700014
  Ph: +91-33-22891126/7/8 • E-mail: kolkata@cbspd.com
• *Mumbai:* 83-C, Dr. E. Moses Road, Worli, Mumbai - 400018
  Ph: +91-9833017933, 022-24902340/41 • E-mail: mumbai@cbspd.com

Representatives:
• Hyderabad: 0-9885175004   • Nagpur: 0-9021734563
• Patna: 0-9334159340   • Pune: 0-9623451994
• Jharkhand: 0-9811541605   • Uttarakhand: 0-9716462459

Printed at:
India Binding House, Noida, UP (India)

Dedicated
to
LORD SRI VENKATESWARA

ABOUT THE BOOK

Dictionary of Medicinal Plants is a compilation of more than 1200 species, known to have medicinal use for man giving full details of their chemical constituents, medicinal uses, and how to cure specific diseases. The plant species mentioned cover geographically many parts of the globe. The genera are arranged alphabetically followed by the species name and family to which it belongs and also common vernacular name.

The book will be of immense use to pharmacists, doctors of homoeopathy, ayurvedic, herbal medicine, and to all students interested in medicinal plants.

ABOUT THE AUTHOR

Dr. A.V.S.S. Sammbamurty is a Senior Reader in Botany at Sri Venkateswara College, Delhi University, teaching Genetics and Taxonomy for the past 3 decades to B.Sc. and B.Sc. (Hons.) students. He did his B.Sc., M.Sc. and Ph.D. from Andhra University, Waltair and published 25 research papers on rice cytogenetics. He participated in several Indian and foreign symposia, seminars and congresses. He is a Fellow of the Genetic Society of India and the Botanical Society of India, and a member of several scientific bodies.

He is an author of the following textbooks meant for B.Sc., B.Sc. (Hons.) and M.Sc. students of Botany : (1) Genetics, (2) A Textbook of Botany, Vol. I, (3) A Textbook of Botany, Vol. II, (4) Dictionary of Botanical Terms, (5) Ecology, (6) Handbook of Genetics, (7) Morphology and Taxonomy, (8) Plants in Medicine and Industry, (9) A Textbook of Modern Economic Botany, (10) Mycology and Plant Pathology, (11) MCQs in Botany for Medical Entrance Tests, (12) Taxonomy of Angiosperms, (13) Dictionary of Medicinal Plants, (14) A Textbook of Economic Botany, (15) A Textbook of Botany (Volumes I, II and III, in Press) and (16) Molecular Genetics (In Press).

Preface

The present compilation of medicinal plants in dictionary form had been conceived by me for a very long time. Medicinal plants are a boon to human kind in a concealed but potent form. The quest for medicinal plants by man had a long antecedent history ever since the dawn of human civilization. Ethnobotany or tribal medicine or birth of Ayurveda dates back to thousands of years. Chemical compounds present in various plants are enormous and man has been tapping the medicinal uses of these compounds by various methods and curing many human diseases. In the present day, though allopathic medicines are greatly used by doctors, still specific diseases are cured only by natural herbal medicines, like jaundice and many more diseases. India and China are known for their antiquity in herbal medicine or ayurvedic medicine. There has been tremendous research going on to tap the natural resources in developing herbal medicine. It is with this purpose that the present compilation has been attempted to present all the available material on medicinal plants, their chemical constituents and their mode of action on man for specific diseases and their curative action. Botanical descriptions of each plant are avoided due to space limitations. Botanical descriptions can be had from different floras and it is only the expert in plant identification can identify a particular species. For medicinal purposes, the exact identification of the species is very important and not bookish knowledge. There is always the problem of identifying plants by Latin names by tribal people. Here only an expert in both botanical knowledge and knowledge in medicinal uses of plants can help for proper identification of the species with medicinal use. Even now, there are several species on earth unidentified for their correct medicinal use.

It is hoped that the present book 'Dictionary of Medicinal Plants' will be useful to doctors, pharmacists, chemists, industrialists, and students of botany to probe further in the identification of medicinal plants, their chemical constituents and their medicinal properties in curing certain specific diseases of man.

The present work, though not comprehensive, tries to present a large number of species available with medicinal use and represents a large extent of **worked out examples**.

A.V.S.S. Sammbamurty

References

1. Anonymous, **Medicinal Plants of India**, Vol. I to IV, CSIR, New Delhi (India), 1980.
2. Andrew Chevallier Mnimh, **The Encyclopaedia of Medicinal Plants**, Dorling Kindersley Ltd., London, 1996, 340 p.
3. Biswas, K., **Common Medicinal Plants of Darjeeling and the Sikkim Himalayas**, Govt. of West Bengal, Commerce and Industries Deptt., Cinebone, 1956, 154 p.
4. Chopra, R.N., Nayar, S.L. and Chopra, I.C., **Glossary of Indian Medicinal Plants**, C.S.I.R., New Delhi (India), 1956, 328 p.
5. Kirthikar, K.K. and Basu, B.D., **Indian Medicinal Plants**, Vol. I and II, 1935.
6. Lewis, W.H. and Elvin Lewis, M.P.F., **Medical Botany - Plants affecting Man's Health**, John Wiley and Sons, New York, 1977, 515 p.
7. Sammbamurty, A.V.S.S. and Subrahmanyam, N.S., **A Textbook of Economic Botany**, Wiley Eastern Ltd. (India), 1989, 875 p.
8. Sammbamurty, A.V.S.S. and Subrahmanyam, N.S., **Medicinal Plants in Industry**, CBS Publishers and Distributors, New Delhi (India), 2000, 288 p.
9. Volak Jan and Studola Jiri, **Illustrated Book of Herbs**, Coxton Editions, London, 1998, 256 p.

Introduction to Medicinal Plants

HERBAL MEDICINE

There are many thousands of medicinal plants in use throughout the world, with a tremendous range of actions and degrees of potency. Most have specific action on particular body systems and are known to be suitable for treating certain type of diseases. Normally plant medicines act on the body as a whole in a systematic way to produce a healthy balanced body system.

The human body is much better suited to treatment with herbal remedies than with the isolated chemical medicines. We have evolved side-by-side with plants over tens of thousands of years and our digestive system and physiology as a whole are geared to digesting and utilizing plant based foods, which often have a medicinal value as well as providing sustenance. The dividing line between 'foods' and 'medicines' may not always be clear. Are lemons, papayas, onions and oats foods or medicines? The answer, very simply, is that they are both. Lemon (*Citrus limon*) improves resistance to infection; papaya (*Carica papaya*) is taken in some parts of the world to expel worms; onions (*Allium cepa*) relieves bronchial infections; and oats (*Avena sativa*) support convalescence. Herbal medicines not only provide nutrients, but when needed they also strengthen and support the action of the digestive system, speeding up the rate of processing food and improving the absorption of nutrients. Once taken in by the body, nutrients and medicines are carried to the body's estimated three trillion cells! The circulatory system has a remarkable ability to adopt to an endlessly shifting pattern of demand.

HERBS AND BODY SYSTEMS

One of the most common ways of classifying medicinal plants is to identify their actions, e.g., whether they are sedative, antiseptic, or diuretic, and the degree to which they affect different body systems. Herbs often have a pronounced action on a particular body system, e.g., a plant that is strongly antiseptic in the digestive tract may be less so in the respiratory tract. Some examples of herbs and their mode of action on the body are given below.

Skin

Antiseptics e.g., tea tree (*Melaleuca alterniolia*) disinfect the skin; **Emollients** e.g., marigold (*Calendula offcinalis*) reduce itchness, redness and soreness; **Astringents** e.g., witch hazel (*Hamamelis virginiana*) tighten the skin; **Depuratives** e.g., Burdock (*Arctium lappa*) encourage removal of waste products; **Healing and vulnerary herbs** e.g., self heal (*Prunella vulgaris*) and comfrey (*Symphytum officinate*) aid the healing of cuts, wounds and abrasions.

Immune System

Immune stimulants e.g., echinacea (*Echinacea sp*) and lapacho (*Tabebuia sp*) encourage the immune system to ward off infection.

Respiratory System

Antiseptics and antibiotics e.g., garlic (*Allium sativum*) help the lungs resist infection. **Expectorants** e.g., elecampane (*Inula helenium*) stimulate the coughing up of the mucous. **Demulcents** e.g., marsh mallow (*Althaea offcinalis*) soothe irritated membranes. **Spasmolytics** e.g., visnaga (*Ammi visnaga*) relax bronchial muscles.

Endocrine Glands

Adaptogens e.g., ginseng (*Panax ginseng*) help the body to adjust to external pressures and stress; **Hormonally active herbs** e.g., agnus castus (*Vitex agnus-castus*) stimulate production of sex and other hormones. **Emmenagogues** e.g., black cohosh (*Cimicifuga racemosa*) encourage or regulate menstruation.

Urinary System

Antiseptics e.g., bachu (*Barosma betalina*) disinfect urinary tubules; **Astringents** e.g., horse tail (*Equisetum arvense*) tighten and protect the urinary tubules; **Diuretics** e.g., corn silk (*Zea mays*) stimulate flow of urine.

Musculo-Skeletal System

Analgesics e.g., yellow jasmine (*Gelsemium sempervirens*) relieve joint and nerve pain; **Anti-inflammatories** e.g., white willow (*Salix alba*) reduce swelling and pain in joints; **Anti-spasmodics** e.g., cinchona (*Cinchona sp*) relax tense and cramped muscles.

Nervous System

Nervines e.g., rosemary (*Rosemarinus officinalis*) support and strengthen the nervous system; **Relaxants** e.g., lemon balm (*Melissa officinalis*) relax the nervous system; **Sedatives** e.g., mistletoe (*Viscum album*) reduce nervous activity; **Stimulants** e.g., kola nut (*Cola acuminata*) increase nervous activity; **Tonics** e.g., oats (*Avena sativa*) improve nerve function and tone and help to restore the nervous system as a whole.

Circulation and Heart

Cardiotonics e.g., damshen (*Salvia militiorrhiza*) vary in action, some slow heart beat rate, while others increase it. Some improve the regularity and strength of the heart's contractions. **Circulatory stimulants** e.g., cayenne (*Capsicum frutescens*) improve the circulation of blood to the extremities. **Diaphoretics** e.g., juhua (*Chrysanthemum X morifolium*) encourage blood flow to the surface of the body, promote sweating and lower blood pressure. **Spasmolytics** e.g., cramp bark (*Viburnum opulus*) relax the muscles, help to lower blood pressure.

Digestive Organs

Antiseptics e.g., ginger (*Zingiber officinalis*) protect against infection; **Astringents** e.g., bistort (*Polygonum bistarta*) tighten up the inner lining of the intestines and create a protective coating over them; **Biffers** e.g., worm wood (*Artemesia absinthium*) stimulate secretion of digestive juices by the stomach and intestines. **Carminatives** e.g., sweet flag (*Acorus calamus*) relieve wind and gripping pain; **Cholagoges** e.g., fringe tree (*Chionanthus virginicus*) improve the flow of bile into the intestines; **Chloretics** e.g., artichoke (*Cynara scolymus*) stimulate secretion of bile by the liver. **Demulcants** e.g., psylline (*Plantago* sp) soothe the digestive system and protect against acidity and irritation. **Hepatics** e.g., bupleurum (*Bupleurum chinense*) prevent liver damage. **Laxatives** e.g., senna (*Cassia senna*) stimulate bowel movements. **Stomachics** e.g., cardamomum (*Eletteria cardamomum*) protect and support the stomach.

ACTIVE CONSTITUENTS

The medicinal effects of certain plants are well known. Senna, for example, has been taken as a laxative for thousands of years, and Aloe vera was known to Cleopatra as a soothing skin remedy. It is only recently, however, that the active constituents responsi- *ble for the medicinal actions of plants have been isolated and observed. Knowing a little about the chemicals contained in plants helps you to understand how they work within the body.*

Mucilage

Found in many plants, mucilage is made up of polysaccharides (large sugar molecules) that soak up water, producing a sticky jelly-like mass. Mucilage lines the mucous membranes of the digestive tract, protecting against irritation, acidity and inflammation. This soothing and protective action appears to extend to other areas, including the mucous membranes of the throat, lungs, kidneys and urinary tubules. Slippery elm (*Ulmus rubra*) is a typical mucilaginous herb.

Phenols

This group of compounds includes salicylic acid – the natural forerunner of aspirin. Salicylic acid is found in many plants, for example, wintergreen (*Gaultheria procumbens*) and white willow (*Salix alba*) another phenol is thymol-a constituent of thyme (*Thymus vulgaris*) phenols are antiseptic and reduce inflammation when taken internally, yet they have an irritant effect when applied to the skin.

Tannins

Tannins are produced to a greater or lesser degree by all plants. The harsh, astringent taste of tannin-laden bark and leaves makes them unpalatable to insects and grazing animals.

Tannins contract the tissues of the body – hence their use to "tan" leather. They draw the tissues closer and improve their resistance to infection. Oak bark (*Quercus robur*) and black catechu (*Acacia catechu*) are both high in tannins.

Coumarins

Coumarins of different kinds are found in many plant species and have widely divergent actions. The coumarins in melilot (*Melilotus officinalis*) thin the blood, while bergapten, found in celery (*Apium graveolens*), is used as a sunscreen, and khellin, found in visnaga (*Ammi visnaga*), is a powerful muscle relaxant.

Anthraquinones

Anthraquinones are the main active constituents in herbs such as senna (*Cassia senna*) and Chinese rhubarb (*Rheum palmatum*), both of which are taken to relieve constipation. Anthraquinones have an irritant laxative effect on the large intestine, causing contractions of the intestinal walls and stimulating a bowel movement approximately 10 hours after being taken. They also make the stool more liquid, easing bowel movements.

Flavonoids

Found in many plants, flavonoids have a wide range of actions. They are anti-inflammatory and are especially useful in maintaining healthy circulation. Rutin, a flavonoid found in plants including buckwheat (*Fagopyrum esculentum*) and lemon

(*Citrus limon*) strengthens capillary walls.

Anthocyanins

These pigments, which give flowers and fruits a blue, purple or red hue, help to keep the blood vessels healthy. Blackberry (*Rubus fruticosus*) and grapes (*Vitis vinifera*) contain appreciable quantities of anthocyanins.

Glucosilinates

Found exclusively in species of the mustard family, glucosilinates have an irritant effect on the skin, causing inflammation and blistering. Applied as poultices to painful or aching joints, they increase blood flow to the affected area, helping to remove the build-up of waste products (a contributory factor in joint problems). Glucosilinates also help to reduce thyroid function. Both radish (*Raphanus sativus*) and mustard (*Sinapsis alba*) contain significant quantities of glucosilinates.

Volatile Oils

Volatile Oils-which are extracted from plants to produce essential oils-are some of the most important plant constituents of all. Tea tree (*Melaleuca alternifolia*), for example, is known to contain over 60 different volatile compounds within its volatile oil, many of them being strongly antiseptic. Some volatile oils contain sesquiterpenes, such as azulene, found in German chamomile (*Chamomilla recutita*). These constituents have anti-inflammatory effect.

Saponins

These are two types of saponins – triterpenoid and steroidal saponins. The latter get their name from their similarity to the human body's own naturally occurring steroid hormones. Many plants containing steroidal saponins have a marked hormonal activity, liquorice (*Glycyrrhiza glabra*) being one of the best known. Triterpenoid saponins, for example those in cowslip root (*Primula veris*), are often strong expectorants, and may also aid in the absorption of nutrients.

Cardiac Glycosides

Found in various medicinal plants, most famously in common foxglove (*Digitalis purpurea*), yellow foxglove (*D. lutea*) and wooly foxglove (*D. lanata*), cardiac glycosides such as digitoxin, digoxin and gitoxin have a strong, direct action on the heart, helping to support its strength and rate of contraction when it is failing. Cardiac glycosides are also significantly diuretic. They help to transfer fluids from the tissues and circulatory system to the urinary tract, thereby lowering blood pressure.

Cyanogenic Glycosides

Though these glycosides are based on cyanide, a very potent poison, they have a helpful sedative and relaxant effect on the heart and muscles in small doses. Wild cherry bark (*Prunus serotina*) and elder (*Sambucus nigra*) both contain cyanogenic glycosides, which contribute to both plants' ability to suppress and soothe dry coughs.

Vitamins

Some plants contain significant levels of vitamins. Watercress (*Nasturtium officinale*), for example, contains an appreciable quantity of vitamin E, and the hips of dog rose (*Rosa canina*) have particularly high levels of vitamin C. Most other medicinal plants contain at least some vitamins. While the content may be small it contributes to overall daily intake for other plants that are rich in vitamins.

Bitters

Bitters are a varied group of constituents linked only by their pronounced bitter taste, The bitterness itself stimulates secretions by the salivary glands and digestive organs. Such secretions can dramatically improve the appetite and strengthen the overall function of the digestive system. With the improved digestion and absorption of nutrients that follow, the body is nourished and strengthened. Many herbs have bitter constituents, notably wormwood (*Artemisia absinthium*) and chiretta (*Swertia chirata*).

Alkaloids

A very mixed group, alkaloids mostly contain a nitrogen-bearing molecule (NH_2) that makes them particularly pharmacologically active. Some are well-known drugs and have a recognized medical use. Vincristine, for example, derived from Madagascar periwinkle (*Vinca rosea*), is used to some types of cancer. Other alkaloids, such as atropine, found in deadly nightshade (*Atropa belladonna*), have a direct effect on the body, reducing spasms, relieving pain and drying up bodily secretions.

Minerals

Some herbs are particularly rich in minerals. Horsetail (*Equisetum arvense*), for example, have high levels of silica. Dandelion (*Taraxacum officinale*) has large quantities of potassium, and unlike other diuretics which flush this mineral out of the body, it helps to maintain high levels of potassium. These plants act as mineral supplements in their own right, while other herbs with a small concentration contribute to overall intake.

Definitions and Terms

Alternative	• Blood purifier; • Cleans and rebuilds the blood, as well as the lymphatic system.
Antiscorbutics	• Blood purifier; • Cleans and rebuilds the blood.
Antispasmodics	• Relieve contractions or spasms.
Anodyne	• Alleviates pain by reducing nerve sensitivity.
Antacid	• Neutralises stomach and intestinal acid.
Antiarthritic	• Relieves problems of arthritic conditions.
Antiseptic	• Inhibits growth of intruding organisms without destroying tissue; • Prevents putrification, cell decay and pus or gangrene.
Antihydropic	• Relieves vomiting.
Antioxidant	• Compounds or substances that bind with free radicals, inhibiting their action and preventing further damage at the molecular level.
Antispasmodic	• Prevents or relieves involuntary muscle contractions; • Relieves nervous irritation.
Aperient	• Mild purgative to bowels.
Aromatherapy	• Use of essential oils to treat cosmetic; psychological or medical concerns.
Aromatics	• Fragrant or spicy-tasting; • Stimulate gastrointestinal mucous membranes.
Astringents	• Promotes firmness; • Induces density of the tissues.
Bitter	• Stimulates gastrointestinal mucous membranes.
Bromelain	• Digestive enzyme with strong anti-inflammatory characteristics; • Also can be used to reduce weight; • Found in high concentrations in pineapple.
Cathartics	• Promote bowel actions; • Clean the liver and gall ducts.

Convulsant	• Cause convulsions;
	• Use stimulant (cayenne, peppermint) before taking convulsant.
Detoxification	• Cleaning can involve the liver and bowel;
	• Generally includes a change in diet, as well as the use of a detoxification program;
	• Deeper cleansing may be accomplished through juicing and fasting;
	• Consult with a physician before attempting any detoxification or fasting program;
	• Naturopathic practitioners consider this the first step in the journey to good health.
Demulcents	• Promote natural mucilaginous shields;
	• Sooths membranes.
Detergents	• Clean wounds, ulcers and boils.
Disinfectant	• Destroys noxious properties of decay.
Diuretics	• Stimulate urine production;
	• Some stimulate kidneys;
	• Others revitalise tissue power.
Diaphoretic	• Promote sweating;
	• Stimulate sebaceous and sudoriferous glands;
	• Lowers body temperature.
Digestive enzymes	• Many different available;
	• Primarily used to help digest foods in cases of hypoacidic stomachs;
	• Should be used for short durations only, during a program to change dietary habits, or during large meals to reduce gas production.
Ecbolic	• Induce abortions.
Expectorant	• Expels (or spit up) mucus.
Emetic	• Expels matter from stomach (aids vomiting).
Emmenagogue	• Promote menstrual flow;
	• Regulate or normalise menstrual flow.
Emollient	• Softens and protects tissues.
Expectorant	• Promotes mucus discharge in lungs.
Fluorine	• Non-essential mineral that supposedly reduces significant incidences of dental cavities;
	• Stored primarily in the teeth and bone;

- Excessive accumulation is called fluorosis, early signs of which may be chalky white patches on the teeth;
- Extreme fluorosis weakens tooth enamel resulting in surface pitting;
- Food sources include fluoridated water, coffee, tea and fish.

Free radicals
- Unstable electrons that knock other electrons off nuclei, creating a domino effect;
- If left unchecked, can lead to unstable or mutated cell growth and destruction of tissue;
- One of the favorite theories of aging at this point;
- Can be initiated by sunlight (and other forms of radiations), polluting of air, food and water; exposure to chemicals, and cigarette smoke.

Glucasamine sulphate
- Used in cases of rheumatoid and osteo arthritis to increase lubrication between the joints, reducing inflammation.

Hepatic
- Influences liver, increasing flow of bile.

Hypnotic
- Promotes sleep;
- Relaxant.

Laxative
- Mild purgative.

Multi-Minerals
- Contain a wide variety of both micro and macro-minerals essential for health and vitamin utilisation.

Multi-Vitamins
- Contain a wide variety of essential vitamins necessary for good health;
- Often used during maintenance programs.

Narcotic
- Powerful anodyne and hypnotic.

Nervine
- Nourish, strengthen and rehabilitate the nerve cells.

Purgative
- Cause discharge from bowel (stronger than cathartic).

Sedative
- Calms/tranquilizes by lowering functional activity.

Stimulant
- Stimulate functional activity and energy in the body;
- They cleanse the body's systems;
- Carry the blood to all body parts;

	• Balances circulation throughout the body.
Stomachic	• Stimulates/tones action of stomach.
Sudorific	• Produces sweating when taken hot;
	• Acts as tonic when taken cold.
Tonics	• Increase vigor, energy and strength to every body organ;
	• Promote elimination;
	• Produces tone in the fiber and tissues.
Trans fatty acids	• Oxidation of fatty acids either by heating or free radical exposure.
Vermicides (Anthelmintic)	• Destroys worms, not necessarily discharging them.
Vermifuge (Anthelmintic)	• Expel worms.
Vulneraries	• Promote healing (wounds and cuts).
Women's tea	• Chinese formulation used to help cases of painful, prolonged or missing periods;
	• It regulates hormonal flow and is most useful in pre-menopausal cases.

Note : The text is meant for informational purposes only and should not be construed as medical diagnosis. Experts in the field of naturopathy hold varying views. The information presented here is based on the most widely available and most commonly accepted research. Knowledge in this field is continually changing as new studies are conducted.

Abbreviations Used

Abortif.	Abortifacient	Essen. oil	essential oil
Absorb.	absorbent	Expect.	expectorant
Alk.	alkaloid		
Alter.	alternative	Febge.	febrifuge
Amenor.	amenorrhoea	Galact.	galactagogue
Amorph.	amorphous	Glucd.	glucoside
Antibil.	antibilious	Gonor.	gonorrhoea
Anthelm.	anthelmintic		
Antid.	antidote	Haemat.	haematuria
Antidysen.	antidysenteric	Haemor.	haemorrhage
Antilith.	antilithic		
Antimal.	antimalarial	Indign.	indigestion
Antiper.	antiperiodic	Inflam.	inflammation
Antiphlegm.	antiphlegmatic	Irrit.	irritant
Antiphl.	antiphlogistic		
Antipyr.	antipyretic	Lactag.	lactagogue
Antiscor.	antiscorbutic	Laxt.	laxative
Antisep.	antispasmodic	Leucor.	leucorrhoea
Antisyp.	antisypillitic		
Aper.	aperient	Mat. Med.	materia medica
Aphrodis.	aphrodisiac	Menor.	menorrhagia
Arom.	aromatic	Mucil.	mucilage
Astrin.	astringent		
		Nutri.	nutritious
Carmin.	carminative	Phlegm.	phlegmatic
Catar.	catarrhal	Purg.	purgative
Cath.	cathartic		
Cholag.	cholagogue	Refrig.	refrigant
Chr.	chronic	Resolv.	resolvent
Constip.	constipation	Restor.	restorative
		Rheum.	rheumatic
Dec.	Deccan	Rubft.	rubefacient
Decoct.	decoction		
Demul.	demulcent	Sialog.	sialogogue
Deod.	deodorant	Stim.	stimulant
Diaphor.	diaphoretic	Stomch.	stomachic
Diar.	diarrhoea	Subst.	substitute
Digest.	digestive	Syn.	synonym
Diur.	diuretic		
Dysen.	dysentery	Tox.	toxic
Dysmen.	dysmenorrhoea		
Dyspep.	dyspepsia	Var.	variety
		Vern.	vernacular
Emmen.	emmenagogue	Vet.	veterinary
Emol.	emollient	Vesic.	vesicant

Contents

ABIES BALSAMEA
(Pinaceae)

BALSAM FIR

Parts used: Oleo-resin, leaves.
Constituents: Balsam fir leaves contain a liquid oleo-resin.
Medicinal actions and uses: Balsam fir is antiseptic and stimulant, and has been used in North America and Europe for catarrh, chest infections such as bronchitis, and urinary tract conditions such as cystitis and frequent urination. Externally, balsam fir was rubbed on the chest or applied as a plaster for respiratory infections. It is little used in herbal medicine today.

ABIES WEBBIANA
(Pinaceae)

Leaves: carminative, expectorant, stomachic, tonic, astringent, used in Asthma, bronchitis.
Juice of leaves: anti per, essential oil.

ABROMA AUGUSTA
(Sterculiaceae)

Root bark – emmen. uterine tonic, in dysmen.
Roots – alkal. and water soluble bases.
Diabetes, profuse urination and frequent urination, dryness of mouth and great thirst.

ABRUS PRECATORIUS
(Leguminosae)

JEQUIRITY

Parts used: Root, leaves, seeds.
Constituents: Jequirity seeds contain abrin, indole alkaloids, and anthocyanins. The root and leaves contain glycyrrhizin and traces of abrin. Abrin is extremely toxic. Glycyrrhizin is expectorant, anti-inflammatory and anti-allergenic.
Medicinal actions and uses: Jequirity seeds have been used medicinally in the past as a contraceptive, abortifacient (to induce a miscarriage) and as a treatment for chronic conjunctivitis. However, they are so poisonous that even external application is no longer justifiable. Even small amounts brought into contact with an open wound can prove fatal. The leaves and roots can be substituted for liquorice (*Glycyrrhiza glabra*), and have been used in the Ayurvedic tradition in the treatment of asthma, bronchitis, and other chest conditions. They have been used in Chinese medicine to treat fever.
Cautions: Never use the seeds. Use the leaves and roots only under professional supervision. Jequirity is subject to legal restrictions in some countries.

ABUTILON ASIATICUM
(Malvaceae)

Leaves: In gonor. applied to ulcers, internally for stones in the bladder and as an eye wash.
Leaves, bark and root – Demulc. diur.

ABUTILON HIRTUM
(Malvaceae)

Leaves	– Demulc.
Bark	– Astr. diur.
Roots	– In fever
Seeds	– Laxat. demulc.
Mucilage	– Asparagin.

ABUTILON INDICUM
(Malvaceae)

KANGHI, INDIAN MALLOW

Parts used: Root, bark, leaves, seeds.
Constituents: Kanghi contains mucilage, tannins and asparagine. Asparagine is diuretic.
Medicinal actions and uses: Also known as Indian mallow, kanghi is used in much the same way as marshmallow (*Althaea officinalis*) one of the main European demulcent herbs. The root, leaves and bark of kanghi are mucilaginous and are used to soothe and protect the mucous membranes of the respiratory and urinary systems. A decoction of the root is given for chest conditions such as bronchitis. The mucilaginous effect benefits the skin; an infusion, poultice or paste made from the powdered root or bark is applied to wounds and used for conditions such as boils and ulcers. The seeds are laxative and "useful in killing threadworms, if the rectum of the affected child be exposed to the smoke of the powdered seeds."
The plant has an antiseptic effect within the urinary tract.
Related Species *A.trisulcatum*, native to Central America, is used to treat asthma in children, and is applied as a poultice for treating cancerous sores and ulcers, especially of the mouth and cervix.

ABUTILON THEOPHRASTII
(Malvaceae)

Leaves – demulc.
Bark – astrn. diur.
Roots – in fever
Seeds – laxat. demulc.
Seeds yield semi-drying oil.

ACACIA ARABICA
(Leguminosae)

BABUL

Part used: Bark.
Constituents: Babul contains tannins, mucilage and flavonoids.
Medicinal actions and uses: Strongly astringent, babul is used to contract and toughen mucous membranes throughout the body in much the same way as does witch hazel (*Hamamelis virginiana*) or oak bark (*Quercus robur*). Babul may be made into a variety of preparations, for instance, a lotion for treating bleeding gums, a gargle for easing the irritation of sore throats, a wash for eczema, an eyewash for conjunctivitis and other eye problems, and a douche for excessive vaginal discharge. It is also taken internally to treat diarrhoea. In Ayurvedic medicine, babul is considered a remedy that is helpful for treating premature ejaculation.

ACACIA CATECHU
(Leguminosae)

BLACK CATECHU, CUTUCH

Parts used: Bark, heartwood, leaves, shoots.
Constituents: The shiny, black-brown extract of leaves and young shoots, which is called "cutch", becomes a brittle solid when dried, and is the form in which black catechu is generally sold. Cutch contains 25-60% tannins, 20-30% mucilage, flavonoids and resins catechin and catechutannic acid.

Medicinal actions and uses: Black catechu is a strong astringent and clotting agent. It helps reduce excess mucus in the nose, the large bowel or vagina. It is also used to treat eczema, haemorrhages, diarrhoea and dysentry. It may be used as an infusion, tincture, powder or ointment. A small piece of cutch dissolved in the mouth is an excellent remedy for bleeding gums and mouth ulcers. The powder and tincture are also applied to infected gums and have been used to clean the teeth. In Ayurvedic medicine, decoctions of the bark and heartwood are used for sore throats.

ACACIA CONCINNA
(Mimosaceae)

Pods – aper. expect, emetic.
Leaves – cath. in billiousness saponin, alkaloid.

ACACIA FERRUGINEA
(Mimosaceae)

Bark and Pods – astrin.
Gum – demulc., emol., nutrient.

ACACIA PENNATA
(Mimosaceae)

Leaf juice – mixed with milk given to infants for indigestion.
Leaves – chewed with sugar and cumin in bleeding gums.
Juice of Bark – antidote for snake poison.
Fruit – used as fish poison.

ACACIA SENEGAL
(Mimosaceae)

Gum – demulc. emol. internally used in inflammation of intestinal mucosa, externally to cover inflamed surfaces, such as burns, sore nipples etc.

ACALYPHA HISPIDA
(Euphorbiaceae)

Flower – Diarch.
Leaves – Beaten up with green tobacco leaf and infusion of rice applied to inverterate ulcers.

ACALYPHA INDICA
(Euphorbiaceae)

Plant – Emetic, expect. used as a substitute for senega, useful in broncht. pneumonia and asthma.
Root – Cath.
Leaves – Laxat, used in scabies, in snake bite. Acalyphine, cynogeneic glucoside, and Triacetone amine, active principle HCN, causes discolourations of blood and gastro intestinal irritation.

ACAMPE PAPILLOSA
(Orchidaceae)

Root – Bitter tonic, used in rheumatism.
Alkaloid and bitter resin.

ACANTHUS ILICIFOLIUS
(Acanthaceae)

Leaves – used as fermentations in rheumatism and Neurologia.
Leaves and tender roots – used in snake bite
Resin and alkaloid
Mangrove plant.

ACANTHUS MOLLIS
(Acanthaceae)

ACANTHIUS, BEAR'S BREECHES
Parts used: Leaves, roots.
Constituents: Acanthus contains large quantities of mucliage and a tannin.

Medicinal uses and actions: The herb's appreciable quantities of mucilage and tannin substantiate its traditional use as a treatment for dislocated joints and burns. These constituents are found in many wound-healing plants, for example comfrey (*Symphytum officinale*) and plantain (*Plantago major*). Acanthus paste applied to a dislocated joint tends to normalise the affected-muscles and ligaments, simultaneously relaxing and tightening them to encourage the joint back into its proper place. The plant's emollient properties are also useful in the treatment of irritated mucous membranes within the digestive and urinary tracts. Acanthus is similar to marshmallow (*Althaea officinalis*) in that it can be used externally to ease irritation, and internally to heal and protect.

Achillea millefolium

ACHILLEA MILLEFOLIUM
Asteraceae (Yarrow)

The non-woody parts of the flowering stems, sometimes only the flowers, free of stalks, are used medicinally. The principal constituent is an essential oil with azulenes that turn blue after distillation. The plant also contains the alkaloids achilleine and stychydrine, tannins and bitter compounds. These constituents give Yarrow antiseptic, stomachic, antispasmodic, astringent and diaphoretic properties and it has a variety of uses both internally and externally. For example, herbalists use an infusion for digestive upsets, diarrhoea, flatulence, menstrual disorders, colds and fevers. Externally a decoction is used to treat slow-healing wounds, skin rashes and eczema, chapped skin and as a gargle and bath preparation. **Yarrow should always be taken in moderation and never for long periods** because it may cause skin irritation.

The fresh leaves – and the flowers – also have many cosmetic uses. The taste is slightly bitter and peppery and young leaves, chopped up, give 'bite' to a mixed salad.

ACHRAS ZAPOTA
(Sapotaceae)

Fruit – preventive against biliousness and febrile attacks.
Bark – tonic, febge.
Seeds – diuretic
Glucoside, tannin, saponin, alkaloid and sapotinin a bitter principle.

ACHYRANTHES ASPERA
(Amaranthaceae)

Plant – Pungent, purg, diur, in dropsy, piles, boils, skin, eruptions, colic, snake-bite.
Infusion of roots – astrin
Seeds – emetic, in hydrophobia.

ACHYRANTHES BIDENTATA
(Amaranthaceae)

Niu Xi

Parts used: Root.
Constituents: *Achyranthes* species contain triterpenoid saponins.
Medicinal actions and uses:. In traditional Chinese medicine, *niu xi* is believed to invigorate blood flow. It is used to stimulate menstruation when a period is delayed or scanty. The herb is also prescribed to ease period pain. *Niu xi* is used to relieve pain in the lower back, especially where the discomfort is attributable to kidney stones. The herb is also taken as a treatment for mouth ulcers, tooth ache, bleeding gums and nose bleeds.

ACONITUM FEROX
(Ranunculaceae)

Extremely poisonous, used in leprosy, fever, cholera, rheumatism.
Toxic, alkaloid, pseudoaconitine.

ACONITUM NAPELLUS
(Ranunculaceae)

Aconite, Monkshood

Part used: Root
Constituents: Aconite contains 0.3-2% terpenoid alkaloids, principally aconitine.
Medicinal actions and uses: Aconite is poisonous in all but the smallest doses, and is rarely prescribed for internal use. More common, it is applied to unbroken skin to relieve pain from bruises or neurological conditions. In Ayurvedic medicine, aconite is used to treat neuralgia, asthma and heart weakness. Aconite is also used extensively in homeopathy as an analgesic and sedative.

ACONITUM PALMATUM
(Ranunculaceae)

Non poisonous root, tonic, antiper, in diarh. and rheumatism.
Rhizome – alkaloid, palmatisine.

ACORUS CALAMUS
Araceae (Sweet Flag)

The rhizomes are used medicinally. When dried they are strongly aromatic and brittle. The constituents include upto 4 per cent of an essential oil with asarone as its chief component, sesquiterpenes, the bitter compounds acorin and acoretin, and tannins. Sweet Flag is used in herbalism as a stomachic and carminative. It is also added to bath preparations to alleviate nervous exhaustion. The essential oil is extracted and combined with other substances in stomachic powders, teas and drops. It is also used in perfumery.

Acorus calamus

ACROCEPHALUS INDICUS
(Labiatae)
Plant – Expect.

ACRONYCHIA PEDUNCULATA
(Rutaceae)
Plant – fish poison
Bark – used as application to sores and ulcers.
Essential oil.

ACTEA SPICATA
(Ranunculaceae)
Root – Nerve sedative, emetic, purg.
Toxic principle, oil of baneberry

ACTINODAPHNE HOOKERI
(Lauraceae)
Infusion of leaves – in urinary disorders and in diabetes
Seed oil – application to sprains.
Amorphous alkaloid, actino-daphinine, reddish brown oil, containing oleates, resinates and trilaurine.

ADANSONIA DIGITATA
(Bombacaceae)
Fruit pulp – aper. demulc. astring. in dysent.
Leaves – used as diaphoret., and as prophylactic against fevers.

ADENANTHERA PAVONIA
(Leguminosae)
Decoction of leaves – in chronic rheumatism, gout, haemat.
Seeds – for boils and inflam.
Seeds contain HCN-glucd, lignocenic acid, alkal.

ADENIA PALMATA
(Passifloraceae)
Root and fruit – poisonous
Juice of leaves and fruits – used externally for skin diseases.

ADHATODA VASICA
(Acanthaceae)
Leaves and Roots – in cough, chr. bronchitis, asthma, phthisis.
Leaves – used in rheumatism, insecticidal.
Flowers, leaves and Roots – antisep.
Leaves – alkaloid, vasicine and small amounts of essential oil.

ADIANTUM CAUDATUM
(Polypodiaceae)
Fronds – for skin diseases, diabetes, cough and fever.

ADIANTUM CAPILLUS-VENERIS
(Polypodiaceae)
MAIDENHAIR FERN
Parts used: Aerial parts.
Constituents: Maidenhair fern contains flavonoids (including rutin and isoquercitin), terpenoids (including adiantone), tannin, and mucilage.
Medicinal actions and uses: Maidenhair fern is still used by Western herbalists to treat coughs, bronchitis, catarrh, sore throat, and chronic nasal catarrh. The plant also has a longstanding reputation as a remedy for conditions of the hair and scalp.

ADIANTUM PEDATUM
(Polypodiaceae)
Plant – pectoral in chr. catarrh, demulc, expect. tonic. astring. emmen.

ADINA CORDIFOLIA
(Rubiaceae)
Bark – Febge, antisep.
Juice – used to kill worms in sores
Bark-tannin, yellow pigment adinin.

ADONIS AESTIVALIS
(Ranunculaceae)

Plant – used as cardiac stim. and diur.
Flowers – laxt., diuret., lithothriptic.
Amorphous glucoside, adonidin.

ADONIS VERNALIS
(Ranunculaceae)

FALSE HELLEBORE,
YELLOW PHEASANT'S EYE

Parts used: Aerial parts.
Constituents: False hellebore contains cardiac glycosides, including adonitoxin.
Medicinal actions and uses: False hellobore contains cardiac glycosides similar to those found in foxglove (*Digitalis purpurea*,). These substances improve the heart's efficiency, increasing its output while at the same time slowing its rate. Unlike foxglove, however, false hellebore's effect on the heart is slightly sedative, and it is generally prescribed for patients with hearts that are beating too fast or irregularly. False hellebore is also recommended as a treatment for certain cases of low blood pressure. In common with other plants containing cardiac glycosides, false hellebore is strongly diuretic and can be used to counter water retention, particularly in cases of poor circulatory function. False hellebore is used in homeopathic medicine as a treatment for angina.

AEGLE MARMELOS
(Rutaceae)

BAEL, BENGAL QUINCE

Parts used: Fruit, leaves, root, twigs.
Constituents: Bael contains coumarins, flavonoids, alkaloids, tannins and fixed oil.
Medicinal actions and uses: The astringent half-ripe bael fruit reduces irritation in the digestive tract and is excellent for diarrhoea and dysentery. The ripe fruit is demulcent and laxative, with a significant vitamin C content. It eases stomach pain and supports the healthy function of this organ. Bael's astringent leaves are taken to treat peptic ulcers. The tree's most unusual application is for earache. A piece of dried root is dipped in the oil of the neem tree (*Azadirachta indica*) and set alight. Oil from the burning end is dripped into the ear.
Pulp of ripe fruit – arom., cooling, laxat.
Unripe or half-ripe fruit – astrin., digest. stomachic. in diar.
Root bark – intermittent fevers. fish

Adonis vernalis

poison. Fruit contains marmalosin, coumarin, alkaloid. Umbelliferone, essential oil consisting of α and β-phellandrene.

AERVA LANATA
(Amaranthaceae)

Plant – anthelm., diur.
Root – demulc., diur. in the treatment of headache.

AERVA TOMENTOSA
(Amaranthaceae)

Decoction of the plant – used to remove swellings.

AESCULUS HIPPOCASTANUM
(Hippocastanaceae)

HORSE CHESTNUT

Parts used: Seeds, leaves, bark.
Constituents: Horse chestnut contains triterpenoid saponins (notably aescin), coumarins and flavonoids. Aescin, the main active constituent, has anti-inflammatory properties. In Germany and other European countries, specialised aescin preparations are used because aescin is not easily absorbed from the gut. Coumarin glycosides – aesculin, aesculoside and fraxins and tannins.

Medicinal actions and uses: Horse chestnut is astringent, anti-inflammatory and helps to tone the vein walls, which, when slack or distended, may become varicose, haemorrhoidal or otherwise problematic. Horse chestnut also reduces fluid retention by increasing the permeability of the capillaries and allowing the reabsorption of excess fluid back into the circulatory system. The herb has been taken internally in small to moderate doses for leg ulcers, varicose veins, haemorrhoids and frostbite, and applied externally as a lotion, ointment or gel. In France, an oil extracted from the seeds has been used as an external treatment for rheumatism. In the US, a decoction of the leaves has been given in cases of whooping cough. A coumarine component (aesculoside) is included in some sun-screen preparations.

AFRAMOMUM MELEGUETA
(Zingiberaceae)

GRAINS OF PARADISE

Parts used: Seeds.
Constituents: The seeds contain a volatile oil (0.3-0.5%), a pungent principle called paradol (related to gingerol in ginger, *Zingiber officinale*,) and tannins.

Aesculus hippocastanum

Medicinal actions and uses: Principally used as a condiment, the seeds also are a stimulant that strengthens and warms the stomach. Like other members of the ginger family, this plant is used to alleviate indigestion, flatulence and bloating (the latter more commonly in livestock). Grains of paradise also help to ease abdominal discomfort due to colic or griping. The seeds can help to reduce or prevent vomiting and to bring relief from nausea. The plant's stimulant properties make it an invigorating herb, especially helpful for those with weak digestions.

Related Species: *Sha ren*, the seeds of the closely related *A. villosum*, are used in Chinese medicine for similar complaints.

AGANOSMA CALYCINA
(Apocynaceae)

Plant – tonic, useful in diseases caused by disordered bile and blood.

AGANOSMA DICHOTOMA
(Apocynaceae)

Plant – Emetic
Leaves – in biliousness
Flowers – in the diseases of eye.

AGARICUS OSTREATUS
(Agaricaceae)

MUSHROOM

Ground to a paste with water and applied to gums in excessive salivation and stomatitis, internally given in dysent. and diarh. used to stop haemorrhage.

AGASTACHE RUGOSA
(Labiatae)

HUO XIANG (CHINESE),
GIANT HYSSOP

Parts used: Aerial parts.

Constituents: *Huo xiang* contains a volatile oil, including methyl chavicol, anethole, anisaldehyde and limonene.

Medicinal actions and uses: The acrid *huo xiang* is considered a warming herb in Chinese herbal medicine. It is employed in situations where there is excessive "dampness" within the digestive system, resulting in poor digestion and reduced vitality. The herb stimulates and warms the digestive tract, relieving symptoms such as abdominal bloating, indigestion, nausea and vomiting. It is commonly used to relieve vomiting and morning sickness.

Huo xiang is used to treat the early stages of viral infections that feature symptoms such as stomachache and nausea. It is combined with Baical skullcap (*Scutellaria baicalensis*) and other herbs to treat malaise, fever, aching muscles and lethargy. A lotion containing *huo xiang* may be applied to fungal conditions such as ringworm.

AGAVE AMERICANA
(Agavaceae)

AGAVE, CENTURY PLANT

Parts used: Sap.

Constituents: Agave sap contains oestrogen-like isoflavonoids, alkaloids, coumarins and vitamins pro A, B^1, B^2, C, D and K.

Medicinal actions and uses: Demulcent, laxative and antiseptic, agave sap is a soothing and restorative remedy for many digestive ailments. It is used to treat ulcers and inflammatory conditions affecting the stomach and intestines, protecting these parts from infection and irritation and encouraging healing. Agave has also been employed to treat a wide range of other conditions, including syphilis, tuberculosis, jaundice and liver disease.

AGRIMONIA EUPATORIA (Rosaceae)

AGRIMONY

Parts used: Aerial parts.

Constituents: Agrimony contains tannins, coumarins, flavonoids, including luteolin, a volatile oil and polysaccharides, tannins, silicic acid, bitter compounds, vitamins.

Medicinal actions and uses: Agrimony has long been used by herbalists to heal wounds because it staunches bleeding and encourages clot formation. An astringent and milder bitter, it is also a helpful remedy for diarrhoea and a gentle tonic for the digestion as a whole. Com-

bined with other herbs such as corn silk (*Zea mays*) it is a valuable remedy for cystitis and urinary incontinence, and has also been used for kidney stones, sore throats, hoarseness, rheumatism and arthritis. All parts of the plant yield a good yellow dye.

AGROPYRON REPENS (Gramineae)

COUCH GRASS

Parts used: Rhizome, seeds, root.

Constituents: Couch grass contains polysaccharides (such as triticin), a volatile oil (mainly agropyrene), mucilage and nutrients. Agropyrene has antibiotic properties. Saponins, sug-

Agrimonia eupatoria

Agropyron repens

ars, (tricticin, inositol, and mannitol.); phenolic glycoside arenein, mucilage and the hydrocarbon agropyrene. **Medicinal actions and uses:** A gentle, effective diuretic and demulcent, couch grass is mostly commonly used for urinary tract infections such as cystitis and urethritis. It both protects the urinary tubules against infection and irritants, and increases the volume of urine, thereby diluting it. It can be taken usually with other herbs, to help treat kidney stones, reducing the irritation and laceration they cause. Couch grass is also thought to dissolve kidney stones (in so far as this is possible), and in any case will help to prevent their further enlargement. Both an enlarged prostate and prostatitis (infection of the prostrate gland) will benefit from a couch grass decoction taken over in course of several months. In German herbal medicine, heated couch grass seeds are used in a hot and moist pack that is applied to abdomen for peptic ulcers. Juice from the roots of couch grass has been advocated for treating jaundice and other liver complaints.

AILANTHUS ALTISSIMA
(Simaroubaceae)

TREE OF HEAVEN, CHUN PI

Parts used: Bark, root bark.
Constituents: The bark contains quassinoids (such as ailanthone and quassin), alkaloids, flavonols and tannins. Quassinoids are intensely bitter, antimalarial and act against cancerous cells.
Medicinal actions and uses: In Chinese herbal medicine, tree of heaven

is used to treat diarrhoea and dysentery, especially if there is blood in the stool. The bark of the tree has been used in Asian and Australian medicine to counter worms, excessice vaginal discharge, gonorrhoea and malaria, and it has also been given for asthma. Tree of heaven has marked antispasmodic properties and acts on the body as a cardiac depressant.

AILANTHUS EXCELSA
(Simaroubaceae)

Bark – arom., used for dyspeptic complaints, tonic, febge, expect. antisp. given in chronic. bronchitis and asthma, used as an astrin. in diarh. and dysent.
Bark and Leaves – tonic, used especially in debility after child birth.

AILANTHUS MALABARICA
(Simaroubaceae)

Bark – carm. tonic. febge.
Resin – in dysen.
Juice of fresh bark – in dysen.
Components – Quassin, ailantic acid.

AJUGA REPTANS
(Labiatae)

BUGLE

Parts used: Aerial parts.
Constituents: Bugle contains iridoid glycosides, including harpagide, which is also found in devil's claw (*Harpagophytum procumbens*).
Medicinal actions and uses: Bugle is bitter, astringent and aromatic, but opinion varies as to its value as a medicine. It has mild analgesic properties, and it is still used occasionally as a wound healer. It is also mildly

laxative and traditionally has been thought to help cleanse the liver.

AJUGA BRACTEATA
(Labiatae)

Plant – Bitter astrin. arom. tonic. used in agnes.

Leaves – in fever as substitute for cinchone.

ALANGIUM SALVIFOLIUM
(Alangiaceae)

Root bark – Purg. anthelm. useful in fever and skin diseases.

Leaves – as poultice in rheumatic pains.

Components – Amorphous alkaloid alangine.

Bark alkaloids – akhar kantine akoline, lamarkine.

Root bark contains two isomeric alkaloids – Alangium A and Alangium B and a third alkaloid alanginine.

Imp. Increases peristaltic movements of intestine, produces irregular respiration, reduces blood pressure, and depresses heart temporarily in low doses.

ALBIZZIA AMARA
(Mimosaceae)

Seeds – astrig. given in piles, diarh. - gonor.

Flowers – externally applied to inflam. boils and ulcers.

Leaves – useful in ophthalmia **Saponin** constituent.

ALBIZZIA CHINENSIS
(Mimosaceae)

Infusion of bark – used as a lotion for cuts, scabies and skin diseases.

Plant – fish poison Gum, saponin.

ALBIZZIA LEBBECK
(Mimosaceae)

Plant – in snake bite and scorpion sting.

Bark and Seeds – astrin. given in piles and diarrh. tonic. restr.

Root Bark – in powder form used to strengthen gums.

Leaves – in night blindness gum, saponin and Tannin.

ALBIZZIA ODORATISSIMA
(Mimosaceae)

Bark – applied externally is considered efficacious in leprosy and invertebrate ulcers.

Leaves – boiled in ghee used as remedy for cough.

Constituent – Gum

ALBIZZIA PROCERA
(Mimosaceae)

Leaves – insecticide, made into poultice applied to ulcers.

ALCEA ROSEA
Malvaceae (Hollyhock)

The flowers of dark, sometimes also double, forms are used medicinally. All parts of the plant must be free of mallow rust (*Puccinia malvacearum*). The constituents include abundant mucilage, the pigment anthocyanin and tannins. Hollyhock has excellent emollient and demulcent properties, which make it a useful treatment for inflammation of the mucous membranes, cough, asthma, chronic gastritis and enteritis, and for constipation. In herbalism it is usually administered in the form of an infusion. Hollyhock is also used in soothing

Alcea rosea

herbal compresses and in bath preparations for skin disorders and cuts and bruises. The fresh leaves have similar beneficial effects. The dark pigment is used to colour medicines and foodstuffs.

ALCHEMILLA VULGARIS (Rosaceae)

LADY'S MANTLE

Parts used: Aerial parts, root.
Constituents: Lady's mantle contains tannins, a glycoside and salicylic acid.
Medicinal actions and uses: Lady's mantle has always been prized as a wound healer. Its astrigency ensures that blood flow is staunched and the first stage of healing soon gets under way. As the name implies, it is a valuable herb for treating conditions suffered by women, and is taken principally to reduce heavy menstrual bleeding, to relieve menstrual cramps and to improve regularity of the cycle. Lady's mantle is also prescribed for fibroids and endometriosis. It has been used to facilitate childbirth, and is thought to act as a liver decongestant. Its astringent properties make it a useful herb for the treatment of diarrhoea and gastroenteritis.

ALCHEMILLEA XANTHOCHOLORA
Rosaceae (Lady's Mantle)

The flowering stems, including the basal leaves, are used medicinally. The constituents include tannins, essential oils, saponins, bitter compounds and salicylic acid. Lady's Mantle is mildly astringent, diuretic and anti-inflammatory and herbalists prescribe it for digestive disorders such as gastritis and enteritis, flatulence and diarrhoea. It also relaxes muscular spasms and regulates the menstrual cycle, particularly during

Alchemillea xanthocholora

the menopause. Externally Lady's Mantle is used in bath preparations for wounds, bruises & skin disorders, and it is a valuable herbal cosmetic.

ALETRIS FARINOSA
(Liliaceae)
STAR GRASS,
TRUE UNICORN ROOT, COLIC ROOT

Parts used: Rhizome, leaves.

Constituents: Star grass contains steroidal saponins based on diosgenin, as well as a bitter principle, volatile oil and a resin.

Medicinal actions and uses: It is difficult to gain a clear picture of star grass's medicinal value. Due to an apparent oestrogenic action, it has been employed in this century chiefly for gynaecological problems, particularly at the menopause. It is also given for period pain and irregular periods. Some authorities hold that it prevents threatened miscarriage. Star grass is also a good digestive herb, proving beneficial in treating loss of appetite, indigestion, flatulence and bloating. The herb has also been employed in the treatment of rheumatism.

ALEURITES MOLUCCANA
(Euphorbiaceae)

Oil from seeds – purg. subst. for castor oil.
Seeds – fixed oil, skin of fruit-essential oil.

ALHAGI PSEUDALHAGI
(Leguminosae)

Plant – Laxt. diurt. expect.
Infusion – diaphor.
Oil from leaves – used for rheumatism.
Flowers – used for piles.

ALLAMANDA CATHARTICA
(Apocynaceae)

Leaves – Cath.
Bark – hydragogue in ascites Alkaloid, glucd.

ALLIARIA OFICINALIS
(Brassicaceae)

Herb and Seeds – esteemed as diur., diaphor., and expect., and is used as external application in gangrenous affections and to promote suppuration, healing of cuts, bruises and others.
Glued myrosin, sinigrin, essential oil.

ALLIUM AMPELOPRASUM
(Liliaceae)

Bulbs – stim., expect. used to hasten suppuration of boils.
Essential oil with allyl-disulphide.

ALLIUM ASCALONICUM
(Liliaceae)

Bulb – aphroids in earache

ALLIUM CEPA
(Liliaceae)
ONION

Parts used: Bulb.

Constituents: Onion contains a volatile oil with sulphurous constituents, sulphur containing compounds such as allicin (an antibiotic) and alliin, flavonoids, phenolic acids and sterols.

Medicinal actions and uses: Onion boasts a long list of medicinal actions – diuretic, antibiotic, anti-inflammatory, analgesic, expectorant, and antirheumatic. It is also beneficial to the circulation. Onions are taken the world over for colds, flu and coughs. Like garlic (A. sativum), onion offsets tendencies to angina, arteriosclerosis, and heart attack. It is also

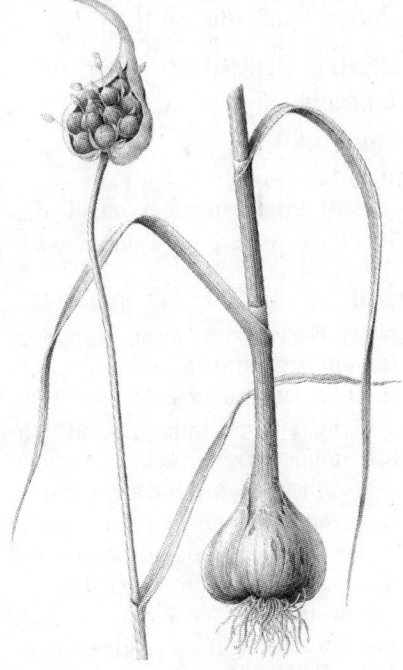

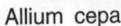
Allium cepa   Allium sativum

useful in preventing oral infection and tooth decay. The warmed juice can be dropped into the ear for earache, and baked onion is used as a poultice to drain pus from sores. Onion has a long-standing reputation as an aphrodisiac, and it is also used cosmetically to stimulate hair growth.

Essential oil and organic sulphides catechol, protocatechinic acid, contains heart stimulant, increases pulse volume and frequency of systolic pressure and coronary flow, stimulates intestinal smooth musculature and uterus, promotes bile production, and reduced blood sugar.

ALLIUM SATIVUM
(Liliaceae)
GARLIC

A bulbous perennial growing to 30 cm - 1 m (1-3 ft), with pale pink or green-white flowers.

Key Constituents:
Volatile oil (alliin, allinase, allicin)
Scordinins
Selenium
Vitamins A, B, C and E

Key Actions:
Antibiotic and antiseptic
Expectorant
Increase sweating
Lowers blood pressure
Reduces blood clotting
Anti-diabetic
Expels worms
Chloretic, expectorant intestinal infections, arteriosclerosis, stimulates bile secretion.
Essential oil containing allyl-propyl disulphide, diallyl-disulphide and two more sulphur compounds. Antisep.

and hyposensitive principle; allicin; allisatin I and allisatin II.

ALLIUM URSINUM
(Liliaceae)

RAMSONS

Parts used: Bulb, aerial parts.
Constituents: Ramsons contain volatile oil, aldehydes, vinyl sulphide and vitamin C.
Medicinal actions and uses: Used mainly as a folk remedy and as a food, ramsons are similar to garlic (*A. sativum*) but weaker in action. They lower high blood pressure and help to prevent arterosclerosis. As ramsons ease stomach pain and are tonic to the digestion, they have been used for diarrhoea, colic, wind, indigestion and loss of appetite. The whole herb is used in an infusion against threadworms, either ingested or given as an enema. Ramsons are also thought to be beneficial for asthma, bronchitis and emphysema. The juice is used as an aid to weight loss. Applied externally, the juice is a mild irritant. It stimulates local circulation, and therefore may be of benefit in treating rheumatic and arthritic joints.

ALLOPHYLLUS SERRATUS
(Sapindaceae)

Root – astrin. employed to check diarh.

ALNUS GLUTINOSA
(Betulaceae)

ALDER

Parts used: bark, leaves
Constituents: Alder contains lignans, tannin (10-20%), emodin (an anthraquinone) and glycosides)
Medicinal actions and uses: The astringent alder is employed most often

Alnus glutinosa

as a mouthwash and gargle for tooth, gum and throat problems. The drying action of a decoction of the bark helps to contact the mucous membrances and reduce inflammation. A decoction may also be used to staunch internal or external bleeding, and to heal wounds. It is also used as a wash for scabies. In Spain, alder leaves are smoothed and placed on the soles of the feet to relieve aching. Leaves are used to help reduce breast engorgement in nursing mothers.

ALOCASIA INDICA
(Araceae)

Leaves – styptic, astrin.
Tuber – in piles, constip. and anasarca.

ALOE BARBADENSIS
(Liliaceae)

ALOE VERA, ALOES

Plant – stomach., purg., emmen. anthelm., in piles and rectal fissures.
Dried juice – cath., given in constipation, for curing pimples on face.
Pulp – in menstrual suppressions
Root – in colic.

Aloin, isobarbaloin, emodin, gum, resin; juice contains antraquinone derivatives like emodin, and chrysophanic acid; whole plant contains uronic acid, oxidase, catalase and sugars.

A perennial with succulent leaves 60 cm (2 ft) long and a spike of yellow or orange flowers.

Key Constituents:
Anthraquinones (aloin, aloe-emodin)
Resins
Tannins
Polysaccharides
Aloectin B
Key actions:
Heals wounds
Emollient
Stimulates secretions of bile
Laxative

ALPINA GALALANGA
(Zingiberaceae)

Rhizomes – used in rheumatism, fever, and catarrh, affections, specially in bronchial catarrh. Stomach. stimulant, aphrodis. carmin. and flavouring agent.
Essential oil.

ALPINIA OFFICINARUM
(Zingiberaceae)

Galangal (Hindi), Gao Liang (Chinese)

A perennial aromatic plant growing to 2 m (6 ft), with white, red lipped flowers and lance-shaped leaves.

Key Constituents:
Volatile oil (about 1%) containing alpha-pinence, cineole, linalool.
Sesquiterpene lactones (galangol, galangin)
Key actions:
Warming digestive tonic
Stimulant

Carminative
Prevents vomiting
Antifungal

ALSTONIA SCHOLARIS
(Apocynaceae)

Fever Bark

Parts used: Stem bark, root bark.
Constituents: The bark of both species contains indole alkaloids. *A. constricta* contains reserpine, a powerful hypotensive.
Medicinal actions and uses: Fever bark has been taken to treat malarial fever (and has been called Australian quinine), but its efficacy against malaria remains unclear. The bark is antispasmodic and lowers blood pressure, and is now used mainly to reduce high blood pressure. Strongly bitter, the bark is also taken to treat diarrhoea.
Echiterine, ditamine, echitamine, echitamidine alkaloids, febrifuge.

ALTERNANTHERA SESSILIS
(Amarnathaceae)

Plant – galact. cholag. febge.
Skin and leaves – used in snake-bite
Young shoots contain protein & iron.

ALTHAEA OFFICINALIS
(Malvaceae)

Marshmallow

Parts used: Root, leaves, flowers.
Constituents: Marshmallow root contains about 37% starch, 11% mucilage, 11% pectin, flavonoids, phenolic acids, sucrose and asparagine.
Medicinal actions and uses: Useful whenever a soothing effect is needed, marshmallow protects and soothes the mucous membranes. The root counters excess stomach acid, peptic ulceration and gastritis. Marshmallow is also mildly laxative and beneficial

Althaea officinalis

ALTHAEA ROSEA
(Malvaceae)
Seeds – demulc. diur. febge.
Roots – astrin. demulc. in dysent.
Flowers – cooling, diur. used in rheumatism.
Seeds contain drying oil.

ALYSICARPUS LONGIFOLIUS
(Fabaceae)
Roots – Subst. for liquorice.

AMANITA PHALLOIDES
(Mushroom)
Can dissolve RBCs and drain off all blood through alimentary canal, causing death like the venom of viper.

AMARANTHUS BLITUM
(Amaranthaceae)
Herb – Cooling, stomach. useful in biliousness, haemorrhagic diathesis, emol.
Plant contain HNO_3, protein and iron.

AMARANTHUS CAUDATUS
(Amaranthaceae)
Plant – Used for purifying blood and in piles and as diur. in strauguary also given in serofula and applied to serofulous sores.

AMARANTHUS TRICOLOR
Plant – Astren. in menor, diarh., dysent. and haemor. from the bowels externally as emol. poultice, as an application in ulcerated conditions of throat, mouth and as a wash for ulcers.
Fatty oils

AMARANTHUS HYPOCHONDRIACUS
(Amaranthaceae)
Amarnath
Parts used: Aerial parts

for many intestinal problems, including regional ileitis, colitis, diverticulitis and irritable bowel syndrome. Taken as a warm infusion, the leaves treat cystitis and frequent urination. Marshmallow's demulcent qualities bring relief to dry coughs, bronchial asthma, bronchial catarrh and pleurisy. The flowers, crushed fresh or in a warm infusion, are applied to help soothe inflamed skin. The root is used in an ointment for boils and abscesses, and in a mouthwash for inflammation. The peeled root may be given as a chewstick to teething babies.
Herbal preparations as an emollient, demulcent, anti tussive and expectorant.

Constituents: Amaranth contains tannins, including a red pigment used to dye foods and medicines.

Medicinal actions and uses: Amaranth is an astringent herb that is used primarily to reduce blood loss and to treat diarrhoea. A decoction of amaranth is taken as a remedy in cases of heavy menstrual bleeding, excessive vaginal discharge, diarrhoea & dysentry. It is also used as a gargle to soothe inflammation of the pharynx and to hasten the healing of mouth ulcers.

AMMANIA BACCIFERA
(Lythraceae)

Leaves – acrid. used to raise blisters, in rheum. pains, fevers and as rubft., in skin diseases.

AMMI MAJUS
(Umbelliferae)

BISHOP'S WEED

Parts used: Seeds

Constituents: The seeds contain furanocoumarins (including bergapten), flavonoids and tannins.

Medicinal actions and uses: Bishop's weed produces strongly aromatic seeds. In an infusion or as a tincture, they calm the digestive system. They are also diuretic, & like visnaga, have been used to treat asthma and angina. Bishop's weed reputedly helps treat patchy skin pigmentation in vitiligo. It has also been used for psoriasis.

AMMI VISNAGA
(Umbelliferace)

VISNAGA, KHELLA

An erect annual growing to 1 m (3 ft.) with leaves divided into wisps and cluster of small white flowers.

Key Constituents:
Khellin (1%)
Visnagin
Khellol glycoside
Volatile oil (0.2%)
Flavonoids
Sterols
Key actions:
Antispasmodic, Anti-asthmatic, Relaxant

AMMOMUM SUBULATUM
(Zingiberaceae)

Seeds – Stomach. useful in neuelogia, used in gonor. as aphrodisiac. antid. to scorpion sting and snake-bite.
Oil from seeds – arom. stim. stomach. applied to eye lids to allay inflammation.
Essential oil.

AMORPHOPHALLUS CAMPANULATUS
(Araceae)

Tuber – Stomach. tonic. restor. carmin. in piles and dysent. when fresh acts as an acrid stim. and expect. and much used in rheumatism. Enzyme.

ANACARDIUM OCCIDENTALE
(Anacardiaceae)

CASHEW

Parts used: Nuts, leaves, bark, root, gum.

Constituents: The gum contains anacardic acid, which is bactericidal, fungicidal, and kill worms and protozoa, phenol cardol, anacardein.

Medicinal actions and uses: Though many parts of the plant are used medicinally, cashew nut is chiefly a food – after removal of its toxic lining. The nut is highly nutritious, containing

45% fat and 20% protein. The leaves are used in Indian and African herbal medicine for toothache and gum problems, and in West Africa for malaria. The bark is used in Ayurvedic medicine to detoxify snake bite. The roots are purgative. The gum is applied externally in cases of leprosy, and for corns and fungal conditions. The oil between the outer and inner shells of the nut is caustic and causes an inflammatory reaction even in small doses. In folk medicine in the tropics, the oil is used very sparingly to eliminate warts, corns, ringworms and ulcers.

Root – Purg.
Kernel – nutri. demulc. emol.
Fruit – antidiarrhoel.

ANACYCLUS PYRETHRUM
(Compositae)

PELLITORY

Parts used: Root, essential oil.
Constituents: Pellitory contains anacycline, inulin and volatile oil.
Medicinal actions and uses: Pellitory root is taken as a decoction or chewed to relieve toothache and increase saliva production. The decoction may also be used as a gargle to soothe sore throats. In Ayurvedic medicine, the root is considered tonic and is used to treat paralysis and epilepsy. The diluted essential oil is used in mouthwashes and to treat toothache.

ANAGALLIS ARVENSIS
(Primulaceae)

SCARLET PIMPERNEL

Parts used: Aerial parts.
Constituents: The herb contains saponins (including anagalline), tannins and cucurbitacins. The latter

are cytotoxic (damaging to cells), cyclamin.
Medicinal actions and uses: Little used by medical herbalists today, scarlet pimpernel has diuretic, sweat-inducing and expectorant properties. As an expectorant, it was used to stimulate the coughing up of mucus and help recovery from colds and flu. It has been used to treat epilepsy and mental problems for 2,000 years but there is little evidence to support its efficacy.
Herb – in gout, cerebral affections, hydrophobia, leprosy, dropsy, epilepsy, marina, as fish poison, snakebite.

ANAMIRTA COCCULUS
(Menispermaceae)

COCCULUS

Parts used: Leaves, fruit.
Constituents: Cocculus contains picrotoxin (up to 5%) and alkaloids. Picrotoxin is a very powerful poison and nerve stimulant. Anamirtin, cocculin.
Medicinal actions and uses: Cocculus fruit is sold commercially as a remedy for parasites. It is mainly applied externally to kill parasites such as lice, and it is also used to treat other skin afflictions. In Ayurvedic medicine, cocculus fruit is classified as astringent, antifungal and anthelmintic (de-worming) and is used for skin ulcers and fungal conditions such as ringworm. The plant is so toxic that it is rarely taken internally; however, it has been used in this way in Indian folk medicine, to contract the uterus after childbirth. The herb is also used in homeopathic medicine as a remedy for heart conditions.

ANANAS COMOSUS
(Bromeliaceae)

PINEAPPLE

Parts used: Fruit, leaves.

Constituents: Pineapple fruit contains bromelain, a protein-splitting enzyme that aids digestion. It has significant levels of vitamins A and C.

Medicinal actions and uses: The sour, unripe fruit improves digestion, increases appetite and relieves dyspepsia. In Indian herbal medicine, it is thought to act as a uterine tonic. The ripe fruit cools and soothes, and is used to settle wind and reduce excessive gastric acid. Its significant fibre content makes it useful in relieving constipation. The juice of the ripe fruit is both a digestive tonic and a diuretic. The leaves are considered to be useful in encouraging the onset of menstrual periods and easing painful ones.

AMPELOCISSUS ARNOTTIANA
(Vitaceae)

Juice of Root – With kernel of coconut, employed as depurative and aper. used as alter. in form of decoct. considered a purifier of blood and diuret. rendering the secretion healthy.

ANAPHALIS NEELGERRYANA
(Compositae)

Leaves – bruised and applied to wounds and cuts as a plaster.

ANASTATICA HIEROCHUNTICA
(Brassicaceae)

Used in difficult labour

ANACHUSA OFFICINALIS
Boraginaceae (Alkanet)

The flowering stems, sometimes also the roots and flowers, are used medicinally. The constituents include silicic acid, the alkaloids cynoglossine and consolidine, mucilage and pigments (anthocyanins). These substances give Alkanet an expectorant action and it is thus used in herbal medicine to treat coughs, bronchitis and other chest and throat infections. The tender young leaves are a rich source of vitamin C and they can be eaten raw in salads or cooked like spinach. The plant's rough surface may, however, irritate the skin.

Anachusa officinalis

ANDIRA ARAROBA
(Leguminosae)

Used in Ringworm
Chrysophanic acid

ANDROGRAPHIS PANICULATA
(Acanthaceae)

Plant – febge. tonic. altern. anthelm. useful in debility. dysen. dyspep.
Infusion of plant – in fever
Roots and leaves – febge, stomach. tonic. alter. anthelm.
Kalmaghin, bitter principle andrographolide.

ANEMONE PULSATILLA
(Ranunculaceae)

PASQUE FLOWER

Parts used: Dried aerial parts.
Constituents: Pasque flower contains the lactone protoanemonin (which on drying forms anemonin), triterpenoid saponins, tannins and volatile oil.
Medicinal actions and uses: Pasque flower is less commonly used now in herbal medicine than in the past, though it is still considered a valuable remedy for cramping pain, menstrual problems and emotional distress. It is considered a specific treatment for spasmodic pain of the reproductive system, both male and female, and is given quite frequently for premenstrual tension and period pain, especially when these are accompanied by nervous exhaustion. In France, it has traditionally been used for treating coughs and as a sedative for sleep difficulties. Pasque flower is also used to treat eye problems such as cataracts. The fresh plant is not used because it is strongly irritant. Pasque flower is one of the most commonly used of all homeopathic remedies.

ANEILEMA NUDIFLORUM
(Commelinaceae)

Herbage – cooked in oil, employed in treatment of leprosy.

ANEILEMA SCAPIFLORUM
(Commelinaceae)

Roots – astrin. tonic, in snake-bite fantile convulsions, asthma and incontinence of urine; dried, powdered and mixed with sugar aphrodis. and with juice of TULSI leaves in spermatorrhoea.

ANEMONE OBTUSILOBA
(Ranunculaceae)

Roots – mixed with milk and given internally for contusions, and used externally as a blistering agent.
Anemonin

ANETHUM GRAVEOLENS
(Umbelliferae)

DILL

Parts used: Seeds, essential oil, leaves.
Constituents: Dill seeds contain upto 5% volatile oil (about half of which is carvone), flavonoids, coumarins, xanthones and triterpenes.
Medicinal actions and uses: Dill has always been considered a remedy for the stomach, relieving wind and calming the digestion. Dill's essential oil relieves intestinal spasms and griping and helps to settle colic, hence it is often used in gripe water mixtures. Chewing the seeds improves bad

Anethum graveolens

Angelica archangelica

breath. Dill makes a useful addition to cough, cold and flu remedies, and is a mild diuretic. Like caraway (*Carum carvi*) it can be used with antispasmodics, such as cramp bark (*Viburnum opulus*) to relieve period pain. Dill increases milk production, and when taken regularly by nursing mothers, helps to prevent colic in their babies.

Carminative, stomachic, tranquilliser, promotes milkflow in nursing mothers. Relieves colic and to induce sleep Dill is extensively used in food industry as a flavouring agent.

ANGELICA ARCHANGELICA
(Umbelliferae)
ANGELICA

Parts used: Root, leaves, stems, seeds.

Constituents: Angelica root contains a volatile oil (consisting mainly of beta-phellandrene), lactones and coumarins. An extract of the root has been shown to be anti-inflammatory, coumarins, glycosides organic acids, bitter compounds, tannins.

Medicinal actions and uses: Angelica is a warming and tonic remedy, having a role to play in a wide range of illnesses. All parts of the palnt will help relieve indigestion, wind and colic. Angelica can also be used in cases of poor circulation, as it improves blood flow to the peripheral parts of the body. It is considered a specific treatment of Buerger's disease, a condition that narrows the arteries of the hands and feet. By improving blood flow and stimulating the coughing up of phlegm, angelica's warm, tonic properties bring relief from bronchitis and debilitating conditions affecting the chest. For respiratory conditions, the roots are most commonly used, but the stems and seeds may be employed as well. Tonic, carminative, stomachic and antispasmodic loss of appetite, flatulence, brochial catarrh. Distilled oil is used in perfumery.

ANGELLICA DAHURICA
(Umbelliferae)

BAI ZHI

Parts used: Root
Constituents: Volatile oil and coumarins, imperatorin, marmesin and phellopterin.
Medicinal actions and uses: Headaches, aching eyes, nasal congestion and toothache warming and tonic; cures sores and boils.

ANGELICA SINENSIS
(Umbelliferae)

CHINESE ANGELICA, DAND CUT (CHINESE)

A stout, erect perennial growing to 2 m, with large bright green leaves and hollow stems.
Key **Constituents:**
Coumarins
Volatile oil (butylidine phthalide, ligustilide, sesqiter-penes, carvacrol)
Vitamin B_{12}
Beta-sitosterol
Key Actions:
Tonic
Blood tonic
Antispasmodic
Sedative
Promotes menstrual flow

ANISOCHILUS CARNOSUS
(Labiatae)

Plant – stim. expect. useful in cough of children

Juice of fresh leaves – cooling, mixed with sugar-candy given for coughs and colds.
Essential oil

ANISOMELES MALABARICA
(Labiatae)

Infusion of leaves – in colic and dyspep. catar. affections and intermittent fevers, given to children in colic, dyspep. and fever arising from teething.
Plant – scorpion sting and snake-bite. Decoction of plant of essential oil distilled from leaves-used externally in rheumatism alkaloid and essent. oil.

ANNONA MURICATA
(Annonaceae)

Fruit – antiscor.
Seeds – emetic. astrin. used as fish poison, insecticidal.
Leaves contain essential oil, muricine and muricinine alkaloids from bark.

ANNONA RETICULATA
(Annonaceae)

Bark – astrin.
Fruit – anthelm. antidysen.
Leaves and Seeds – insecticidal
Bark-alkaloid anonaine.

ANNONA SQUAMOSA
(Annonaceae)

CUSTARD APPLE

Parts used: Leaves, bark, fruit, seeds.
Constituents: Custard apple contains fruit sugars and mucilage.
Medicinal actions and uses: In the West Indies, the young shoots are used with peppermint (Metha × piperita) to relieve colds and chills. In Cuban medicine, the leaves are taken to reduce uric acid levels. The leaves, bark

and unripe fruit are all strongly astringent and are used to treat diarrhoea and dysentery. The crushed seeds are mixed with an inert powder and employed as an insecticide.

ANOGEISSUS LATIFOLIA
(Combretaceae)
Bark – bitter. astring.
Plant – scorpion-sting and snake-bite
Tannin, gum like gum arabic.

ANTENNARIA DIOICA
Asteraceae (Mountain Everlasting)

The flowerheads are used medicinally. The coloured flowers have a stronger action than the white ones and they are separated from the others before drying. The dried herb has a pleasant aroma and a bitter taste. The constituents include essential oils, tannins, bitter compounds and organic pigments which give Mountain Everlasting antidiarrhoeal, mild diuretic, expectorant and choleretic properties. In herbal medicine it is used for gastrointestinal infections, bilious conditions and bronchitis. It is also an ingredient of tea mixtures.

ANTHOCEPHALUS INDICUS
(Rubiaceae)
Bark – Tonic, febge, astrin. in snake-bite.
Decoction of leaves – used as gargle in cases of aphthae and stomatitis. Cinchotannic acif like principle.

ANTHOXANTHUM ODORATUM
(Poaceae)

The grass is specially provocative of hay fever and hay asthma, a medicinal tincture from it is sniffed well into the nose and throat for immediate relief in attack.

ANTHEMIS COTULA
(Compositae)

MAYWEED, STINKING MAYWEED

Parts used: Flowers, leaves.
Constituents: Mayweed contains sesquiterpene lactones (including anthecotulide).
Medicinal actions and uses: Although it is similar in appearance to Roman and German chamomile, mayweed is far less effective as a medicine. It has been used as an antispasmodic and to induce menstrua-

Antennaria dioica

tion, and was traditionally employed for supposedly hysterical conditions relating to the uterus. It is rarely used in contemporary herbal medicine.

ANTHRISCUS CEREFOLIUM
(Umbelliceae)

CHERVIL

Parts used: Aerial parts.
Constituents: Chervil contains a volatile oil, coumarins and flavonoids.
Medicinal actions and uses: Chervil is a good remedy for settling the digestion. It is also used to "purify the blood" and to help lower blood pressure, as well as being considered a diuretic. Juice from the fresh plant is applied to various skin conditions, including wounds, eczema and abscesses.

ANTIARIS TOXICARIA

Sap – arrow poison.
Seeds – febge, in dysen.
Glucosides, α-antiarin, β-antiarin, γ-antiarin, antiaresin, toxicarin.
Possesses strong digitalis-like action on the heart.

ANTHYLLIS VULNERARIA
Fabaceae (Kidney Vetch)

The flowers are used medicinally but old ones are not dried because they turn brown and disintegrate. The constituents include tannins, saponins, mucilage, sugars, and organic pigments (flavonoids), and these give the plant astringent, antiseptic and laxative actions. An infusion is used in herbalism as a general tonic: externally a decoction is used in compresses or bath preparations for treating inflamed wounds, ulcers and eczema, and in gargles and mouth-

Anthyllis vulneraria

washes. It can be used as a substitute for ordinary tea mixed with the leaves of Wild Strawberry (*Fragaria esca*) and Raspberry (*Rubus daeus*) and the flowers of Blackthorn (*Prunus spinosa*).

APHANAMIXIS
POLYSTACHYA
(Meliaceae)

Bark – astrin. used in spleen and liver diseases, tumours, abdominal complaints.
Seed oil – used in liniment in rheumatism.

APAHNES ARVENSIS
(Rosaceae)

PARSLEY PIERT

Parts used: Aerial parts.
Constituents: Parsley piert contains tannins.
Medicinal actions and uses: Astringent, diuretic and demulcent, parsley

piert is used to treat kidney and bladders problems, and is frequently used in the treatment of bladder stones (gravel), which cause irritation and obstruct urine flow. Best taken in an infusion, the herb is also a useful remedy for cystitis and recurrent urinary infections. It is combined with marshmallow (*Althacea officinalis*) in cases where a demulcent action is needed.

APIUM GRAVEOLENS
(Umbelliferae)
CELERY, SMALLAGE

A biennial with a ridged shiny stem, glossy leaves and small flowers, growing to about 50 cm.

Key Constituents:
Volatile oil (1.5-3%) containing linonene (60-70%) phthalides and beta-selinene.
Coumarins
Furancomarins (bergapten)
Furanoids (aplin)

Key Actions:
Antirheumatic
Carminative
Antispasmodic
Diuretic
Lowers blood pressure
Urinary antiseptic

APOROSA LINDLEYANA
(Euphorbiaceae)
Decoction of Root – given to jaundice, fever, headache, seminal loss and insanity.

AQUILARIA AGALLOCHA
(Thymelaeaceae)
Wood – Stim. carm. tonic. aphrodis. astrin. in diarrah. and vomiting. in snake bite.
Essential oil.

ARACHIS HYPOGAEA
Fruit and oil – astrin. to the bowels.
Unripe nuts – lactag.
Oil – aper. emol. used as a substitute of olive oil.
Nutmeal – arachin, con-arachin, fat, protein, vitamins B_1 and B_2, nicotinic acid, vitamin E, vita B_C (pyridoxin), lecithin.

ARALIA RACEMOSA
(Araliaceae)
AMERICAN SPIKENARD
Parts used: Root.
Constituents: American spikehead contains a volatile oil, tannins and diterpene acids.
Medicinal actions and uses: Many of American spikenard's current uses come directly form Native American precedents. The herb encourages sweating, is stimulant and detoxifying. It is taken for rheumatism, asthma and coughs. Applied externally as a poultice, American spikenard is used to treat a number of different skin conditions, including eczema.

ARBUTUS UNEDO
(Ericaceae)
STRAWBERRY TREE
Parts used: Leaves, fruit.
Constituents: Strawberry tree contains up to 2.7% arbutin, methylarbutin and other hydroquinones, a bitter principle, and tannins. Arbutin is powerfully antiseptic in the urinary system.
Medicinal actions and uses: Strawberry tree is valued as an astringent and antiseptic herb. Its antiseptic ac-

tion within the urinary tract makes it a useful remedy for treating cystitis and urethritis. Strawberry tree's astringent effect has been put to use in the treatment of diarrhoea and dysentary. Like many other astringent plants, it makes a gargle that is helpful for sore and irritated throats.

ARCTIUM LAPPA
(Compositae)

Burdock, Nio Bang Zi (Chinese)

A biennial, with stems that grow to 1.5 m, reddish purple flowerheads and hooked bracts.

Key Constituents:

Bitter glycosides (artiopicrin)
Flavonoids (arctin)
Tannins, mucilage, resin
Polyacetylenes
Volatile oil

Inulin (up to 45%)
Sesquiterpenes

Key Actions:

Cleansing
Mild diuretic
Antibiotic
Antiseptic
Diaphoretic, hypoglycaemic, chloretic used in the mixtures. Decoction used externally for bathing wounds, ulcers and eczema.

ARCTIUM TOMENTOSUM
Asteraceae (Wolly Burdock)

The roots of one-year or over-wintering plants, collected before flowering, are used medicinally. The pharmacological investigation of burdocks is not yet complete, but the chemical composition of Wolly Burdock's seems to be the same as Greater Burdock's –

Arctium lappa

Arctium Tomentosum

both contain a large amount of inulin. The presence of inulin – a polysaccharide composed of units of fructose – gives burdocks hypoglycaemic properties. Woolly Burdock is mostly used in herbalism to treat skin diseases and as a hair oil. Sometimes the fresh leaves, fresh root or just the juice from the root are used – they promote bile secretion, urine flow & sweating.

ARCTOSTAPHYLOS UVA-URSI
(Ericaceae)

Uva-Ursi, Bearberry

Parts used: Leaves, berries.
Constituents: The leaves of uva-ursi contain hydroquinones (mainly arbutin, up to 17%), tannins (up to 15%), phenolic glycosides and flavonoids. Arbutin and other hydroquinones have an antiseptic effect in the urinary tract.
Medicinal actions and uses: Uva-ursi is one of the best natural urinary

Arctostaphylos uva-ursi

antiseptics. It has been used extensively in herbal medicine to disinfect and astringe the urinary tract in cases of acute and chronic cystitis and urethritis. However, it is not a suitable remedy if there is a simultaneous infection of the kidneys.

ARDISIA SOLANACEA
(Myrsinaceae)

Roots – febge. in diarh. rheumatism.

ARECA CATECHU
(Arecaceae)

Nut – aphrodis. useful in urinary disorders, astrin. anthelm. nerve tonic, emmen. in veterany medicine as verminfuge for tapeworm in snakebite.
Choline, isoguvocine, alkaloids-arecaine, arecaidine, arecoline, guvacine, α-catechin, arecodine, arecaine, action on heart and respiration. Nicotine-like action.

ARENARIA RUBRA
(Carophyllaceae)

Sandwort, Sand Spurrey

Parts used: Aerial parts.
Medicinal actions and uses: Sandwort is a diuretic herb that is thought to relax the muscle walls of the urinary tubules and bladder. Sandwort is most commonly taken in the form of an infusion to treat kidney stones, acute and chronic cystitis, and other conditions of the bladder.

ARENARIA SERPHYLLIFOLIA
(Caryophyllaceae)

Herb – used in china for bladder diseases, considered valuable for calculus troubles and acute and chr. cystitis.

ARENGA OBTUSIFOLIA
(Arecaceae)

Plant – used in fish poison
Fruits – anticoagulant

ARENGA PINNATA

Root – in bronch. stomach.
Juice of outer fleshy covering of the fruit – corrosive, fish poison.

ARGEMONE MEXICANA
(Papaveraceae)

MEXICAN POPPY

Parts used: Aerial parts, latex, seeds.
Constituents: Mexican poppy contains isoquinoline alkaloids similar to those in the opium poppy (*Papaver somniferum*).
Medicinal actions and uses: The fresh latex of Mexican poppy contains protein-dissolving constituents, and is used to treat warts, cold sores and blemishes on the lips. The whole plant acts as a mild painkiller. An infusion of the seeds – in small quantities – is used in Cuba as a sedative for children suffering from asthma. In greater quantities, the oil in the seeds is purgative. The flowers are expectorant and are good for treating coughs and other chest conditions.
Yellow juice of plant – for dropsy, jaundice, and cutaneous affections.
Oil – purgative, skin infections.
Seeds – antidote to snake poison.
Alkaloids – berberine, protopine.

ARGYREIA AGGREGATA
(Convolvulaceae)

Leaves – made into paste applied externally in cough and quinsy.

ARGYREIA CUNEATA
(Convolvulaceae)

Leaves – used in diabetes. Leaves contain glycoside; oral administration of a milk extract of the leaves for 3-5 days brings about a significant remission of the characteristic symptoms of diabetes.

ARGYREIA SPECIOSA
(Convolvulaceae)

Root – alter. tonic, useful in rheumatism and diseases of the nervous system.
Leaves – antiphl. used as emol. poultices for wounds and externally in skin diseases.
Fatty acid.

ARIKURYROBA SCHIZOPHYLLA
(Arecacheae)

Juice of unripe fruit – used for inflam. of eyes in Brazil.

ARISAEMA CONSAGUINEUM
(Araceae)

TIAN NAN XING

Parts used: Dried rhizome.
Constituents: *Tian nan xing* contains triterpenoid saponins and benzoic acid.
Medicinal actions and uses: In Chinese herbal medicine, *tian nan xing* is thought to encourage the coughing up of phlegm. The dried rhizome is used principally for chest problems. When prescribed internally it is always combined with fresh ginger root (*Zingiber officinale,*). The fresh rhizome is only ever used externally, for ulcers and other skin conditions.

ARISAEMA SPECIOSUM
(Araceae)

Root – antid. to snake poison
Tuber – given to sheep as remedy for colic and to kill worms which infest cattle.

ARSTIDA ADSCENSCIONIS
(Poaceae)

In Madagascar an ointment consisting of lard and the ashes of flowers is used topically for itch and Ringworm.

ARISTOLOCHIA BRACTEATA
(Aristolochiaceae)

Plant – purg. anthelm. emmen.
Juice of leaves – applied to foul and neglected ulcers.
Bruised leaf – mixed with castor oil applied to eczema
Decoction of Root – used for expelling roundworms.
Volatile substance and alkaloid.

ARISTOLOCHIA CLEMATITIS
(Aristolochiaceae)

Parts used: Root, aerial parts.
Constituents: Birthwort contains aristolochic acids, a volatile oil and tannins aristolochic acid aristolochine alkaloid.
While stimulating white blood cell activity, aristolochic acid is also carcinogenic and damaging to the kidneys.
Medicinal actions and uses: Little used today, birthwort was formerly used to treat wounds, sores and snake bite. It has been taken after childbirth to prevent infection and is also a potent menstruation-inducing herb and a (very dangerous) abortifacient. A

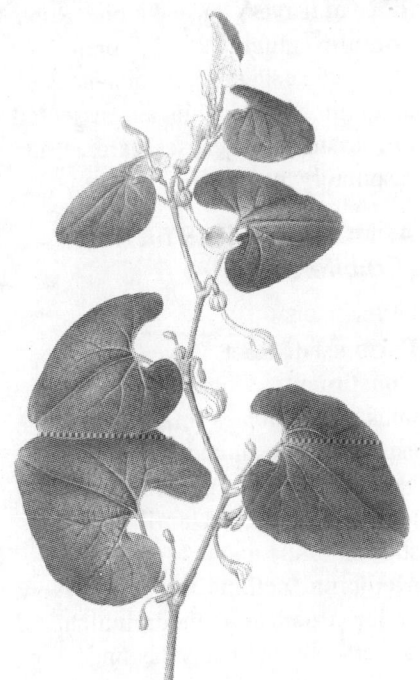

Aristolochia clematitis

decoction was taken to encourage healing of ulcers. Birthwort has also been used for asthma and bronchitis. Aristolochic acid is cytotoxic and in sufficient doses it can cause haemorrhage, miscarriage in pregnant women, permanent damage to the liver and kidneys, and even cardiac and respiratory failure. Birthwort is also anti-inflammatory and a decoction has been used externally to treat wounds, ulcers, eczema and other skin complaints, there are, however, many safer remedies and its use in these ways is not recommended.

ARISTOLOCHIA INDICA
(Aristolochiaceae)

Root – tonic, stim. emmen. emetic, in fevers. in powder form given in honey for leucoderma.

Juice of leaves – in snake-bite. Roots contain glucd. bitter principle glucosidic called isoaristolochic acid, allantoin, aristolochin alk. essential oil. aristolochin causes cardiac and respiratory paralysis.

ARMORACIA RUSTICANA
(Cruciferae)

HORSERADISH

Parts used: Root, leaves.

Constituents: Horseradish root contains glucosilinates (mainly sinigrin), asparagine, resin and vitamin C. On being crushed, sinigrin produces allyl isothiocyanate, an antibiotic substance.

Medicinal actions and uses: Now undervalued as a medicinal herb, horseradish has many healing properties. It strongly stimulates the digestion, increasing gastric secretions and appetite. It is a good diuretic and promotes perspiration, making it useful in fevers, colds and flu. It is also expectorant and mildly antibiotic, and can be of use in both respiratory and urinary tract infections. A sandwich of freshly grated root is a home remedy for hay fever. Externally, a poultice of the root can soothe chilblains.

ARNICA MONTANA
Asteraceae (Arnica)

All parts of the plant are of medicinal value, but the flowerheads, and sometimes the rhizomes, are mostly used. Only the ray – and disc-florets – without the involucre and receptacle – are processed. Among the constituents of the florets are traces of an essential oil, organic pigments (carotenoids),

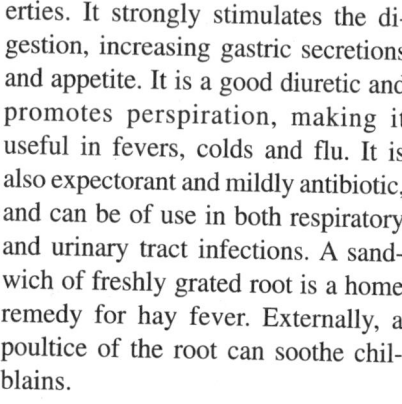

Armoracia rusticana

Arnica montana

the bitter compound arnicin, the saponin arnidendiol, arnisterol and the flavonoid glyosides isoquercetin and astragalol. The rhizomes contain tannins, up to 6.3 per cent of an essential oil and resin. Arnica has tonic, anti-inflammatory and vasodilating properties but it is also a severe irritant of the internal organs and is now rarely prescribed for internal use. **It should be taken internally only under strict medical supervision.** Externally a diluted infusion or decoction is used for bathing superficial wounds, bruises and sprains and as a component of mouthwashes and gargles. **Repeated external use can cause skin irritation**. In homeopathy a tincture is used for trauma.

ARTABOTRYS ODORATISSIMUS (Annonaceae)

In Malay archipelago the decoction of leaves is given for cholera.
Essential oil is perfumery alkaloids artabotrine and shaveoline, toxic.

ARTEMISIA ABROTANUM (Compositae)

SOUTHERNWOOD
Parts used: Aerial parts.
Constituents: Southernwood contains a volatile oil, abrotanin and tannins, absinthol, tannins.
Medicinal actions and uses: Southernwood is a bitter tonic. It strengthens and supports digestive function by increasing secretions in the stomach and intestines. An infusion of southernwood has been given to children as a treatment for worms, but this is not recommended without

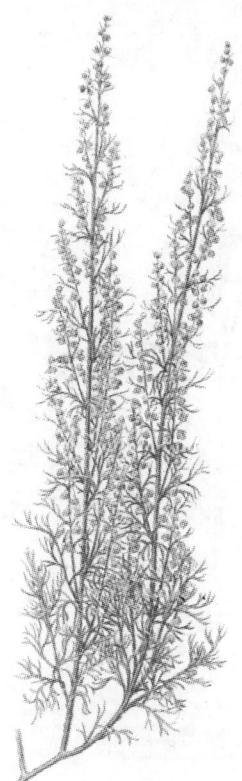

Artemisia abrotanum

professional supervision. Like other *Artemisias*, southernwood stimulates menstruation and is commonly taken to encourage the onset of irregular or absent periods. Stomachic, diuretic, anthelmentic, diaphoretic. Externally used in bath preparations as hair rinse, to treat frost bite and cuts.

ARTEMISIA ABSINTHIUM Asteraceae (Wormwood)

The flowering stems are used medicinally. The constituents include an essential oil with thujone and thujole, a bitter compound (absinthin), organic acids and tannins. The herb has a very bitter taste. It has tonic, stomachic,

Artemisia absinthium

choleretic, carminative, antiseptic and anthelmintic properties. It was once used for many disorders but nowadays it is mostly used, on its own or in tea mixtures, for various digestive upsets. **Taken over a long period, however, Wormwood becomes habit-forming and will eventually cause serious brain damage. It is thus advisable to take Wormwood internally only under the supervision of a qualified medical or herbal practitioner.** The neurotoxic agents are thought to be thujone and thujole. The use of Wormwood in alcoholic beverages is now prohibited by law. Externally a decoction is used as a gargle and in compresses for bruises. The essential oil recovered from the fresh plant is used in homeopathic tinctures.

ARTEMISIA ANNUA (Compositae)

QING HAO, CHINESE WORMWOOD

A perennial growing to about 1 m, with green feathery leaves covered in fine hairs.

Key **Constituents:**
Volatile oil (abrotamine, beta-bourbonene)
Sesquiterpene lactone (artemisinin)
Vitamin A
Key Actions:
Bitter
Reduces fever
Antimalarial
Antibiotic

ARTEMISIA CAPILLARIS (Compositae)

YIN CHEN HAO

Parts used: Aerial parts.
Constituents: *Yin chen hao* contains a volatile oil and coumarins. The volatile oil is antifungal.
Medicinal actions and uses: *Yin chen hao* is an effective remedy for liver problems, being specifically helpful for treating hepatitis with jaundice. Traditional Chinese medicineh holds that it is bitter and cooling, clearing "damp heat" from the liver and gall ducts and relieving fever. *Yin chen hao* is also anti-inflammatory and diuretic. It was formerly applied in the fom of a plaster for headaches.

ARTEMISIA CINA (Compositae)

LEVANT WORMWOOD

Parts used: Flowerheads.
Constituents: Levant wormwood contains santonin (a sesquiterpene

lactone), artemisin and a volatile oil (with up to 80% cinole). Santonin is directly toxic to roundworms and, to a lesser extent, threadworms.

Medicinal actions and uses: Used almost exclusively to expel worms, Levant wormwood is strongly bitter and aromatic and has a tonic and stimulant effect on the digestion. The dried flowerheads are occasionally mixed with honey to disguise their bitterness.

ARTEMISIA DRACUNCULUS
Asteraceae (Tarragon)

Although Tarragon is mainly used for flavouring food, the flowering stems are occasionally used for medicinal purposes. The constituents include an essential oil with estragole and phelandrine, bitter compounds and tannins. Tarragon is used in an infu-

sion, as a powder or tincture to stimulate the appetite, to aid digestion and as a general tonic. The distilled essential oil is used in the manufacture of some toilet preparations.

Tarragon is widely used in the food and canning industries. For home cooking no garden should be without it. Tarragon vinegar is easily prepared by steeping fresh leaves just before flowering in a good, white wine vinegar for two or three weeks, then straining the product into small bottles. Separate leaves can be used in salads, sauces, stews and pickles.

ARTEMISIA VULGARIS
Asteraceae (Mugwort)

The flowering stems are used medicinally. The main constituents are an essential oil, with cineole and thujone, tannins and bitter compounds. Like other species of *Artemisia* Mugwort is used as a general tonic, as an appe-

Artemisia dracunculus

Artemisia vulgaris

tizer and stomachic and as a seasoning, but its action is not as strong. It is a traditional treatment for nervous disorders, insomnia and gynaecological complaints. The essential oil extracted from the fresh leaves is used in preparations as an expectorant, diuretic and anthelminthic. Because Mugwort contains the neurotoxic substance thujone **care should always be taken over the dosage and duration of treatment**.

In cooking Mugwort is a useful condiment for roast meat, especially pork and mutton and fat poultry such as goose and duck. It is an effective moth repellent.

In addition, it can be taken to encourage the elimination of worms. Mugwort also increases bile flow and mildly induces the onset of menstruation.

ARTOCARPUS HETEROPHYLLUS
(Moraceae)

Leaves – used in skin diseases, antid. to snake bite.
Root – used internally in diarh.
Juice of plant – applied to glandular swellings and abscesses to promote suppuration.
Unripe fruit – astrin.
Ripe fruit – laxat.
Wood yields colouring matter morin and cyanomaclurin, bark contains tannin, steroketone and artosterone from latex which has highly androgenic properties.

ARTOCARPUS HIRSUTUS
(Moraceae)

Dry leaves and Juice – together with zedoary and camphor used as application to bubos and swelled testicles. Seeds have fixed oil.

ARUM MACULATUM
Araceae (Lords–and–Ladies)

The rhizome contains the toxic glycoside aronin, also saponins, starch and calcium oxalate, which gives the plant its sharp burning taste. It is diuretic and strongly purgative. **Lords and ladies should never be collected and used for self-medication.** Symptoms of poisoning are severe vomiting, abdominal cramps and diarrhoea.

Arum maculatum

ARUNDO DONAX
(Poaceae)

Decoction of rhizomes – emol. diur, said to stimulate menstrual discharge and diminish the sectretion of milk, alkaloid gramine, doaxine donaxarine, raises blood pressure.

ASARUM EUROPAEUM
Aristolochiaceae
(Asarabacca)

The rhizome and the leaves have medicinal actions. The principal constituents are an essential oil from which asarone (known as Asarabacca camphor) is crystallised, also starch, resin and flavonoids. The dried herb has a bitter taste. It is a strong emetic, diuretic and purgative. A pinch of the powdered herb in snuff mixtures induces sneezing and a copious flow of mucus. **It is dangerous to use Asarabacca internally except under strict medical supervision**: strong doses can cause haemorrhaging and in pregnant women, miscarriage.

Asarum europaeum

ASCLEPIAS CURASSIVA
(Asclepiadaceae)

Root – emetic, purg., remedy in piles and gonor.
Juice of leaves – anthelm. sudorific. for arresting haemorrhages and gonor.
Plant – in phthisis, poisonous. glucd. asclepuadin, vincetoxin.

ASCLEPIAS TUBEROSA
(Asclepiadaceae)
Pleurisy Root
Part used: Root.
Constituents: Pleurisy root contains cardenolides and flavonoids. It is oestrogenic.
Medicinal actions and uses: Though its most specific usage is relieving the pain and inflammation of pleurisy, pleurisy root has other applications. It is useful for hot, dry and tight conditions in the chest. It promotes the coughing up of phlegm, reduces inflammation and in addition, helps reduce fevers by stimulating perspiration. The root is also taken for the treatment of chronic diarrhoea and dysentery.

ASPARAGUS ADSCENDENS
(Liliaceae)
Roots – demulc. galact. tonic. useful in diarh. and general debility. Asparagin.

ASPARAGUS GONOCLADUS
(Liliaceae)
Root – aphrodis. boiled with oil applied to cutaneous diseases, given in gonor.

ASPARAGUS OFFICINALIS
(Liliaceae)
Asparagus
Parts used: Root, shoots.

Constituents: Asparagus contains steroidal glycosides (asparagosides), bitter glycosides, asparagine and flavonoids. Asparagine is a strong diuretic.

Medicinal actions and uses: Asparagus is a strong diuretic that is useful for a variety of urinary problems, including cystitis. It is also useful in the treatment of rheumatic conditions, helping to hasten the "flushing" of waste products accumulated in the joints out of the body in the urine. Asparagus is also bitter, mildly laxative and sedative.

ASPARAGUS RACEMOSUS (Liliaceae)

Root – refrig. demulc. diur. aphrodis. antis. alter. anti diar. antidysent. galact. and as demulc. in veternary medicine.

ASPHODELUS TENUIFOLIUS (Liliaceae)

Seeds – diur. applied externally to ulcers and inflammed parts.

ASPERULA ODORATA (Rubiaceae)

Sweet Woodruff

Parts used: Aerial parts.
Constituents: Sweet woodruff contains iridoids, coumarins (0.6%), tannins, anthraquinones and flavonoids. The flavonoids act on the circulation and are diuretic.
Medicinal actions and uses: Sweet woodruff is considered tonic, with significant diuretic and anti-inflammatory effects. Its coumarin and flavonoid constituents make it helpful in treating varicose veins and phlebitis. It has been used as an antispasmodic, and it is given to children and adults for insomnia.

ASPIDOSPERMA QUEBRACHO-BLANCO (Apocynaceae)

Quebracho

Part used: Bark.
Constituents: Quebracho contains indole alkaloids (including yohimbine) and tannins.
Medicinal actions and uses: With its antispasmodic effect on the bronchial tubes, quebracho is used threapeutically to treat asthma and emphysema. It is also a tonic and reduces fever. This herb is astringent and has been used externally on wounds and burns.

ASPLENIUM ADIANTUM-NIGRUM (Polypodiaceae)

Plant – bitter. diur. laxat. useful in ophthalmia, diseases of spleen, jaundice, produces sterility in women.
Decoct. or syrup of fronds – used in Europe as expect. pectoral and emmen.

ASRANTIA MAJOR Apiaceae (Great Masterwort)

The rhizomes and flowering stems have medicinal action. Their main constituent is an essential oil that acts as a stomachic. In herbal medicine the dried herb is used in an infusion or as a powder to promote the flow of digestive juices and thus stimulate the appetite. Great Masterwort is also included in diuretic tea mixtures.

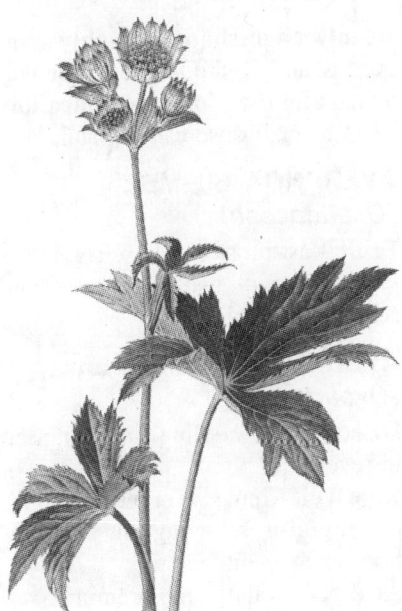

Asrantia major

ASTER AMELLUS
(Asteraceae)

The chinese use the *roots* for coughs, pulmonary affections, in haemorr. and malaria.

ASTERACANTHA LONGIFOLIA
(Acanthaceae)

Decoct. of roots – diur.
Seeds – given for gonor. and with milk sugar in spermatorrhoea.
Leaves, roots and Seeds – diur. employed for jaundice, dropsy, rheumatism, anasarca and diseases of the urinary tract.
Phytosterol, essential oil, mucilage, pot. salts, semi-drying oil, diastase, hipase, and protease. alkal.

ASTRAGALUS MEMBRANACEUS
(Leguminosae)

ASTRAGALUS, MILK VETCH, HUANG QI (CHINESE)

A perennial growing to 40 cm, with hairy stems and leaves divided into 12-18 pairs of leaflets.

Key Constituents:
Asparagine
Calcyosin
Formononetin
Astragalosides
Kumatakenin
Sterols

Key Actions:
Adaptogenic
Immune stimulant
Diuretic
Vasodiator
Antiviral

ATLANTIA MONOPHYLLA
(Rutaceae)

Oil from berries – used externally in chronic rheumatism and paralysis.
Root – antisp. stim.
Leaves – in snake-bite
Essential oil.

ATRACTYLODES MACROCEPHALA
(Compositae)

BAI ZHU

Part used: Rhizome.
Constituents: *Bai zhu* contains a volatile oil (0.35-1.35%), which includes atractylol, and the lactones atractylenolide II and III. Atractylol has a liver protective activity.
Medicinal actions and uses: *Bai zhu* has traditionally been used as a tonic, building *qi* and strengthening the spleen. The rhizome has a sweet, pungent taste, and is used to relieve fluid retention, excessive sweating, and digestive problems such as diarrhoea

and vomiting. Combined with Baical skullcap (*Scutellaria baicalensis*,) it is employed to prevent miscarriage.

ATROPA BELLADONNA
Solanaceae (Deadly Nightshade)

The dried leaves and root are used medicinally. They include various potent alkaloids, such as hyoscyamine, atropine, belladonnine and scopolamine. These substances particularly affect the parasympathetic nervous system and are used together or in isolated form in proprietary and homeopathic medicines prescribed by medical practitioners as antispasmodics, sedatives and analgesics and as antidotes to various poisons. Atropine is used in ophthalmology.

Atropa belladonna

Deadly Nightshade is highly dangerous and it should never be collected and used for self-medication. Even a small dose can be fatal.

AVERRHOA BILIMBE
(Oxalidaceae)

Fruit – astrin. stomach. refrig. in the form of curry useful in piles and scurvy. Juice **pH 4.47.**

AVERHOA CARAMBOLA
(Oxalidaceae)

Dried fruits – cooling. antiscor. used in fevers.
Ripe fruit – remedy for bleeding piles and useful in relieving thirst and febrile excitement.
Acid pot. oxalate and vitamin A.

AVICENNIA OFFICINALIS
(Verbenaceae)

Bark – astrin.
Root – aphrodis.
Unripe seeds – used as poultice to hasten suppuration of boils and abscesses.
Tannins
Mangrove plant.

AVICENNIA TOMENTOSA
(Verbenaceae)

Root – aphrodis.
Bark – astrin.
Unripe seeds – used as poultice to hasten suppuration of boils and abscesses.
Lapachol.
Mangrove plant.

AVENA SATIVA
(Gramineae)

OATS

Parts used: Seeds, straw (dried stems).

Avena sativa

depressant, gently raising energy levels and supporting an over-stressed nervous system. Oats are used to treat depression and nervous debility, as well as the exhaustion that results from multiple sclerosis, chronic neurological pain and insomnia. Oats are thought to stimulate sufficient nervous energy to help relieve insomnia, and mental fatigue.

Oats are one of the principal herbal aids to convalescence after a long illness. Externally, the grain is emollient and cleansing, and a decoction strained into a bath can help soothe itchiness and eczema.

AZADIRACHTA INDICA
(Melicaeae)

NEEM

Parts used: Bark, leaves, twigs, seeds, sap.

Constituents: Neem contains meliacins, triterpenoid bitters, tannins and flavonoids.

Medicinal actions and uses: Considered a pharmacy in its own right in India, every part of the neem tree may be used medicinally. The bitter, astringent bark is applied as a decoction for haemorrhoids. The leaves are traditionally steeped for malaria, peptic ulcers and intestinal worms. Neem juice (expressed from the leaves), infusion or ointment, is applied externally to ulcers, wounds, boils and eczema. The twigs are used to clean and teeth, firming up the gums and preventing gum disease. Neem oil, expressed from the seeds, is commonly used as a hair dressing. Strongly antifungal and antiviral, it

Constituents: Oats contain saponins, alkaloids, sterols, flavonoids, silicic acid, starch, proteins, (including gluten), vitamins (especially B vitamins) and minerals (especially calcium).

Medicinal actions and uses: Oats are best known as a nutritious cereal, but they benefit the health in many other ways. Oat bran lowers cholesterol, and an oat-based diet may improve stamina, Oats, and oat straw in particular, are tonic when taken medicinally. Nerve tonic.

Oat straw is prescribed by medical herbalists to treat general debility and a wide variety of nervous conditions. The grains and straw are mildly anti-

prevents lice and other infestations. This oil is also used to treat leprosy, and may be used as a vehicle for other active ingredients. The sap is another traditional external remedy for leprosy. The seeds are spermicidal. Used in snake bite and scorpion sting.

Berries – purg. emol. anthelm.

Oil – Stim. antisep. alter. in rheumatism and skin diseases.

AZIMA TETRACANTHA
(Salvadoraceae)

Root – diur. in rheumatism, given in dropsy.

Root Bark – in rheumatism.

Leaves – Stim. given with food as remedy for rheumatism.

Juice of leaves – to relieve cough of phthesis and asthma.

Bark – expect.

BACOPA MONNIERI
(Scrophulariaceae)

WATER HYSSOP, BRAHMI (HINDI)

Parts used: Aerial parts.

Constituents: Water hyssop contains steroidal saponins, including bacosides. Brahmi resembles strychnine, alk. herpestine.

Medicinal actions and uses: In India, water hyssop is used principally for nervous system disorders such as neuralgia, epilepsy and mental illness, but it is also employed for a wide range of other disorders, including indigestion, ulcers, wind and constipation, asthma and bronchitis, and infertility. In China, it is taken as a yang tonic for impotence, premature ejaculation, infertility and rheumatic conditions. In Indonesia, the plant is a remedy for filariasis (a tropical disease caused by worms). In Cuba, water hyssop is used as a purgative, and a decoction of the whole plant is taken as a diuretic and laxative. The expressed juice is mixed with oil and applied as a rub for arthritic pain.

Plant – nerve tonic, used in asthma, epilepsy, insanity, hoarseness, diur. aper.

Skin and leaves – in snake-bite.

BALANITES AEGYPTIACA
(Simarubiaceae)

Bark, unripe fruit and leaves – purg., anthelm.

Seeds – expect. given cough and colic.

Plants – in snake-bite.

Bark – used as anthelm. for cattle and its juice as fish prison.

Saponin,   tetraglucoside   of sapanogenin.

BALIOSPERMUM MONTANUM
(Euphorbiaceae)

Seeds – purg. used externally as stim. and rubft. and in snake-bite

Root – cath. used in dropsy, anasarca and jaundice.

Decoct. of leaves – in asthma.

Oil from seeds – hydrogogne cath. external application in rheumatism.

BALLOTA NIGRA
(Labiatae)

BLACK HOREHOUND

Parts used: Aerial parts.

Constituents: Black horehound contains   diterpenoids,   including

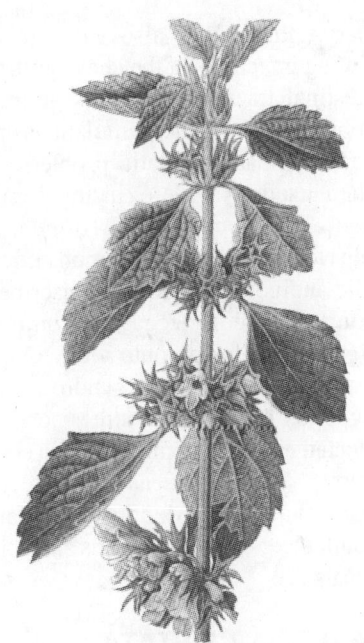

Ballota nigra

marrubiin, which is also a constituent of white horehound (*Marrubium vulgare*;). tannins, essential oils, organic acids and pectin.

Medicinal actions and uses: While it has long been considered a remedy for convulsions, low spirits and menopausal problems, black horehound is not as commonly used today as it was in the past. Authorities differ over whether there is any substance to claim for its earlier applications. The herb is currently used by Anglo-American herbalistts as an anti-emetic-preventing or reducing nausea or vomiting. It is perhaps most useful when nausea arises from disorders of the inner ear (such as Meniere's disease) as opposed to those of the digestive system. Black horehound is thought to be mildly sedative and antispasmodic, and is occasionally

taken for arthritis and gout. Black horehound may be substituted for white horehound, but its medicinal effect is inferior.

BAMBUSA ARUNDINACEA
(Gramineae)

SPINY BAMBOO

Parts used: Root, leaves, sprouts.
Constituents: Spiny bamboo, juice contains high levels of silica.
Medicinal actions and uses: The various parts of spiny bamboo are used in many different ways in Indian and Ayurvedic medicine. The root is considered astringent and cooling, and is used to treat joint pain and general debility. The leaves are used to stimulate menstruation, and, being antispasmodic, to help relieve period pain. The leaves also are used to tone and strengthen stomach function, are taken to expel worms, and have the reputation of being aphrodisiac. The young sprouts are eaten to relieve nausea, indigestion and wind, and a poultice of the sprouts is applied to help drain wounds that have become infected. The juice is rich in silica, and aids in the strengthening of cartilage in conditions such as osteoarthritis and osteoporosis.
Leaves – emmen. used in hematesis and vet. practice. coughs and colds. Tonic, useful in snake bite.
Chemical Constituents: cholin, betain, nuclease, urease, proteolytic enzyme, diastatic, emulsifying enzyme, 'bangsolochin', ta bashir'; young shoots contain cynogenic glucd, and are poisonous, oxalic acid, young shoots benzoic acid, reducing sugar, resins and waxes, young shoots

lethal to mosquito larvae. Sprout juice – HCN, benzoic acid.

BANISTERIOPSIS CAAPI
(Malpighiaceae)

AYAHUASCA

Part used: Bark.

Constituents: Ayahuasca contains beta-carboline alkaloids (including harmine, harmaline and delta-tetrahycroharmine) which stimulate hallucinations.

Medicinal actions and uses: Though known as a powerful hallucinogen, ayahuasca is also a medicine, being used as a remedy to cure a range of diagnosed conditions. However, ayahuasca is taken by the healer rather than by the patient. In the shamanistic societies of the Amazon, ayahuasca allows the healer to communicate with the spirit world where illness arises, interceding on behalf of the ill person and the community to restore health and harmony to all – quite unlike the individualised approach of Western medicine. Beyond its ability to affect mood, the bark is emetic and purgative. At low doses it is used as a mild detoxifier.

BAPTISIA TINCTORIA
(Leguminosae)

WILD INDIGO

Parts used: Roots, leaves.

Constituents: Wild indigo contains isoflavones, flavonoids, alkaloids, coumarins and polysaccharides. The isoflavones are oestrogenic.

Medicinal actions and uses: Wild indigo is a strong antiseptic and immune-stimulant. It is considered particularly effective for upper respiratory infections such as tonsillitis and pharyngitis, and is also valuable in treating infections of the chest, gastrointestinal tract and skin. Its antimicrobial and immune-stimulant properties combat lymphatic problems – when used with detoxifying herbs such as burdock (*Arctium lappa*, it helps to reduce enlarged lymph nodes. Wild indigo is frequently prescribed along with echinacea (*Echinacea angustifolia*) for chronic viral conditions or chronic fatigue syndrome. A decoction of the root soothes sore or infected nipples and infected skin conditions. When used as a gargle or mouthwash, the decoction treats mouth ulcers, gum infections and sore throats.

BARLERIA CRISTATA
(Acanthaceae)

Roots and leaves – used to reduce swelling.
Infusion – given in cough.
Plant – in snake-bite.

BARLERIA PRIONITIS
(Acanthaceae)

Juice of leaf – used in catar. affections of children, which are accompanied by fever and much phlegm.
Dried bark – in cough.
Juice of bark – in anasarca.
Leaves – chewed to relieve toothache.
Paste of root – applied to disperse boils and glandular swellings.

BAROSMA BETULINA
(Rutaceae)

BUCHU

A bushy shrub growing to 2 m, with stemless, slightly leathery leaves dotted with oil glands.

Key **Constituents:**
Volatile oil (1.5-2.5%), including
pulegone, menthone, diesphenol.
Sulphur compounds
Flavonoids (diosmin, rutin)
Mucilage
Key Actions:
Urinary antiseptic
Diuretic
Stimulant
Uterine stimulant

BARRINGTONIA ACUTANGULA
(Lecythidaceae)

Powdered seed – emetic, expect. and
as snuff in headache.
Bark, root and seed – fish poison.
Leaves and roots – bitter tonic.
Root – cooling, aper.
Juice of leaves – in diarh.
Glued saponin, barringtonin, tannin.

BARRINGTONIA RACEMOSA
(Lecythidaceae)

Root – deobstruent, cooling.
Fruit – in cough, asthma and diarh.
Kernels of Drugs – with milk given
in jaundice, and other bilious diseases.
Seeds – in colic and ophthalmia.
Seeds and bark – Vermifuge, fish
poison, tonic and insecticidal.
Glucd, saponin, tannin, toxic princi-
ples.

BASELLA RUBRA
(Basellaceae)

Leaves – demulc. diur. useful in
gonor. and balanites.
Juice of leaves – used in urticaria,
and in cases of constipation particularly in children and pregnant women.

Vitamins A and B, protein, calcium,
iron.

BAUHINIA PURPUREA
(Caesalpiniaceae)

Bark – astrin. in diar.
Root – carmi.
Flowers – laxat.
Tree yields gum; bark contains tan-
nin; seeds contain non-drying oil.

BAUHINIA RACEMOSA
(Caesalpiniaceae)

Gum – used medicinally
Decoct. of leaves – in headaches and
malaria.
Bark – astrin. in diur. and dysent.

BAUHINIA TOMENTOSA
(Caesalpiniaceae)

Decoct. of root bark – given in
inflamm. of liver, anthelmen.
Buds and young flowers – in
dysentric infections.
Fruit – diarh.
Plant – used in snake-bite and scorpion sting.

BAUHINIA VAHLII
(Caesalpiniaceae)

Seeds – tonic, aphrodis.
Leaves – demulc. mucilaginous, gum,
tannins.

BAUHINIA VARIEGATA
(Caesalpiniaceae)

Bark – alter. tonic, astrin. useful in
skin diseases, ulcers in scrofula.
Dried buds – used in dysent. and
piles, diarh. and worms.
Decoct. of root – in dyspep.
Root – antid. to small poison.
Gum, Tannins.

BELAMCANDA CHINENSIS
(Iridaceae)

Roots – aper. resolv. antid. to snake poison.

Pulp of stem – Considered stomach.

Rhizome – important drug in chinese materia medica and chief remedy for tonsillitis and given in chest and liver complaints and added to tonics.

Glucd shekanin isolated from the rhizomes of this plant and tectoridin from *Iris trctorum* are identical. Crystalline glucd. belamcandin isolated from roots.

BELLIS PERENNIS
Asteraceae (Daisy)

The flowerheads are used medicinally. The main constituents are saponins, an essential oil, tannins, mucilage, flavones and a bitter compound, all of which give Daisy astringent and expectorant properties. It has a beneficial effect on gastritis, enteritis and diarhhoea, and infections of the upper respiratory tract. In herbal medicine it is usually used as an infusion. Daisy also makes an attractive addition to tea mixtures. Externally it is used in compresses and bath preparations to treat skin disorders, wounds and bruises. A decoction from the fresh leaves is used for the same purposes.

Young fresh leaves can be eaten raw in salads or added to soups.

BENINCASA HISPIDA
(Cucurbitaceae)

WAX GOURD, PETHA

Parts used: Fruit rind, fruit, seeds.

Constituents: Wax gourd contains, saponins and guardinine.

Medicinal actions and uses: In Chinese herbal medicine, a decoction of wax gourd seeds is used to "drain dampness" and "clear heat". It is given for chest conditions and vaginal discharge. In combination with Chinese rhubarb (*Rheum palmatum*), it is prescribed for intestinal abscesses. In Ayurvedic medicine, the seeds are used to treat coughs, fever and excessive thirst, and to expel tapeworms. The fruit is classified as cooling, diuretic and laxative. It is thought to act as an aphrodisiac and is used for peptic ulceration and debility. In an ancient Indian recipe, the juice from the fruit is mixed with lime juice (*Citrus aurantiifolia*) to prevent or stop bleeding.

BERBERIS AQUIFOLIUM
(Berberidaceae)

ORGEON GRAPE

Part used: Root.

Constituents: Oregon grape contains isoquinoline alkaloids (including berberine, berbamine and hydrastine).

Bellis perennis

These alkaloids are strongly antibacterial and are thought to reduce the severity of psoriasis.

Medicinal actions and uses: Oregon grape is chiefly used for gastritis and general digestive weakness, to stimulate gallbladder function, and to reduce catarrhal problems (mainly of the gut). It also treats eczema, psoriasis, acne, boils and herpes, and skin conditions linked to poor gallbladder function.

BERBERIS VULGARIS
(Berberidaceae)

BARBERRY

Parts used: Stem bark, root bark, berries.

Constituents: Barberry contains isoquinoline alkaloids, including berberine and berbamine. Berberine is strongly antibacterial and amoebicidal and stimulates tannins, resins and organic pigments.

Medicinal actions and uses: Barberry acts on the gallbladder to improve bile flow and ameliorate

conditions such as gallbladder pain, gallstones and jaundice. Its strongly antiseptic property helps amoebic dysentery, cholera and other similar gastro intestinal infections.

The bark is astringent, antidiarrhoeal and healing to the intestinal wall – in short, barberry has a strong, highly beneficial effect on the digestive system as a whole. Like Oregon grape (*B. aquifolium, see preceding entry*) and goldenseal (*Hydrastis canadensis*), barberry helps chronic skin conditions such as eczema and psoriasis. The decoction makes a gentle and effective wash for the eyes, although it must be diluted sufficiently before use.

Rich in vitamin C, sugars and pectin and they make refreshing herbal teas. They relieve biliousness and used as laxative. Disorders of kidney, liver and gall bladder, stomach. antipyret.

BERGENIA LIGULATA
(Saxifragaceae)

Root – tonic used in fever, diar. and pulmonary affections, antiser. bruised and applied to boils and ophthalmia.

Root contains gallic acid, tannic acid, glucose, mucilage, wax. etc.

BERGIA ODORATA
(Elatinaceae)

Used for cleaning teeth and applied to broken bones.

Leaves – rubbed down in water ued as poultice in sores.

BETA VULGARIS
(Chenopodiaceae)

RED BEET, WHITE BEET

Part used: Root.

Berberis vulgaris

Constituents: White beet contains betain, which promotes the regeneration of liver cells, and the metabolism of fat cells. Red beet contain betanin – an anthocyanin similar to those found in red wine – which is partly responsible for red beet's immune-enhancing effect. Zinc, vitamin B_1 & C, iron.

Medicinal actions and uses: White beet acts to support the liver, bile ducts and gall bladder, influencing fat metabolism and helping to lower blood fat levels. Red beet juice is thought to stimulate the immune system. However, it must be taken in very large quantities – at least a litre a day according to one authority. Red beet juice is prescribed by herbalists as part of a cancer-treatment regime.

Seeds – cooling, diaphor.

Betula pendula

BETULA PENDULA
Betulaceae (Silver Birch)

The young leaves are used medicinally. When dried they are aromatic and have a bitter taste. Among the constituents are saponins, tannins, traces of an essential oil, resin and bitter compounds. These substances give Silver Birch mild diuretic and disinfectant properties, but they do not irritate the kidneys. For this reason Silver Birch leaves are a basic ingredient of herbal tea mixtures used for urinary infections and for kidney stones. The herb is also diaphoretic, particularly in combination with flowers of lime (*Tilia*) and it eases rheumatic pain. Externally the leaves make invigorating bath preparations. Dry distillation of fresh birch wood yields birch tar, which is used in smoothing ointments for skin ailments.

BIDENS PILOSA
(Asteraceae)

Infusion of plant taken in Malay for coughs.

In Brazil leaves used as styptic and as vulnerary, applied to foul ulcers and swollen glands.

In Gold Coast, the juice of the leaves is squeezed into the eyes or the ears to cure eye complaints or ear complaints.

BIDENS TRIPARTITA
(Compositae)

BUE MARIGOLD

Parts used: Aerial parts.

Constituents: Bur marigold contains flavonoids, xanthophylls, volatile oil, acetylenes, sterols and tannins.

Medicinal actions and uses: Bur marigold is little used medicinally today, but it was once esteemed as a medicine. Astringent and diuretic, it may be employed to treat bladder and kidney problems. It also has a longstanding reputation for quickly staunching blood flow, and can be used for uterine haemorrhage and conditions producing blood in the urine. Bur marigold's astringency is beneficial in counteracting peptic ulceration, diarrhoea and ulcerative colitis. When used to treat digestive tract ailments, it is usually combined with a herb that reduces flatulence, such as ginger (*Zingiber officinale*).

BIOPHYTUM SENSITIVUM
(Geraniaceae)

Leaves – diur.
Powdered Seeds – applied to wounds.
Decoct. of roots – in gonor. and lithiasis.
Ash – Stomach.

BISCHOFIA JAVANICA
(Euphorbiaceae)

Juice of leaves – considered cure for sores.
Leaves contain vitamin C, twig bark contains tannin, seeds contain drying oil.

BIGNONIA CATALPA
(Bignoniaceae)

INDIAN BEAN TREE
Parts used: Bark, fruit.

Constituents: The bark contains catalpine, and oxylenzoic and protocatechetic acids.

Medicinal actions and uses: The mildly sedative and narcotic bark is used to treat asthma, whooping cough and other spasmodic coughs in children. The distilled water of the fruit, in combination with herbs commonly used to treat eye problems, such as eyebright (*Euphrasia officinalis*), and rue (*Ruta graveolens*), makes an effective eyewash for conjunctivitis and other eye infections.

BIXA ORELLANA
(Bixaceae)

ANNATTO
Parts used: Seeds, leaves, root.
Constituents: The seed pulp contains carotenoid colouring principles, bixin, faty oil, bixol.

Medicinal actions and uses: In the Caribbean, annatto leaves and roots are used to make an astringent infusion that is taken to treat fever, epilepsy and dysentery. The infusion is also taken as an aphrodisiac. The leaves alone make an infusion that is used as a gargle. The seed pulp reduces the severity of blistering when applied immediately to burns. Taken internally, the seed pulp acts as a general antidote for poisoning.

BLEPHARIS EDULIS
(Acanthaceae)

Seeds – resolv. diur. aphrodis. expect.
Crystalline bitter principle; seeds contain allantoin, bitter glucd, blepharin, catechol, tannins, saponin and glucose.

BLUMEA BALSAMIFERA
(Asteraceae)

Warm infusion – sudorific
Decoct. – Expectorant
Plant – fish poison.
Camphor, leaves-essential oil, containing a camphor known as Ngai-camphor and a glucd. Injection of extract lowers blood pressure; used in treatment of excitement, insomnia, and hypertension.

BLUMEA DENSIFLORA
(Asteraceae)

Leaves – occasionally used as sudorific.
Juice of leaves – insect-repellent.
Essential oil, camphor.

BLUMEA ERIANTHA
(Asteraceae)

Juice of plant – Carmin.
Warm infusion – sudorific.
Cold infusion – diur. emmen.
Essential oil, camphor-like smell.

BLUMEA LACERA
(Asteraceae)

Plant – bitter, antipyret.
Juice of leaves – anthelm. astrin. febge. stim. diur.
Root – in cholera
Essential oil, camphor

BOERHAAVIA DIFFUSA
(Nyctaginaceae)

Root – diur. laxat. expect. in asthma, stomach. in oedema, anemia, jaundice, ascites, anasarca, scanty urine, and internal inflammation, antidote to snake venom.
Alkaloid – punarnavine raises blood pressure.

BONNAYA REPATANS
(Scrophulariaceae)

Herb – applied externally for worms in the skin.

BORAGO OFFICINALIS
(Boraginaceae)

BORAGE

Parts used: Aerial parts, flowers, seed oil.
Constituents: Borage contains mucilage, tannins and pyrrolizidine alkaloids, which in isolation and toxic to the liver. Saponins, mucilage, silicic acid.
Medicinal actions and uses: With its high mucilage content borage is a demulcent herb and soothes respiratory problems. Its emollient qualities make it helpful for sore and inflamed skin – prepared either as freshly squeezed juice, in a poultice or as an infusion. The flowers encourage

Borago officinalis

sweating and the leaves are diuretic. The seed oil is particularly rich in polyunsaturated fats, and is superior in this respect to evening primrose oil (*Oenothera biennis*). Borage seed oil is used to treat premenstrual complaints, rheumatic problems, eczema and other chronic skin conditions. Anti-inflammatory, diuretic, diaphoretic, demulcent, tonic, bronchitis, colds.

BORASSUS FLABELLIFER
(Arecaceae)

Root – cooling, restor.
Juice of plant – diur. stim. antiphlegm., useful in inflammatory affections and dropsy.
Pulp – demulc. nutri.
Nutritive value of the sap called TODDY depends on the small amount of sugar and yeast in it and the latter is a good source of vitamin B complex.

BORRERIA HISPIDA
(Rubiaceae)

Decoct. of Root – alter.
Seeds – Stim.
The vapour is inhaled to kill tooth worms.

BOSWELLIA SERRATA
(Burseraceae)

Gum – diaphor. diur. astrin. emmen. in rheumatism, nervous and skin diseases.
Essential oil, gum resin.

BOTRYCHIUM LUNARIA
(Ophioglossaceae)

Plant – used in dysen. ruptures for healing wounds, a good vulnerary.

BRAGNATIA WALLICHI
(Aristolochiaceae)

Roots – in cholera, diar., dysen.
Plant – mixed with oil and made into an ointment, said to be beneficial for carbuncles and inveterate ulcers.
Roots contain isoaristolochic acid.

BRASSICA CAMPESTRIS
(Brassicaceae)

Tuberous roots and Seeds – considered antiscor.
Seeds yield oil of colza which is official in Sweden as oleum nape.

BRASSICA CAMPESTRIS VAR. RAPA
(Brassicaceae)

Seeds – mixed with hot water form an efficient-counter-irritant poultice.
Oil – Combined with camphor forms an efficacious embrocation in muscular rheumatism, stiff neck etc., it is used in dengue fever with benefit, and is rubbed on the chest in broncht. Roots and leaves considered stomach in Indo-China. Oil contains glycerides of erucic acid.

BRASSICA CERNUA
(Brassicaceae)

Decoct. of Seeds – in lumbago, cough and indign.
Leaves – used in Indo-China as antidysent. diaphor.
Fresh leaves contain oxalic acid and calcium.

BRASSICA INTEGRIFOLIA
(Brassicaceae)

Seeds – Warming, sudorific, used in spasmodic, neuralgic, and rheum. affections.

Oil – used as an embrocation, applied to skin in eruptions and ulcers.
Oil contains glycerides of erucic acid.

BRASSICA NIGRA
Brassicaceae (Black Mustard)

The seeds are used medicinally. Their constituents include glycoside-bonded oil of mustard (up to 35 per cent), the enzyme myrosinase, an alkaloid (sinapine), mucilage and protein. These substances give Black Mustard rubefacient and irritant properties and it is used medicinally to improve the blood supply to the skin and to the lungs, pleura and kidneys. Plasters and poultices made from mustard powder ease rheumatic and arthritic pain, muscular spasms, strained muscles and congested lungs,; bath preparations have the same effect. **All Black Mustard preparations irritate the skin.**

Brassica nigra

BRASSICA NAPUS VAR. CHINENSIS

Plant – antiscorb. arthritic, resolv.
Seeds – stim. stomach. laxat.

BRASSICA OLERACEA
(Crucuferae)
CABBAGE

Parts used: Leaves.
Constituents: Cabbage is rich in vitamins A, B_1, B_2 and C.
Medicinal actions and uses: Cabbage's best-known medicinal use is an a poultice – the leaves of the wild or cultivated plant are blanched, crushed or chopped, and applied to swellings, tumours and painful joints. Wild cabbage leaves eaten raw or cooked aid digestion and the breakdown of toxins in the liver – so the Romans' eating it to ease a hangover was in fact quite justified. Cabbage is also detoxifying and helpful in the longterm treatment of arthritis. The high vitamin C content of cabbage has made it useful in the prevention of scurvy.

BREYNIA PATENS
(Euphorbiacaee)

Plant – astrin.
Juice of stem – used in conjuctivitis.

BREYNIA RHAMNOIDES
(Euphorbiaceae)

Bark – astrin.
Dried leaves – smoked like tobacco in swelled uvula and tonsils.

BRIDELIA MONTANA
(Euphorbiaceae)

Plant – anthelm.
Root and Bark – astrin.
Tannin.

BRIDELIA RETUSA
(Euphorbiaceae)

Roots and Bark – astrin.
Bark – with gingile oil used as liniment in rheumatism.
Bark contains Tannin.

BUPLEURUM CHINESE
(Umbelliferae)

BUPLEURUM, HARE'S EAR ROOT <
CHAT HU (CHINESE)

A perennial growing to 1 m high, with sickle-shaped leaves and clusters of small yellow flowers.
Key **Constituents:**
Bupleurumol
Triterpenold saponins-saikosides (saikonina)
Flavonoid (rutin)
Key Actions:
protects liver
Anti-inflammatory
Tonic
Antiviral

BRYONIA ALBA
Cucurbitaceae (Black Berried Bryony)

The roots of Black-berried Bryony

Bryonia Alba

contain several glycosides, tannins, an alkaloid, (bryonicine) and resin. Together, these substances have strong purgative, diuretic, emetic, anti-inflammatory and antirheumatic actions. Tinctures of the fresh root are used in homeopathy for fevers, rheumatism, constipation and other disorders. **Neither Black-berried Bryony nor White Bryony should ever be collected and used for self-medication.**

BRYONIA DIOICA
(Cucurbitaceae)

WHITE BRYONY

Part used: Root.
Constituents: White bryony contains cucurbitacins, glycosides, volatile oil and tannins. The cucurbitacins kill cells and so act on tumours.
Medicinal actions and uses: A powerful cathartic and purgative, bryony is used with great caution in herbal medicine today. It is principally prescribed for painful rheumatic conditions. It may be taken internally, or applied as a counter-irritant, causing swelling and increased blood flow to the area. White bryony is also given for other inflammatory conditions such as duodenal ulcers, asthma, bronchitis and pleurisy, and may be used to reduce high blood pressure. The whole herb has an antiviral effect.

BRYONOPSIS LACINIOSA
(Cucurbitaceae)

Plant – bitter, aper. tonic, used in bilious attack, in fevers with flatulence.
Leaves – applied to inflam. Bitter principle Bryonin.

BUTEA MONOSPERMA
(Fabaceae)

PALAS, FLAME OF THE FOREST,
BENGAL KINO

Parts used: Bark, flowers, leaves, gum, seeds.

Constituents: All parts of the tree except the seeds contain tannins. Butin, glucd, butrin glucd, proteolytic, lipolytic enzymes. Gum.

Medicinal actions and uses: The gum that oozes from incisions made in palas bark is known as Bengal kino. Mildly astringent, it is used as a substitute for the kino derived from bastard teak (*Pterocarpus marsupium*). Bengal kino is taken as a decoction or a tincture for acid indigestion, diarrhoea and dysentery, and used as a gargle form sore throats and as a douche for vaginitis. A decoction of the astringent leaves and flowers is taken for diarrhoea, heavy menstrual bleeding and fever, and is applied to haemorrhoids and skin conditions. A decoction of leaves, bark or flowers is also thought to be aphrodisiac, while the flowers are believed to have contraceptive effect. The seeds are purgative, and are mainly used externally to treat herpes and ringworm.

BUXUS SEMPERVIRENS
Buxaceae (Box)

The leaves contain various alkaloids (for example, buxine), an essential oil and tannins, which give Box purgative, diaphoretic and antipyretic properties. **It is dangerous if taken internally and it should never be collected and used for self-medication.** In homeopathy a tincture prepared from fresh leaves is prescribed for fever, rheumatism and urinary tract

Buxus sempervirens

infections. Symptoms of poisoning are vomiting, abdominal pain and bloody diarrhoea.

CACCINIA GLAUCA
(Boraginaceae)

Plant – alter. tonic., diur., demuk. used in syphitis and rheumatism. occurs in – Baluchistan.

CADABA FARINOSA
(Capparidaceae)

Leaves and Roots – purg. anthelm. antisyp. deobstment, emmen. aper., prescribed in decoct. in uterine obstructions.

Leaves – used as poultice for sores.

CADABA TRIFOLIATA
(Cappaaridaceae)

Roots and leaves – purg. emmen. anthelm. antiphl. and indign. of children.
Leaves – employed in preparation of medicated oils.

CAESALPINIA BONDUCELLA
(Leguminosae)

NIKKAR NUT

Parts used: Seeds.
Constituents: The seeds contain a fixed (25%), a bitter principle (bonducin) and tannins, phytosterin, saponin, fatty oil.
Medicinal actions and uses: Nikkar seed are used to treat fevers and are taken as a tonic and aphrodisiac. In India, they are often mixed with black pepper (Piper nigra) for medicinal use. The seeds are also taken for inflammatory conditions such as arthritis. Roasted nikkar seeds are used in the treatment of diabetes. The oil extracted from the seeds is used in cosmetic preparations to soften the skin.

CAESALPINIA CORIARIA
(Caesalpiniaceae)

Powder of pods – astrin. antiper. tonic.
Decoction of pods – used for treatment of bleeding piles.
Bark – antiper. used in chronic fevers.
Pod – rich in Tannin.

CAESALPINIA SAPPAN
(Caesalpiniaceae)

Decoction of wood – considered emmen. useful in diarh. and dysen.

given internally in certain skin diseases.
Brasinin, essential oil, δ-γ-phellandrene, Tannins.

CAJANUS CAJAN
(Fabaceae)

Seeds and Leaves – made into a paste, which is warmed, and applied over the mammae to check secretion of milk.
Seeds – in snake bite seeds contain two globulins, cajanin and coreajanin.

CALCAMINTHA ASCENDENS
(Lamiaceae)

CALAMINT

Parts used: Aerial parts.
Constituents: Calamint contains a volatile oil (about 0.35%) consisting mainly of pulegone.
Medicinal actions and uses: Calamint stimulates sweating, and hence helps lower fevers. It also settles wind and indigestion. It is expectorant, and is a good cough and cold remedy. This range of applications makes it a good medicinal herb for mild respiratory infections. It should preferably be mixed with others herbs such as yarrow (Achillea millefolium) and thyme (Thymus vulgaris).

CALAMINTHA CLINOPODIUM
(Lamiaceae)

Plant – astrin. carmin. and heart tonic. Rhizome contains stachyose.

CALAMUS ROTANG
(Arecaceae)

Root – given chronic fevers, antid. to snake venom.

Leaves – in diseases of the blood, in biliousness.

Wood – vermifuge.

CALAMUS TRAVANCORICUS (Arecaceae)

Tender leaves – used in biliousness, worms, dyspep. in ear diseases, considered anthelm.

CALENDULA OFFICINALIS
Asteraceae (Pot Marigold)

Either the whole flowerheads or just the ray-florets are used medicinally. Among the constituents are an essential oil, pigments (carotenoids), bitter compounds, saponins, flavonoid glycosides, mucilage and resin bitter substances calendulin, salicylic acid sesquiterpene alcohols, esters of lauric, myristic, palmitic, stearic, and pentadecyclic acids. These give Pot Marigold vulnerary, anti-inflammatory, choleretic and antispasmodic properties. It is not often used internally nowadays, but extracts, tinctures and ointments are sometimes used externally to heal stubborn wounds, bed sores, persistent ulcers, varicose veins, bruises, gum inflammations and skin rashes. It is an excellent mouth wash after tooth extraction. Pot Marigold is probably more often used in complexion creams and lotions for cleansing, softening and soothing the skin. In the pharmaceutical industry the bright-orange pigments in the flowers are used to make medicinal preparations more attractive.

CALLICARPA ARBOREA (Verbenaceae)

Bark – arom. bitter, tonic., carmin.

Decoction of bark – applied to cutaneous diseases.

CALLICARPA LANATA (Verbenaceae)

Decoction of bark and root – useful in fever, hepatic obstruction and skin diseases.

Root – in cutaneous affections.

Leaves – boiled in milk used as a wash for aphthae of the mouth.

CALLICARPA LONGIFOLIA (Verbenaceae)

Root, leaves and bark – useful in the treatment of spine.

Decoct. of leaves – prescribed for colic and fevers.

Decoct. of roots – for diarh. and syphilis.

Leaves – fish poison.

CALLIGONUM POLYGONOIDES (Polygonaceae)

Roots – Bruised and boiled, in com-

Calendula officinalis

bination with catechu, used as gargle for sore gums.
Flowers are rich in protein.

CALONYCTION MURICATUM
(Convolvulaceae)

Seeds – bitter, purg., used as substitute for *Ipomoea hederaceae*, and in powder form as febge.
Resin and fixed oil.

CALOPHYLLUM APETALUM
(Guttiferae)

Resin – antiphl. anodyne
Oil of Seeds – used in leprosy and cutaneous affections.
Dried seeds kernels yields oil in the bitter taste.

CALOPHYLLUM INOPHYLLUM
(Guttiferae)

Oil of Seeds – Specific for skin diseases, and for application of rheumatism.
Bark – astrin. in internal haemorrh.
Gum – emetic, purg.
Juice – purg.
Leaves – fish poison.
Kernel yields oil, Bark-tannins, leaves contain saponin and hydro-cyanic acid.

CALLUNA VULGARIS
Ericaceae (Heather)

The flowers alone or the flowering stems are used medicinally. The constituents include the flavonoid glycosides quercitrin and myricitrin, tannins, silicic acid and resin. These give Heather anti-inflammatory, diuretic and mild sedative properties. It is used in combination with other herbal preparations to treat diarrhoea,

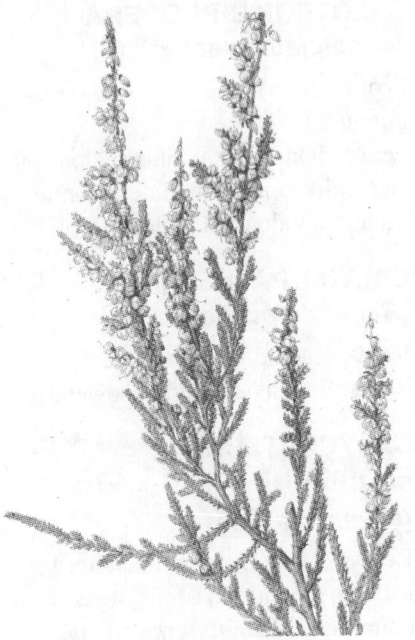

Calluna vulgaris

rheumatic pain, colds and coughs, urinary infections, and it is effective against kidney diseases and enlarged prostrate gland.

CALOTROPIS GIGANTEA
(Ascelpiadaceae)

Root bark – in dysen. substitute for ipecacuanha, diaphoret. expect. emetic, in form of paste applied to elephantiasis.
Tincture of leaves – irrit. in combination with *Euphorbia nerifolia* used as purg.
Powdered flowers – in colds, coughs, asthma and indigestion.
Bitter resins akundarin, calotropin; latex contains uscharin, calotoxin, calaetin, α-calotropeol, β-calotropeol, β-anyrin and calcium oxalate, glutathione, prokoclastic enzyme, giganteol, isogiganteol. Latex strong irritant to the skin and mucous membrane.

CALOTROPIS PROCERA
(Asclepiadaceae)

Properties same as *Calotropis gigantea*).
Leaves and stalks contain calotropin and calotropagenin; latex contains uscharin, calotoxin, calactin.

CALTHA PALUSTRIS
(Ranunculaceae)

Roots – aerial, poisonous.
Roots contain helleborin and veratrin.

CALYCOPTERIS FLORIBUNDA
(Combretaceae)

Leaves – bitter, astrin. anthelm. laxt. in colic, ground and administered with butter as cure for dysen. and malaria, external application for ulcers.
Root – in snake-bite
calycopterine, leaves contain tannin.

CALYSTEGIA SEPIUM
Convolvulaceae (Hedge Bindweed)

The flowering stems have medicinal

Calystegia sepium

action but are nowadays rarely used as a herbal remedy. The constituents include tannins, the glycoside jalapin and mucilage. These give Hedge Bindweed purgative, choleretic and mild diuretic properties. The purgative action is unpredictable and therefore potentially dangerous. **hedge Bindweed should thus not be used for self-medication**.

CAMELLIA SINENSIS
(Theaceae)

Tea

Parts used: Leaves, buds.
Constituents: Tea contains xanthines, caffeine (1-5%), theobromine, tannins.
Medicinal actions and uses: Tea is useful in the treatment of infections of the digestive tract. In Ayurvedic medicine, tea is considered astringent, sweat-inducing, digestive, and a nerve tonic, and is used for eye problems, haemorrhoids, fever and tiredness. Tea leaves may be used externally to soothe insect stings, swellings and sunburn.
Leaf contains carotene, riboflavin, nicotinic acid, pantothenic acid, ascorbic acid, malic acid, oxalic acid, queratin, theophylline, hypoxanthine, adenine, gums, dextrins and inositol.

CANANGA ODORATA
(Annonaceae)

Ylang-Ylang

Parts used: Flowers, essential oil.
Constituents: The essential oil contains linalool (11-30%), safrole, eugenol, geraniol, and sesquiterpenes (including 15-25% germacrene).

Medicinal actions and uses: The flowers and essential oil are sedative and antiseptic. The oil has a soothing effect, and its main therapeutic uses are to slow an excessively fast heart rate and to lower blood pressure. With its reputation as an aphrodisiac, ylang-ylang may be helpful in treating impotence.

Oil – used as an application in cephalogia, ophthalmia and gout. Cananga oil.

CANARIUM BENGALENSE
(Burseraceae)

Leaves and Bark – used externally for rheum, swellings.
Resin.

CANARIUM COMMUNE
(Burseraceae)

Resin – in form of ointment applied to indolent ulcers, substitute for Mistura amygdalae.
Fruit – laxat.
In Cambodia tuber considered stim. bechic, diaphor. and styptic.
Essential oil from oleoresin contains anethole and terpenes protein, fat, ash.

CANARIUM STRICTUM
(Burseraceae)

Gum – used with gingili oil in rheum. pains in chronic skin diseases.
Essential oil.

CANAVALIA ENSIFORMIS
(Fabaceae)

Fruits – if eaten, create abdominal complaints, hernia, and colic.
Cystin, tyrosin, tryptophan and alkaloid, seed-agrinine canavalin, consanavalin A concanavaline B,

arginine, desaminocanavanine, choline, trigonellene, betonicine, cavicine and kitogine.

CANELLA WINTERANA
(Canellaceae)

CANELLA, WILD CINNAMON
Part used: Bark.
Constituents: Canella contains about 1% volatile oil (including eugenol, alpha pinene and caryophyllene), alpha aldehydes (including canellal), resin and mannitol.
Medicinal actions and uses: Canella is cytotoxic (kills cells), antifungal and repels insects. It is also strongly aromatic, stimulant and antiseptic. Canella is often used in the West Indies and Latin America as a substitute for cinnamon (*Cinnamomum verum*,). The infusion is drunk for its pleasant flavour and tonic effect (the bark is thought a sexual stimulant). Canella is also used for stomach problems, indigestion and puerperal fever, an infection that develops after childbirth.

CANNA ORIENTALIS
(Cannaceae)

Root – Diaphor. diur. in fevers and dropsy, demulc. stim.
Stalks – Cut into pieces and boiled with rice water and pepper given to cattle as antid. for effects produced by eating poisonous grasses.

CANNABIS SATIVA
Cannabaceae (Hemp)

The leaves and flowering stems of the female plant that contain the dark resin cannabinone are the medicinally active parts. They are dried to yield

Cannabis sativa

the drug known as hashish, cannabis, bhang or marijuana, which is smoked, drunk or chewed for hallucinogenic effects. Hemp also has several medicinal uses, but is its less widely used by medical practitioners than its once was. The main constituents are tetrahydrocannabinol, cannabinol and cannabidiol. These give Hemp sedative, analgesic and antispasmodic properties and it has been used in medicines for insomnia, depression, neuralgia, migraine, asthma and as a local anaesthetic in dentistry. Nowadays a promising application of Hemp is as an anti-emetic is cancer therapy. The plant provides effective treatment for glaucoma; in which pressure within the eye is abnormally high, and is hypotensive, lowering blood pressure. Marijuana relieves asthma, menstrual pains, the pain of childbirth, and of arthritis and rheumatism, and may have-value as an antidepressant. It encourages and induces sleep.

CANSCORA DECUSSATA
(Gentianaceae)
Plant – laxat. alter. nerve tonic
Fresh juice of plant – prescribed in insanity, epilepsy and nervous debility.

CANTHIUM DICOCCUM
(Rubiaceae)
Bark – used in fever
Leaves contain hydrocyanic acid

CAPPARIS DECIDUA
(Capparidaceae)
Top Shoots and Young Leaves – used as a plaster for boils and swellings in powder form used to raise blisters and to relieve toothe ache and as antid. to poison.
Bark – acrid. laxat. diaphor. alexeteric, anthelm. useful in cough, asthma and inflamm.
Fruit – astrin. useful in cardiac troubles and biliousness.
Root and Root bark – Pungent, bitter, given in intermittent fevers and rheumatism.

CAPPARIS SPINOSA
(Capparaceae)
Caper

Parts used: Root bark, bark, flower buds.
Constituents: Caper contains capric acid.
Medicinal actions and uses: The unopened flower buds are laxative and, if prepared correctly with vinegar, are thought to ease stomach pain. The bark is bitter and diuretic and can be taken immediately before meals to increase the appetite. The root bark is

purifying and stops internal bleeding. It is used to treat skin conditions, capillary weakness and easy bruising and is also used in cosmetic preparations. A decoction of the plant is used to treat vaginal thrush.

anthelm, emmen, analgesic, in rheum, paralysis, enlarged spleen, and tubercular glands.

Flower buds contain rutin glucd, ruticacid, pectic acid, saponin.

CAPPARIS ZEYLANICA
(Capparidaceae)

Root bark – sedative, stomach, antihridrotic, bitter, cholag, and in cholera.

Leaves – Counter-irrit. and as cataplasm in boils, swellings and piles.

Contains an alkaloid, a phytosterol, a mucilaginous substance and water soluble acid.

CAPSELLA BURSA-PASTORIS
Brassicaceae (Sheperd's Purse)

The flowering stems are used medicinally. They should always be free of a parasitic fungus, *Cystopus candidus*, which produces a white coating on the plant. The constituents include the amines choline and acetylcholine, an alkaloid (bursine), a flavonoid glycoside (diosmin), organic acids and tannins. These substances give Shepherd's Purse astringent, haemostatic, vasoconstricting and diuretic properties. It is used to check gastric, uterine and pulmonary bleeding, to treat gastritis and enteritis, and urinary and kidney disorders. It affects the smooth muscles in the uterus and it is thought to assist contraction of the womb during childbirth. Any in-

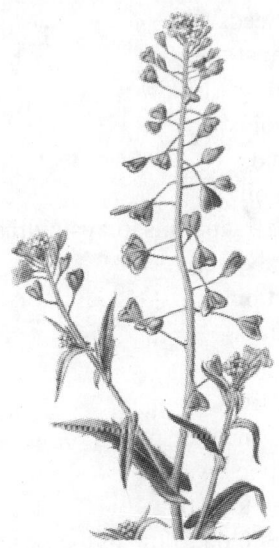

Capsella Bursa-Pastoris

ternal use of Shepherd's Purse should be in moderation and preferably supervised by a qualified medical or herbal practitioner; **large doses are poisonous**. Externally Shepherd's Purse is used in compresses to bath wounds, eczema and other skin disorders.

CAPSICUM ANNUUM
(Solanaceae)

Fruit – Stim. externally as ruebft. used in putrid sore throat and scaralatina, also in ordinary sore throat, hoarseness, dyspep, and yellow fever, in diarh, and piles, in snakebite.

Capsicum, capsaicin, solanine

CAPSICUM FRUTESCENS
(Solanaceae)
CAYENNE, CHILLI

A perennial, spiky shrub growing to 1 m, with scarletered conical fruits with seeds.

Fruits, Seeds
Key Constituents:
Capsaicin (0.1-1.5%)
Carotenoids
Flavonoids
Volatile oil
Steroidal saponins (capsicidins in seeds only)
Key Actions:
Stimulant
Tonic
Carminative
Relieves muscle spasms
Antispeptic sweating
Increases sweating
Increases blood flow to the skin
Analgesic

CARALLIA BRACHIATA
(Rhizophoraceae)

Fruits – used in treatment of contagious ulcers.
Bark – for itch.

CARALLUMA EDULIS
(Asclepiadaceae)

Plant – cooling, alter. anthelm. in leprosy and diseases of blood.

CARAPA GRANATUM
(Meliaceae)

Fruit – used as a cure for swellings of the breast and elephantiasis.
Bark – astrin. used for dysent. diarh. other abdominal troubles and as febge.
Seed Kernels – bitter tonic.
Seed ash – mixed with sulphur and coconut oil applied as ointment for itch.
Tannin and oil.

CARDAMINE PRATENSIS
(Brassicaceae)

Plant – stim. diaphor. diur.
Flowering tops – in epilepsy seeds contain myronic acid and an oil similar to mustard oil.

CARDANTHERA ULIGINOSA
(Acanthaceae)

Juice of leaves – mixed with salt used as blood purifier in Malabar.

CARDARIA DRABA
(Brassicaceae)

Plant – antiscor.
Seeds – used as cure for flatulency, fish poison. Young leaves contain HCN.

CARDIOSPERMUM HALICACABUM
(Menispermaceae)

BALLOON VINE

Parts used: Root, leaves, seeds.
Constituents: Most *Cardiospermum* species contain cyanogenic glycosides, saponin.
Medicinal actions and uses: Brings on delayed menstruation and to relieve backache and arthritis. The leaves stimulate local circulation and are applied to painful joints to help speed the clearing of toxins. The seeds are also thought to help in the treatment of arthritis. The plant as a whole has sedative properties.
Caution: Do not take during pregnancy.
Plant – in rheumat. stiffness of limbs, snake-bite.

Root – Diaphor. diur. aper. laxat. Rubft. emmen. Rheum. nervous diseases.
Leaf juice – cure for earache.

CARDUUS MARIANUS
(Compositae)

MILK THISTLE, MARY THISTLE
A spiny biennial, growing to 1.5 m, with white-veined leaves and purple flowerheads.
Key Constituents:
Flavonlignans (1-4%) (silymarin)
Bitter principles
Polyacetylenes
Key Actions:
Protects the liver
Stimulates secretion of bile
Increases breast-milk production
Antidepressant

CAREYA ARBOREA
(Lecythidaceae)

Bark and fruit – astrin. demulc.
Flowers and juice of fresh bark – given with honey as demulc. in coughs and colds.
Bark – used as antipyr. anti-pruritic in eruptive fevers, particularly in small-box and in snake-bite.
Root, bark and leaves – fish poison. Tannin.

CARICA PAPAYA
(Caricaceae)

PAPAYA
Parts used: Fruit, latex, leaves, flowers, seeds.
Constituents: Papaya fruit contains carpaine, carposide, papain, sucrose, invert sugar, a resinous substance, papain, malic acid, salts of tartaric acid and citric acids; pectins, carotenoid pigments, vitamins, sulphur containing carpasemine; a glucoside carposide and an alkaloid carpaine found in leaves; leaves contain vitamins C and E, enzymes papain and chymopapain from latex.
Medicinal actions and uses: Papaya's main medicinal use is as a digestive agent. The leaves and the fruit can both be used (the unripe fruit is especially effective). The latex from the trunk of the tree is applied externally to speed the healing of wounds, ulcers, boils, warts and cancerous tumours. The seeds are used as a gentle purgative for worms. The latex has a similar but more violent effect. The flowers may be taken in an infusion to induce menstruation, and a decoction of the ripe fruit is helpful for treating persistent diarrhoea and dysentery in children. The ripe fruit is mildly laxative and the leaves are used to dress wounds.

CARISSA CARANDAS
(Apocynaceae)

Fruit – antiscor.
Unripe fruit – astrin.
Ripe fruit – cooling acid.
Root – bitter, stomach, anthelm.
Decoct. of leaves – given at the commencement of remittant fever.
Alkaloid, Salicylic acid.

CARTHAMUS OXYACANTHA
(Asteraceae)

Oil for Seeds – used as a dressing for bad ulcers and as a remedy for itch. Fatty oil.

CARTHAMUS TINCTORIUS
(Compositae)

SAFFLOWER, HONG HUA (CHINESE)

Parts used: Flowers, seeds, seed oil.
Constituents: Safflower contains carthamone, lignans and a polysaccharide.
Medicinal actions and uses: In Chinese herbal medicine, the flowers are given to stimulate menstruation and to relieve abdominal pain. The flowers are also used to cleanse and heal wounds and sores and to treat measles. In the Anglo-American herbal tradition, the flowers are also given as a treatment for fever and skin rashes. The unpurified seed oil is purgative.

CARUM BULBOCASTANUM
(Apiaceae)

Uses same as *carum carvi*.
Essential oil, aldehydes.

CARUM CARVI
Apiaceae (Caraway)

The fruits are used medicinally. Their constituents include an essential oil (3-5%) with carvone and limonene, proteins, starch, sugar, tannins and fatty oil. These substances give Caraway stomachic, antispasmodic, carminative, galactagogic, antiseptic and anthelmintic properties. The dried fruits are used in powdered form, chewed whole or they are crushed and made into an infusion. They are safe for children. The distilled essential oil (caraway oil) is also an effective treatment but large doses are harmful to the liver and kidneys.
Caraway (fruits or oil) is widely used in flavouring food products such as

Carum carvi

bread, meat, cheese, pickles and sauces. It is also an ingredient of some alcholic beverages such as Kummel and gin. The fleshy root can be eaten as a vegetable and the young leaves added to salads and soups.
In addition, the seeds are diuretic, expectorant and tonic and are frequently used in bronchitis and cough remedies, especially those for children. Caraway has a reputation for increasing breast-milk production. The diluted essential oil is a useful remedy for scabies.

CARYOTA URENS
(Arecaceae)

Nut – acrid, cooling, to allay thirst and fatigue, used as an application to the head in cases of hemicramia.

Freshly drawn toddy – laxt. palm juice contains sucrose, reducing sugar, aectic acid.

CASEARIA ESCULENTA
(Samydaceae)

Root – astrin. cath. promotes action of liver.
Decoct. of Root – used in diabetes and piles.
Roots contain a resin, tannic acid, colouring matter, starch, tannin in bark, cathartic acid.

CASSIA ABSUS
(Caesalpiniaceae)

Leaves – bitter, astrin. as cough remedy.
Seeds – astrin. cath. for ring-worm, skin affections, in conjuctivitis and ophthalmia.
Seeds kernels contain chaksine, and isochaksine depressant for heart and nerves.

CASSIA ALATA
(Caesalpiniaceae)

Leaves – used for ringworm, in snake-bite.
Decoct. of leaves and flowers – used internally in broncht. and asthma and for washing eczematous patches.
Plant contains chrysophanic acid.

CASSIA ANGUSTIFOLIA
(Caesalpiniaceae)

Leaves and fruits – laxat. purg.
Glucoside kampferin, anthra-quinone, essential oil, chrysophanic acid, isorhemnetin calcium oxalate, glucosides semnoside A and B.

CASSIA AURICULATA
(Caesalpiniaceae)

Bark and Root – astrin.

Root – used in skin diseases.
Leaves and fruits – anthelm.
Seeds – in ophthalmia and conjuctivitis, in diabetes and chylous urine. Tannin.

CASSIA FISTULA
(Caesalpiniaceae)

Root bark, Seeds and Leaves – Laxt.
Fruit – Cath. applied in rheum. and snake-bite.
Seeds – emetic.
Root – astrin., tonic, febge. purg.
Juice of leaves – in skin diseases.
Leaves contain anthraquinone derivatives and very little tannin. Phlobaphenes pulp contains rhein, major anthraquinine derivative, volatile oil, and three waxy substances and a resinous substance.

CASSIA OCCIDENTALIS
(Caesalpiniaceae)

Plant – febge. purg. diur. tonic.
Leaves, roots and seeds – purg.
Seeds and Leaves – used externally in skin diseases, antiper.
Root – in snake bite
Emodin, oxymethyl-anthraquinones, toxalbumin, tannic acid, mucilage, fatty oil, chrysarobin, fatty oil.

CASSIA SENNA
(Caesalpiniaceae)

Senna (Alexandrian Senna)

A small perennial shrub growing to 1 m, with straight woody stem and yellow flowers.
Key Constituents:
Antrhaquinine glycosides (sennosides)
Naphthalane glycosides
Mucilage

Flavonoids
Volatile oil
Key Actions:
Stimulant
Laxative
Cathartic

CASSIA TORA
(Caesalpiniaceae)

Decoction of leaves – laxat.
Leaves and seeds – in skin diseases, ringworm and itch.
Root – in snake bite
Emodin, glued and a pleasant smelling fixed oil.

CASSYTHA FILIFORMIS
(Lauraceae)

Plant – tonic, alter., in bilious affections, chr. dysent., urethritis and skin diseases, insecticidal, powdered and mixed with gingile oil, it is used as hair tonic; mixed with butter and ginger it is used for cleansing inveterate ulcers.
Juice of plant – mixed with sugar considered as specific in inflammed eyes.
Alkaloid, laurotetamine, which produces cramps and in large doses even death, dulcitol.
A parasitic leafless twiner.

CASTANEA SATIVA
(Fagaceae)

SWEET CHESTNUT

Parts used: Leaves, bark.
Constituents: Sweet chestnut contains tannins, plastoquinones and mucilage.
Medicinal actions and uses: An infusion of sweet chestnut leaves is taken to treat whooping cough,

bronchitis and bronchial catarrh. The preparation tightens the mucous membranes and inhibits racking coughs. A decoction of leaves or bark is also valuable as a gargle for sore throats, and may be taken for diarrhoea. The leaves are used in the treatment of rheumatic conditions, to ease lower back pain and also to relieve stiff joints or muscles.

CATANOSPERMUM AUSTRALE
(Fabaceae)

Pods – astrin.
Unripe seeds – poisonous to cattle causing intense gastroenteritis, but ripe ones are harmless.
Leaves contain saponins.

CASUARINA EQUISETIFOLIA
(Causarinaceae)

Bark – astrin. useful in diarh. and dysent.
Decoction of leaves – used in colic
Bark contains colouring matter, casuarin and tannin.

CATHA EDULIS
(Celastraceae)

KHAT, CATHA

Parts used: Leaves, twigs.
Constituents: Khat contains alkaloids similar to those in *Ephedra* species – norpseudoephedrine (up to 1%) and ephedrine, tannins and a volatile oil. Ephedrine-type alkaloids strongly stimulate the central nervous system, are anti-allergenic and suppress the appetite. Alk. cathine related to bezedrine.
Medicinal actions and uses: Mainly used as a social drug, kaht is also chewed fresh or taken in an infusion

to treat ailments such as malaria. In Africa, it is taken in old age, stimulating and improving mental function. Khat is used in Germany to counter obesity.
Astrin. relief in cough, asthma and other diseases of chest.

CAULOPHYLLUM THALICTROIDS
(Berberidaceae)
BLUE COHOSH, SQUAW ROOT, PAPOOSE ROOT

A perennial growing to 1 m, with large 3-lobed leaves, purple blue flowers and deep blue berries.
Key Constituents:
Alkaloids (caulophylline, laburnine, magnoflorine)
Steroidal saponins (caulosapogenin)
Resin
Key Actions:
Antispasmodic
Diuretic
Promotes menstrual flow
Uterine tonic
Antirheumatic
Increases sweating
Anti-inflammation

CAYRATIA CARNOSA
(Vitaceae)
Root – astrin. ground with black pepper applied to boils.
Leaves – made into poultice employed in the treatment of yoke-sores on the necks of bullocks.
Stems, leaves and roots contain hydrocyanic acid.

CAYRATIA PEDATA
(Vitaceae)
Leaves – astrin. refrig. used for ulcers.

Leaf decoct – used to check uterine reflexes.

CEANOTHUS AMERICANUS
(Rhamnaceae)
NEW JERSEY TEA
Parts used: Root, root bark, leaves.
Constituents: New Jersey tea contains tannins, alkaloids, resin and a coagulant.
Medicinal actions and uses: Being astringent, expectorant and antispasmodic. New Jersey tea is used for sore throats, bronchitis, asthma and coughs. Like other plants containing appreciable amounts of tannins, it has been employed to treat diarrhoea and dysentery. It is also thought to be sedative and to lower blood pressure.

CEDRELA TOONA
(Meliaceae)
Bark – astrin. tonic, antiper, in chr. infantile dysen. external application for ulcers.
Flowers – emmen.
Bitter substance, Red colouring matter nyctanthin, essential oil.

CEDRUS DEODARA
(Pinaceae)
CEDAR
Parts used: Leaves, wood, essential oil.
Constituents: The volatile oil contains Cedrene (50%), atlantol and atlantone (*C. atlantica* only).
Medicinal actions and uses: Cedar of Lebanon is antiseptic and expectorant, acting to disinfect the respiratory tract. In Indian herbal medicine, Himalayan cedar leaves are used to

treat tuberculosis. The heartwood is also given as a decoction for feverish chest ailments such as acute bronchitis, and for insomnia and diabetes. Cedarwood essential oil is generally distilled from the Atlas or African cedar (*C. atlantica*, native to Morocco) and the red cedar (*Juniperus virginia*, native to North America). The oil is strongly antiseptic, astringent, diuretic, expectorant and sedative. Diluted and massaged into the skin, it treats catarrh, chest infections and cystitis. It is also used to treat skin wounds and ulcers. In Ayurvedic medicine, cedarwood essential oil is prescribed for syphilis and leprosy.

CEIBA PENTANDRA
(Bombacaceae)

Gum – tonic, alter., astrin. laxat. in bowel complaints.
Young leaves – emol.
Roots – diur. in scorpion-sting.
Unripe fruit – astrin. demulc.
Juice from the roots – cure for diabetes.
Seeds resemble cotton seed in compositions contain little or no gossypol, oil.

CELASTRUS PANICULATUS
(Celastraceae)

Bark – abortif.
Seeds – bitter, laxat. emetic, stim. aphrodis. in rheum. leprosy, gout, various fevers, paralysis.
Oil from seeds – in beri beri, powerful stim.
Alkaloids, glucosides, colouring matter, seeds yield oil, formic, acetic and benzoic acids, fleshy arils yield fat, celastrine, alk. pariculative.

CELOSIA ARGENTEA
(Amaranthaceae)

Seeds – in diarh. aphrodis. useful in blood diseases and month sores, for clearing the vision, and for diseases of the eye.

CELOSIA ARGENTEA *VAR.* CRISTATA
(Amaranthaceae)

Flowers – astrin. used in diarh. and excessive menstrual discharges.
Seeds – demulc. useful in painful micturition, cough and dysent.
Plant yields butanin, a nitrogen containing anthocyanin, seeds yield fatty oil.

CELSIA COROMANDELIANA
(Scrophulariaceae)

Juice of leaves – sedative, astrin. in diarh. and dysent.
Plant Juice – presented in skin eruptions and fevers.

CELTIS AUSTRALIS
(Ulmaceae)

SOUTHERN NETTLE TREE
Parts used: Leaves, fruit.
Constituents: Southern nettle contains tannins and mucilage.
Medicinal actions and uses: Due to their astringent properties, both the leaves and fruit of southern nettle tree may be used as a remedy. While the fruit is considered more effective, particularly before it has fully ripened, a decoction of both is taken to reduce heavy menstrual and intermenstrual uterine bleeding. The fruit and leaves may be used to astringe the mucous membranes in peptic ulcers, diarrhoea and dysentery.

CENTAUREA CYANUS
Asteraceae (Cornflower)

The florets are used medicinally. They contain organic pigments (anthocyanins) – principally centaurin or cyanidin – a glycoside (cichoriin), saponins, mucilage and tannins. These substances give Cornflower weak diuretic, astringent and tonic properties. In herbal medicine it is used as an infusion on its own or added to tea mixtures for digestive and gastric disorders. The flowers give teas and potpourris a pleasing colour. Externally Cornflower is used in bath preparations or compresses to treat wounds and skin ulcers. It is also an ingredient of hair tonics and is used as an eye wash. The isolated blue pigment is used in the pharmaceutical industry to colour certain medicine and in cosmetic preparations. Cuicin is slightly antibiotic. Sesquiterpene lactones, acetylenes and coumarins.

CENTAURIUM ERYTHRAEA
Gentianaceae (Common Centaury)

The flowering stems are used medicinally. The main components are the glyycosidic bitter compounds gentiopicrin and erythrocentaurin. The other constituents include traces of an essential oil and tannins. The bitter substances stimulate salivary and gastric secretions and Common Centaury is a popular digestive tonic and stomachic. In herbal medicine it is used in the form of an infusion for loss of appetite. Digestive disorders and biliary conditions. A tincture prepared from the fresh plant is used in homeopathy for the same ailments. Common Centaury is also increas-

Centaurea cyanus

Centaurium frythraea

ingly being used in making bitter herbal wines and liqueurs.

CENTELLA ASIATICA
(Umbelliferae)

GOTU KOLA (HINDI) INDIAN PENNYWORT

A perennial, herbaceous creeper, growing to 50 cm, with fan-shaped leaves.

Key Constituents:
Diterpenoid saponins (asiaticocide, brahmoside, thankuniside)
Alkaloids (hydrocotyline) Sitosterol essential oil fatty oil tannin
Bitter principles (Vellarin)

Key Actions:
Tonic
Antirheumatic
Mild diuretic
Sedative
Peripheral vasodilator

CENTIPEDA MINIMA
(Asteraceae)

Powdered leaves and Seeds – induce sneezing and a snuff made from them is used for colds in the head.

Herb – boiled and made into thick paste applied to checks to cure tooth ache.

Infusion of plant – useful in opthalmia.

Seeds – vermifuge.

Essential oil, amorph. bitter substance, alkaloid, glycoside, saponin, biter acidic principle myriosynin.

CENTRANTHERUM ANTHELMINTICUM
(Asteraceae)

Seeds – anthelm. in skin diseases tonic, stomach, diur. employed for destroying pediculi, in scorpion-sting. Bitter principle, resins.

CEPHAELIS IPECACUANHA
(Rubiaceae)

IPECACUANHA

Parts used: Root, rhizome.

Constituents: Ipecacuanha contains isoquinoline alkaloids, tannins and glycosides. The alkaloids are expectorant and at a larger dose, cause vomiting and diarrhoea. They are also strongly amoebicidal.

Medicinal actions and uses: Ipecacuanha is still used in both conventional and herbal medicine, and is listed in most national pharmacopoeias. One of the surest of emetics, even moderate doses will stimulate vomiting until the contents of the stomach are cleared it is particularly useful for drug overdose. At a lower dose, ipecacuanha is a strong expectorant. It is commonly found in many patent cough medicines, and is used in the treatment of bronchitis and whooping cough. Ipecacuanha is also still used for amoebic dysentery.

Alkaloids – emetine, cephaeline, psychotrine, emetamine, methyl psychotrine.

CERATONIA SILIQUA
(Leguminosae)

CAROB

Parts used: Fruit, bark.

Constituents: The fruit contains up to 70% sugars, fats, starch, proteins, vitamins and tannins.

Medicinal actions and uses: Carob pods are nutritious and, due to their high sugar content, sweet-tasting and mildly laxative. However, a decoction of the pulp is also antidiarrhoeal, gently helping to cleanse and also to relieve irritation within the gut.

These appear to be contradictory effects, but carob is an example of how the body responds to herbal medicines in different ways, according to how the herb is prepared, and according to the specific medical problem. The bark is strongly astringent and a decoction of it is taken to treat cases of diarrhoea.

CERATOPHYLLUM DEMERSUM
(Ceratophyllaceae)

Plant – cooling, antiper. useful in biliousness, in scorpion-sting.
Hairs contain myrophyllin.

CERBERA MANGHAS
(Apocynaceae)

Bark – purg.
Nut – narcotic, poisonous
Fruit – employed to kill dogs
Plant – fish poison
Glucoside aerberin, bitter substance odollin, oil, cerebroside, – digitalis like action; poisonous glycoside thevetin present in latex.

CEREUS GRANDIFLORUS
(Cactaceae)

Fresh young shoots – used as cardiac stim. and as partial substitute for digitalis.
Liquid extract or tincture used in cases of dropsy and various cardiac affections.

CERIPOS TAGAL
(Rhizophoraceae)

Plant – astrin.

Decoction of bark – haemostatic.
Decoction of shoot – used as subst. for quinine on African coast.
Mangrove.

CEROPEGIA BULBOSA
(Asclepiadaceae)

Tuberous roots – tonic, digest. Alkaloid aeropegine is the bitter principle of the root.

CETRATIA ISLANDICA
Parmeliaceae (Iceland Moss)

LICHEN
The thallus is used medicinally. It is effective when fully grown but still green. The main constituents are abundant mucilage with lichenin and isolichenin, bitter organic acids, carbohydrates, traces of iodine and vitamin A. The mucilage has expectorant and antitussive properties and Iceland Moss is thus beneficial for coughs, hoarseness, whooping cough, tuberculosis and bronchial asthma. The bitter compounds stimulate the appetite and gastric secretions. Iceland Moss is contained in some proprietary medicines.
Iceland Moss can be ground and made into flour for baking bread and cakes.

Cetratia Islandica

When gently boiled and cooled it makes nutritious jelly. Strongly demulcent, Iceland moss soothes the mucous membranes of the chest, counters catarrh and calms dry and paroxysmal coughs, being particularly helpful as a treatment for elderly people. Iceland moss is also very bitter and, within the gut, has both a demulcent and bitter tonic effect.

It is thus of value in all kinds of chronic digestive problems, for instance irritable bowel syndrome. Iceland moss also gently expels worms.

CHAMAEMELUM NOBILE
(Compositae)
ROMAN CHAMOMILE

Parts used: Flowers, essential oil.
Constituents: Roman chamomile contains up to 1.75% volatile oil (including tiglic and angelic acid esters and chamazulene), sesquiterpene lactones, flavonoids, coumarins and phenolic acids.

Chamaemelum nobile

Medicinal actions and uses: A remedy for the digestive system, Roman chamomile is often used interchangeably with German chamomile (*Chamomilla recutita*). However, an infusion of Roman Chamomile has a more pronounced bitter action than its German namesake. It is an excellent treatment for nausea, vomiting, indigestion and loss of appetite. It is also sedative, antispasmodic and mildly analgesic, and will relieve colic, griping and other cramping pains. By stimulating digestive secretions and relaxing the muscles of the gut, it helps normalise digestive function. Roman chamomile may also be taken for headaches and migraine, even by children. Its marked anti-inflammatory and anti-allergenic properties make it helpful when applied to irritated skin.

CHAMENERION ANGUSIFOLIA
(Onagraceae)
ROSE BAY WILLOW HERB

Parts used: Aerial parts.
Constituents: Rose bay willow herb contains flavones and tannins.
Medicinal actions and uses: Demulcent and astringent, rose bay willow herb treats diarrhoea, mucous colitis and irritable bowel syndrome. It has also been made into an ointment to soothe skin problems is children. Rose bay willow herb has been used in Germany and Austria to treat prostrate problems.

CHAMALIRIUM LUTEUM
(Liliaceae)
HELONIAS, FALSE UNICORN ROOT, BLAZING STAR

A herbaceous saponins (up to 9%) Glycosides (chamaelirin, helonin)
Key Actions:
Uterine and ovarion tonic
Promotes menstrual flow, diuretic.

CHAMOMILLA RECUTITA
Asteraceae (Scented Mayweed)

The flowerheads are used medicinally. Their constituents include up to 1 per cent of an essential oil with azulene (chamazulene) that turns blue on distillation, and bisabolol and farnesene; also flavonoid and coumarin glycosides, mucilage and fatty acids. These substances give Scented Mayweed anti-inflammatory, antiseptic, carminative, diaphoretic, sedative and antispasmodic properties. It is one of the most widely used herbal medicines, particularly for children's ailments. It is also a common treatment for teething problems in homeopathic medicine. In herbal medicine an infusion is used for colds and influenza, indigestion, diarrhoea, urinary infections, insomnia and 'nerves'. Externally it is used in compresses and bath preparations and in eye and mouth washes.

CHAMOMILLA SUAVEOLENS
Asteraceae (Pineapple Weed)

The flowerheads are used medicinally. The constituents include an essential oil (but there is less than in Scented Mayweed or Chamomile, *Chamaemelum nobile*), tannins, glycosides and a bitter compound. Pineappleweed has the same uses as Scented Mayweed except that is is not anti-inflammatory. It is now rarely used in herbal medicine but an infusion is effective for influenza and digestive disorders and it is anthelminthic. Externally it can be used as

Chamomilla recutita

Chamomilla suaveolens

a mouth rinse and gargle and in treatments for skin disorders.

CHASALIA CHARTACEA
(Rubiaceae)

Decoct. of root – used in rheum. pneumonia, head disorders ear and eye diseases and sore throat.
Roots and leaves – used in external applications for wounds, ulcers and headache.

CHEILANTHES TENUIFOLIA
(Polypodiaceae)

The Santals prescribe a preparation from the roots for sickness attributed to witchcraft or the evil eye.

CHEIRANTHUS CHEIRI
(Cruciferae)

WALLFLOWER

Parts used: Leaves, flowers.
Constituents: The herb contains cheiranthin and other cardioactive glycosides, cheirolin, cheirimine, essential oil.
Medicinal actions and uses: Although wallflower was formerly used as a diuretic, there was no understanding of its powerful effect on the heart. In small doses it is cardiotonic, supporting a failing heart in a manner similar to foxglove (*Digitalis purpurea*). In more than small doses it is toxic, and is therefore rarely used in herbal medicine.
Flowers – cardiac, emmen. used in paralysis and impotence.
Dried petals – arom., stim.

CHELONE GLABRA
(Scrophulariaceae)

BALMONY

Parts used: Aerial parts.

Constituents: Balmony contains resins and bitters.
Medicinal actions and uses: A strongly bitter remedy, balmony is principally used to treat gallstones and other gallbladder problems. It stimulates bile flow and has a mildly laxative action. It can be taken to relieve nausea and vomiting, intestinal colic, and to expel worms. It may also be antidepressant. Balmony is a suitable remedy for children.

CHELIDONIUM MAJUS
Papaveraceae (Greater Celandine)

The flowering stems are used medicinally. The constituents include up to 4 per cent of alkaloids (including chelidonine, chelerythrine, sanguisorbine and berberine) bound to chelidonic and other acids; also traces of an essential oil, saponins and

Chelidonium majus

carotenoid pigments. Greater Celandine has antispasmodic, mild sedative and antiseptic properties and it is used in medical practice to alleviate stomach, gallbladder and intestinal pains. It is also a vasodilator and hypertensive. A tincture is used in homeopathy and in folk medicine the ointment prepared from Greater Celandine is used to treat chronic eczemas. The alkaloid chelidonine affects cell division (as does colchicine) and Greater Celandrine is one of many plants being investigated for potential anti-cancer properties. Strong doses of Greater Celandine cause severe gastro-enteritis, violent coughing and breathing difficulties. **All internal use should thus be under strict medical supervision. The use of fresh juice to remove warts is also dangerous.**

CHENOPODIUM AMBROSIODES
Chenopodiaceae (American Wormseed)

The flowering stems of American Wormseed are used medicinally. The constituents include an essential oil with ascaridole as the main component, saponins, tannins and bitter compounds. An infusion prepared from American Wormseed has anthelmintic, antispasmodic, stomachic and tonic properties. It has also been recommended for asthma, nervous disorders and for menstrual disorders. Nowadays, however, the main use of the plant is as a source of the essential oil (chenopodium oil), which is obtained by distillation from the fresh plants and is used particularly against roundworm and hookworm. Ameri-

Chenopodium ambrosiodes

can Wormseed and the oil from it are poisonous in large doses and they should be used only under strict medical supervision. Ascaridol is a powerful worm-expellent.

CHIMAPHILA UMBELLATA (Ericaceae)
PIPSISSEWA

Parts used: Leaves.

Constituents: Pipsissewa contains hydroquinones (including arbutin), flavonoids, triterpenes, methyl salicylate and tannins. The hydroquinones have a pronounced disinfectant effect within the urinary tract.

Medicinal actions and uses: Astringent, tonic and diuretic, pipsissewa is mainly used in an infusion for urinary tract problems such as cystitis and urethritis. It has also been prescribed for more serious conditions such as gonorrhoea and kidney stones. By increasing urine flow, it stimulates the removal of waste products from the body, and is therefore of benefit in the treatment of rheumatism and gout. The fresh leaves may be applied externally to rheumatic joints or muscles, as well as to blisters, sores and swellings.

CHIONACHNE KOENIGII
(Poaceae)

Plant – laxat. aphrodis. useful in burning sensations, stranguary, phthisis, vesical calculi, diseases of blood, biliousness, haemorrhagic diathesis.

CHIONANTHUS VIRGINICUS
(Oleaceae)
Parts used: Root bark, bark.
Constituents: Fringe tree contains a saponin (chionanthin) and a glycoside (phyllirine).
Medicinal actions and uses: The root bark is a liver tonic, stimulates bile flow and acts as a mild laxative. It is prescribed mainly for gallbladder pain, gallstones, jaundice and chronic weakness. While it appears to be of benefit to liver and gallbladder function, there is as yet no research to substantiate its effects. The root bark also appears to strengthen function in the pancreas and spleen. Anecdotal evidence indicates that it may substantially reduce sugar levels in the urine. Fringe tree also stimulates the appetite and digestion and is an excellent remedy for chronic illness, especially where the liver has been affected. For external use, the crushed bark may be made into poultice for treating sores and wounds.

CHLORIS VIRGATA
(Poaceae)
Decoction of plant or roots used by the Xosas of S. Africa as an addition to baths for the treatment of colds and Rheum.

CHLOROPHORA EXCELSA
(Moraceae)
Milky Juice – applied for itch.
Bark – used in Africa as an ingredient of applications for swellings.

CHLOROPHYTUM ARUNDINACEUM
(Liliaceae)
Root – tonic.

CHLOROXYLON SWIETENIA
(Rutaceae)
Bark – astrin.
Leaves – applied to wounds and prescribed in rheumatism.
Plant – irrit.
Alkaloid chloroxylonine, chloroxyline, a powerful irrit. Tannins. essential oil.

CHONDRODENDRON TOMENTOSUM
(Menispermaceae)
PAREIRA
Parts used: Root, stem.
Constituents: Pareira contains alkaloids, including delta-tubocurarine and L-curarine. Tubocurarine is a potent muscle relaxant.
Medicinal actions and uses: Pareira notoriety as a poison hinges on the effect   bloodstream. Provided there are no cut and sores in the mouth, the plant is resonably safe taken orally as a medicinal remedy. The bitter and

slightly sweet-tasting roots and stems are mildy laxative, tonic and diuretic, and also act to induce menstruation. The plant is chiefly used to relieve chronic inflammation of the urinary tubules. In Brazil, it is also used for snake bite, for which purpose an infusion of the root is taken internally while the bruised leaves are applied externally.

CHONDRUS CRISPUS (Gigartinaceae)

CARRAGHEEN, IRISH MOSS

Part used: Whole herb.

Constituents: Carragheen contains large amounts of polysaccharides, proteins (upto 10%), amino acids, iodine and bromine. The polysaccharides become jelly like and demulcent when the plant is immersed in water makes a valuable nutrient in convalescence. Applied externally, this emollient herb soothes inflamed skin. Carragheen also acts to thin the blood.

Medicinal actions and uses: A useful demulcent and emollient, carragheen is mainly taken for coughs and bronchitis. Its expectorant effect encourages the coughing up of phlegm and it soothes dry and irritated mucous membranes. It is of value for acid indigestion, gastritis and urinary infections. Mucilaginous in texture and slightly salty in taste, carragheen.

CHROZOPHORA PROSTRATA (Euphorbiaceae)

Ashes of Root – given to children for cough.

Leaves – considered depurative

Seeds – used as purg.

CHRYSANTHEMUM CINERARIIFOLIUM
Asteraceae (Dalmatian Pyrethrum)

The flowerheads have the active ingredients pyrethrins and cinerins, plus an essential oil and glycosides. Pyrethrins and cinerins are contact insecticides used externally to kill insects and other pests living on the skin of man and animals as well as those that are harmful to plants. These substances are important insecticides because they are non-toxic to mammals and they do not accumulate in the environment or in the bodies of animals. They act by paralysing the nervous system of insects and the

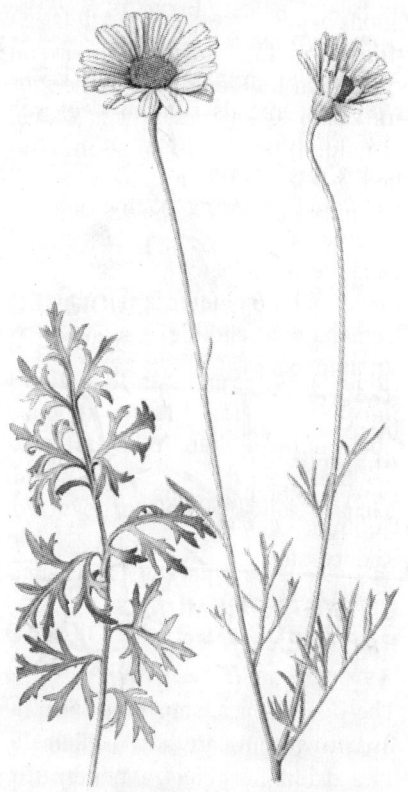

Chrysanthemum cinerariifolium

animals do not become resistant to them. The actions of pyrethrins and similar substances are still the subject of research. Pyrethrum is not used medicinally.

CHRYSANTHEMUM CORONARIUM
(Asteraceae)

Plant – in conjunction with black pepper given in gonor.
Flowers – used as subst. for chamomile, an arom. bitter and stomach.
Bark – purg. used in syphilis. Adenine, chlonine.

CHRYSANTHEMUM MORIFOLIUM
(Asteraceae)

JU HUA (CHINESE, FLORISTS' CHRYSANTHEMUM

A perennial growing to about 1.5 m, with flowerheads composed of yellow ray florets.

Key Constituents:
Alkaloids including stachydrine
Volatile oil
Sesquiterpene lactones
Flavonoids, including apigenin
Betanine and choline
Vitamin B_2
Key Actions:
Increase sweating
Antiseptic
Lowers blood pressure
Cooling
Reduces fever

CHRYSANTHEMUM PARTHENIUM
Asteraceae (Feverfew)

The flowerheads, sometimes also the flwoering stems, are used medicinally. The dried herb has a penetrating aroma and must always be stored well

Chrysanthemum parthenium

away from other herbs. It contains an essential oil with camphor (so-called chamomile camphor), bitter compounds, tannins and mucilage. Herbalists use an infusion as a stomachic, mild sedative, disinfectant, antispasmodic and mild anthelmintic. Feverfew is also receiving increasing attention from the medicine profession as a safe herbal treatment for migraine. In Britain a clinical trial is in progress and scientists have isolated new substances from the plant, which seem to oppose the actions of prostaglandins. Prostaglandins are naturally occurring substances in the body that may play a part in causing migraine. Externally a strong infusion of Feverfew soothes swellings and open wounds and is a mouth rinse after tooth extraction.

CHRYSANTHEMUM VULGARE
Asteraceae (Tansy)

The flowerheads and leaves are used medicinally. They should always be stored well away from other herbs. The principal constituent is an essential oil (0.2-0.6 per cent) with the poisonous thujone. The other constituents include bitter compounds (tanacetins), tannins and organic acids. Tansy has a variety of medicinal uses but is is now mainly used as an anthelmintic – but there are safer remedies. An infusion or powder has been used for this purpose. The essential oil obtained by distilling the fresh flowering stems is used in homeopathy for worms and as an emmenagogue. **Tansy should be used internally only under strict medical supervision**; large doses are powerful irritants and may also cause kidney and brain damage. Externally Tansy is applied to swellings, bruises and varicose veins.

Chrysanthemum vulgare

CHRYSOBALANUS ICACO
(Rosaceae)

Bark, roots and leaves – astrin.
Fruits – used for diarh. and other bowel complaints.
Seeds contain fatty oil.

CHRYSOPOGON MONTANUS
(Poaceae)

Seeds – a popular vermifuge

CICCA ACIDA
(Euphorbiaceae)

Fruit – astrin.
Root and Seed – cath.
Leaves and Roots – used as antid. to viper venom.
Root contains tannin, saponin, gallic-acid and a crystalline substance.

CICER ARIETINUM
(Fabaceae)

Acid exudation – astrin. used in dyspep. constip. and snake-bite.
Oxalic, acetic, malic acids; arginine, tyrosine, lysine, cystine, tryptophan; carotenoids oil soluble vitamins A, D and E; Lecithin, phytin, saponin; biochanin A, B and C isolated from sprouting gram.

CICHORIUM ENDIVIA
(Asteraceae)

Plant – used as resolv. cooling medicine and in bilious complaints.
Root – tonic, demulc. in dyspep. and fever.
Fruit – cooling, in fever, headache, bilious complaints and jaundice.
Bitter substance.

CICHORIUM INTYBUS
Asteraceae (Chicory)

The roots of wild plants are used medicinally. The dried root has a pungent and bitter taste. The constituents include the bitter compounds lactucin and intybin, inulin (up to 58 per cent), tannins, sugars and vitamins. These substances give Chicory aperitif, stomachic, tonic, hyposglycaemic, mild diuretic and laxative properties. In herbal medicine a decoction is used for liver disorders, gallstones and kidney stones, and for inflammations of the urinary tract.

Cicuta virosa

Cichorium intybus

CICUTA VIROSA
Apiaceae (Cowbane)

The rhizomes contain poisonous acetylinic compounds and an essential oil that has a narcotic effect but is not toxic. These substances affect the central nervous system and Cowbane was once used for treating various brain disorders and spasms of the smooth muscles. Nowadays it is usually regarded as far too toxic for internal use. **Cowbane should never be collected and used for self-medication**. The poison acts quickly on the nervous system causing salivation, vomiting, abdominal cramps, widely dilated pupils, delirium and violent convulsions. Death occurs due to paralysis and respiratory failure.

CIMICIFUGA RACEMOSA
(Ranunculaceae)

BLACK COHOSH, SQUAW ROOT

A herbaceous perennial growing to about 2.5 m, with creamy-white flower spikes.

Key Constituents:
Triterpene glycosides (acetin, cimicifugoside)
Isoflavones (formononetin)
Isoferulic acid
Salicylic acid
Key Actions:
Promotes menstrual flow
Antirheumatic
Expectorant
Sedative

CINCHONA CALISAYA
(Rubiaceae)

CHINCHONA, PERUVIAN BARK

An evergreen tree reaching 25 m, with reddish bark and leaves that grow to 50 cm.
Key Constituents:
Alkaloids (up to 15%), mainly quinoline alkaloids (quinine, quinidine) and indole alkaloids (cinchonamine) Bitter triterpenic glycosides (quinovin)
Tannins
Quinic acid
Key Actions:
Bitter
Reduces fever
Antimalarial
Tonic
Stimulates the appetite
Antispasmodic
Astringent
Antibacterial

OTHER SPECIES OF CINCHONA WITH SIMILAR EFFECTS

C. calisaya
C. ledgeriana
C. officinalis
C. robusta
C. succirubra

CINERARIA MARITIMA
(Compositae)

SILVER RAGWORT

Part used: Whole plant.
Constituents: Contains pyrrolizidine alkaloids (including jacobine) and tannins. Pyrrolizidine alkaloids in isolation are highly toxic to the liver.
Medicinal actions and uses: The juice of the whole plant is mainly used for conjunctivitis and other eye problems. Applied to the eyes, it has a mildly irritating effect that increases blood flow to the area, helping to strengthen resistance & clear away infection. It is also used for weak eyesight, and for the early stages of cataracts.

CINNAMOMUM CAMPHORA
(Lauraceae)

CAMPHOR

Parts used: Stems, root, wood, leaves, twigs, volatile oil.
Constituents: The plant contains a volatile oil comprising camphor, safrole, eugenol and terpineol. It also contains lignans. Camphor is irritant and antiseptic; safrole is thought to be carcinogenic. A white crystalline substance derived from the stems, root and other parts of the tree, also called camphor, has powerful antiseptic, stimulant and antispasmodic properties.
Medicinal actions and uses: Camphor is most commonly applied externally as a counter-irritant and analgesic liniment to relieve arthritic & rheumatic pains, neuralgia and back pain. It may also be applied to skin problems, such as cold sores & chilblains, and used as a chest rub for bronchitis and other chest infections. Though the oil has been taken for various complaints, internal use is not advised.

CINNAMOMUM TAMALA
(Lauraceae)

Bark – arom. in gonor.
Leaves – stim. carmin. used in rheum. colic. diarh. and in scorpion sting. Essential oil, d-α-phellandrene eugenol, cinnamic aldehyde.

CINNAMOMUM VERUM
(Lauraceae)

CINNAMON, DALCINI (HINDI)

An evergreen tree growing to 8-18 m, with soft, reddish brown bark and yellow flowers.
Key Constituents:
Volatile oil up to 4% (Clinnamaldehyde 65-75%, eugenol 4-10%)
Tannins (condensed)
Coumarins
Mucilage
Key Actions:
Warming stimulant
Carminative
Antispasmodic
Antiseptic
Antiviral

CINNAMOMUM ZEYLANICUM
(Lauraceae)

Bark – arom. astrin. stim. carmin. useful for checking nausea and vomiting.
Essential oil, eugenol, source of clove oil.

CISSAMPELOS PARIERA
(Menispermaceae)

Root – bitter, antiper. diur. purg. stomach, in dyspep. diarh. dropsy, cough and urinary troubles like cystitis, in snake-bite.
Leaves – external application for itch.

Alkaloids, – seperine, berberine, cissampeline, pelosine, saponin, hyatin, hyatinin and a queteitol and sterol from roots.

CISSUS QUADRANGULARIS
(Vitaceae)

Leaves and young shoots – alter. stomach. used in powder form in digestive troubles.
Juice of stem – used in irregular menstruation and scurvy.
Stem – given internally and applied topically for fracture of bones; beaten into a paste given for asthma.
Contains calcium oxalate, carotene, ascorbic acid.

CITRULLUS COLOCYNTHIS
(Cucurbitaceae)

Fruit and Seed – purg.
Root – purg, used in ascites, jaundice, urinary diseases and rheum.
Fruit and root – antidote to snake poison.
Bitter substance colocynthin, colocynthitin, α-elastrin, hentriacontane and saponins, fixed oil, phytosterolin, 2 phyto-sterols, 2 hydrocarbons, alkaloid, glucoside and tannin, fatty acids.

CITRULLUS VULGARIS
(Cucurbitaceae)

WATERMELON

Parts used: Fruit, seeds.
Constituents: Watermelon contains cirtrullin and arginine, both of which are thought to increase urea production in the liver, so increase the flow of urine. Carotene, lycopine, mannitol, oil, provitamin A, vitamin C, pectin, seeds rich in enzyme urease.

Medicinal actions and uses: Watermelon is best known as a thirst-quenching fruit that comes into season when temperatures are at their hottest. In traditional Chinese medicine it is used precisely to counter "summer heat" patterns – characterised by excessive sweating, thirst, raised temperature, scanty urine, diarrhoea and irritability or anger. Watermelon fruit and juice soothe these symptoms, increasing urine flow and cleansing the kidneys. The fruit's refreshing properties extend to the digestive system, where it clears wind. Watermelon may also be used in the treatment of hepatitis. In hot, stifling weather it is helpful for those suffering from bronchitis or asthma. The cooling fruit pulp may be applied to hot and inflamed skin and to soothe sunburn. The seeds can be mashed and used to expel worms.

CITRUS AURANTIUM
(Rutaceae)
BITTER ORANGE

Parts used: Fruit, peel, leaves, flowers, seeds, essential oil.

Constituents: Bitter orange peel contains a volatile oil with limonene (around 90%), flavonoids, coumarins, triterpenes, vitamin C, carotene and pectin. The flavonoids are anti-inflammatory, antibacterial and antifungal. The composition of the volatile oils in the leaves, flowers and peel varies significantly. Linalyl acetate (50%) is the main constituent in oil from the leaves (petitgrain) and linalool (35%) in oil from the flowers (neroli). The unripe fruit of the bitter orange contains cirantin, which reportedly is contraceptive.

Medicinal actions and uses: The strongly acidic fruit of the bitter orange stimulates the digestion and relieves flatulence. An infusion of the fruit is thought to soothe headaches, calm palpitations and lower fevers. The juice helps the body eliminate waste products, and being rich in vitamin C, helps the immune system ward off infection. If taken in excess, however, its acid content can exacerbate arthritis. In Chinese herbal medicine, the unripe fruit, known as *zhi shi*, is thought to "regulate the *qi*", helping to relieve flatulence and abdominal bloating, and to open the bowels. The essential oils of bitter orange, especially neroli, are sedative. In Western medicine, these oils are used to reduce heart rate and palpitations, to encourage sleep and to soothe the digestive tract. Diluted neroli is applied as a relaxing massage oil.

CITRUS BERGAMIA
(Rutaceae)
BERGAMOT

Part used: Essential oil.

Constituents: Bergamot contains a volatile oil including linalyl acetate (30-60%), limonene (26-42%) and linalool (11-22%), bergapten and a diterpene.

Medicinal actions and uses: Bergamot is little used in herbal medicine, but it can be used to relieve tension, relax muscle spasms and improve digestion.

CITRUS LEMON
(Rutaceae)
LEMON

A small, evergreen tree growing to about 7 m, with light green, toothed leaves.

Key Constituents:
Volatile oil (about 2.5% of the peel), limonene (up to 70%), alpha-terpinene, alpha-pinene, bete-pinene, citral. coumarins
Bioflavonoids
Vitamin A, B_1, B_2, B_3, and C (40-50 mg per 100 g of fruit)
Mucilage
Key Actions:
Antiseptic
Antirheumatic
Antibacterial
Antioxidant
Reduces Fever

CITRUS LIMETTIOIDES
(Rutaceae)

Fruit – refrig. in fever and jaundice
Oil from peel – pinene, limonene, linalool, linalyl acetate.

CITRUS LIMON
(Rutaceae)

Rind of the ripe fruit – stomach, carmin.
Juice of ripe fruit – antiscorb. refrig. in scurvy, in rheum. dysent. diarh.
Oil from peel – α-limonene, d-α-pinene, carphene, linalool, essential oil, hespiridin.

CITRUS MAXIMA
(Rutaceae)

Fruit – nutri. cardiotonic, refrig.
Leaves – useful in epilepsy, chorea and convulsive cough.
Naringin, oil from peel-α-limonene, α-pinene, linalool, geraniol.

CITRUS MEDIA
(Rutaceae)

Root – anthelm. in constip. useful in vomiting, urinating calculus.

Flowers and buds – stim. astrin.
Ripe fruit – stim. tonic
Juice – refrig. astrin.
Oil from peel – limonene, dipentene, citrol.

CITRUS PARADISI
(Rutaceae)

Juice – used for building up resistance to common colds and wound infections.
Vitamin C, B_1 and pectin oil from peel - limonene, volatile constituents, sesquiterpenes, aldehydes, geraniol, cardinene, citrol etc.

CITRUS RETICULATA
(Rutaceae)

Fruit – laxat. aphrodis. astrin. tonic, relieves vomiting.
Flowers – stim.
Oil from peel – d-limonene, terpene, carene, linalool.

CITRUS SINENSIS
(Rutaceae)

Fruit – purifies blood, allays thirst in fevers, cures catarrh. improves appetite.
Juice – useful in biliousness and diarh.
Fresh rind – rubbed on the face as remedy for acne.
Fruit antiscor, vitamin, oil from peel d-limonene, decyclicaldehyde, linalool, d-terpineol, neroli oil, petitgrain oil, glycoside hersperidin.

CLAVICEPS PURPUREA
Clavicepitaceae (Ergot Fungus)

Ergot contains more than a dozen potent alkaloids, most of them deriva-

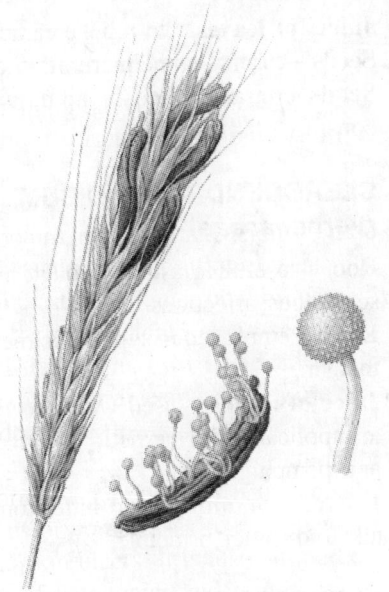

Claviceps purpurea

Clematis recta

tives of lysergic acid, among them ergometrine, ergocitrine, ergocornine, ergotamine and ergotoxin. Ergometrine is the most important of these substances. It is extracted and used in pharmaceutical preparation principally to assist women in childbirth and in the treatment of migraine. Ergomatine is also used to treat migrane. **These medicines are available only in prescription.**
Ergomentinine, ergosine, ergosinine, ergocrystine, ergocrystinine.

CLEMATIS RECTA
Ranunculaceae (Erect Clematis)

The flowering stems have medicinally active constituents, among them glycosides, saponins and other, so far unidentified, substances. In homeopathy tinctures of the fresh plant are used to alleviate rheumatic pains, migraine and headache, and to treat varicose veins, slow-healing wounds and skin ulcers. The fresh leaves may cause stubborn eczemas and irritate the eyes. The toxicity of the plant is slightly lessened by drying. If eaten the plant can cause enteritis, severe abdominal cramps and diarrhoea. For these reasons **Erect clematis should never be collected and used for self-medication.**

CLEISTRANTHUS COLLINUS
(Euphorbiaceae)

Plant – astrin. Extremely poisonous **Extract of leaves, roots and specially fruits** – Violent, gastro intestinal irrit.
Root, leaf and bark – fish poison Saponin, glucoside, oduvin tannins.

CLEMATIS VITALBA
(Ranunculaceae)
TRAVELLER'S JOY

Parts used: Leaves.

Constituents: Traveller's joy contains protoanemonin and saponins. Protoanemonin is caustic and irritant.

Medicinal actions and uses: The leaves of traveller's joy irritate the skin, causing it to redden and blister, but they are also strongly analgesic. Applied to arthritic joints, they help relieve pain and encourage the removal of waste products. The plant is also diuretic, and in the past was taken internally to counter urinary problems. However, the mature plant is now known to be toxic and should not be ingested. The juice is reputed to relieve headaches and migraine if sniffed, but as this might destroy the mucuous.

C. gouriana
C. nepantensis
C. smilacifolia   All are poisonous
C. triloba
C. wightiana

CLEOME BRACHYCARPA
(Capparidaceae)
Plant – bitter, good for scabies, rheum., and inflam.
Leaves – in leucoderma.

CLEOME CHELIDONNI
(Capparidaceae)
Root – considered vermifuge in indo China.
Infusion of plant – used in gingivitis and skin diseases.

CLEOME ICOSANDRA
(Capparidaceae)
Leaves – rubft. vesic. sudorific external application for wounds and ulcers.

Juices of leaves – to relieve earache
Seeds – carmin. anthelm. rubft. vesic.
Seeds contain viscosic acid, visconsin.

CLERODENDRUM INDICUM
(Verbenaceae)
Root – useful in asthma, cough and scrofulous affections.
Resin – employed in syphilictic rheumatism.
Juice of leaves – used with ghee as an application to herpetic eruptions and pemphigus.
Leaves – vermifuge and bitter tonic alkaloid, bitter principle.

CLERODENDRUM INERME
(Verbenaceae)
Juice of leaves – alter. febge.
Leaves – in form of poultice used to resolve buboes.
Juice of root – alter.
Root – by boiling is oil a liniment obtained which is useful in rheum. amorphous bitter principle, resin, gum.

CLERODENDRUM INFORTUNATUM
(Verbenaceae)
Leaves and roots – employed externally for tumours and contain skin diseases.
Leaves – subst. for chiretta as tonic and antiper.
Fresh juice of leaves – vermifuge, used as bitter tonic and febge. in malaria especially of children.
Leaves and flowers – in scorpion-sting.
Sprouts – in snake-bite.

CLERODENDRUM PHLOMIDIS
(Verbenaceae)

Root – bitter tonic, given in convalescence of measles.
Juice of leaves – alter. given in neglected syphilic compliments.

CLERODENDRUM TRICHOTOMUM
(Verbenaceae)

CHOU WU TONG

Parts used: Leaves.
Constituents: *Chou wu tong* contains clerodendrin, acacetin and meso-inositol.
Medicinal actions and uses: In Chinese herbal medicine, *chou wu tong* is prescibed for joint pain, numbness and paralysis, and occasionally for eczema. Traditionally regarded as a plant that "dispels wind-dampness", it is now also being used to help lower blood pressure. The plant is mildly analgesic and, when used with the herb *Siegesbeckia pubescens*, is anti-inflammatory.

CLERODENDRUM SERRATUM
(Verbenaceae)

Root – in febrile and catar. affections useful in malaria.
Leaves – used for fevers, boiled with oil and butter made into an ointment useful in cephalogia and opthalmia also used in snake-bite.

CLITORIA TERNATEA
(Fabaceae)

Seeds – purg. aper.
Root – bitter, cath. purg. diur.
Root bark – diur. laxt.
Plant – used in snake poison.

Seeds contain a fixed oil and a bitter resinous principle, both seeds and root-bark contain tannin.

CNICUS BENEDICTUS
Asteraceae (Blessed Thistle)

The flowering stems and the leaves are used medicinally. The herb gets its bitter taste from the bitter compound cnicin. It also contains traces of an essential oil, tannins, abundant mucilage and minerals. Blessed Thistle has diaphoretic, choleretic, carminative, tonic and antiseptic properties. It is contained in several propriatary medicines and a tincture is used in homeopathy. In herbal medicine small amounts of Blessed Thistle are used to treat digestive disorders and lack of appetite, and to promote the flow of gastric secretions and bile. **Large doses irritate the mouth,**

Cnicus benedictus

digestive tract and kidneys and may cause vomiting and diarrhoea. Internal use of Blessed Thistle should therefore be professionally supervised. The herb should never be taken during pregnancy.

Blessed Thistle is also used to make bitter liqueurs and was once used in beer making. The tender young shoots can be eaten like artichokes and the leaves can be added to salads.

CNIDIUM MONNIERIE
(Umbelliferae)

SHE CHUANG ZI

Parts used: Seeds, essential oil.
Constituents: The volatile oil of *she chuang zi* contains pinene, camphene, bornyl isovalerate and isoborneol.
Medicinal actions and uses:The seeds of *she zuang zi* is most often prescribed externally as a lotion, powder or ointment for skin conditions such as eczema, ringworm and scabies. The seeds are considered particularly helpful for problems affecting the genital area, for example vaginitis and vaginal discharge. *She zuang zi* is also taken internally for impotence, and infertility in both men and women, often in combination with schisandra (*Schisandra chinensis*).

COCCINIA INDICA
(Cucurbitaceae)

Juice from leaves and roots – used in diabetes.
Leaves – applied externally in eruptions of skin.
Plant – internally in gonor.
Enzyme amalyse, hormone and traces of an alkaloid.

COCCULUS HIRSUTUS
(Menispermaceae)

Root – refrig. laxat. sudorific, alter. useful in chr. rheumatism and venereal diseases.
Juices of leaves – when mixed with water forms a jelly which is taken as a cooling medicine for gonor. and used externally for eczema, prurigo and impetigo.

COCCULUS LAURIFORMIS
(Menispermaceae)

Jungle tribes of Malay peninsula use the plant as arrow poison and darts. Toxic alkaloid, coclaurine; bark and leaves contain coclaurine with a curare-like action.

COCHLEARIA ARMORACIA
(Brassicaceae)

Herb – stim. diaph. diur. digest. used as counter irrit. in lumbago.
Root contains a pungent, acrid, vesicating volatile oil; pungency due to glycoside sinigrin and enzyme myrosin, root rich source of vitamin C.

COCHLEARIA OFFICINALIS
(Brassicaceae)

SCURVY GRASS

Parts used: Leaves, aerial parts.
Constituents: Scurvy grass contains glucosilinates, a volatile oil, a bitter principle, tannin, vitamin C and minerals.
Medicinal actions and uses:Besides having a high vitamin C content, scurvy grass has antiseptic and mild laxative actions. The young plant, which has a general detoxicant effect

and contains a wid range of minerals, is taken as a spring tonic. Like watercress (*Nasturtium officinale*), it has diuretic properties and is useful for any condition in which poor nutrition is a factor. It can be used in the form of a juice as an astringent mouthwash for mouth ulcers and can also be applied externally to spots and pimples.

COCHLOSPERMUM RELIGIOSUM
(Cochlospermaceae)

Gum – sweetish, cooling, sedative, used in coughs and gonor.
Dried leaves and flowers – stim. Tree yields a gum.

COCOS NUCIFERA
(Arecaceae)

Fruit – Sweet, aphrodis. diur.
Oil – local application in alopecia and in loss of hair afterfevers and debilitating diseases.
Water of unripe fruit – cooling, useful in thirst, fever and urinary disorders.
Root – diuretic, astrin. and used in uterine diseases.
Flowers – astrin.
Enzyme invortin, oxydase, catalase; milk-histidine, arginine, lysine, tyrosine, tryptophan, proline, leucine, alanine; oil-contains lauric, myrstic and fatty acids, mixed glycerides, phytosterol and squalene, vitamins B group.

CODIAEUM VARIEGATUM
(Euphorbiaceae)

Pounded leaves – applied as poultice on the abdomen of children suffering from urinary troubles.

CODONOPSIS PILOSULA
(Campanulaceae)

CODO NOPSIS, DANG SHEN (CHINESE)
A twinig perennial growing to 1.5m with oval leaves and pendulous green and purple flowers.
Key Constituents:
Triterpenoid saponins
Sterins
Alkenyl & alkenyl glycosides
Polysaccharides
Tangshcnoiside I
Key Actions:
Adaptognic
Stimlant
Tonic

COFFEA ARABICA
(Rubiaceae)

COFFEE
Parts used: Seeds.
Constituents: Coffee contains 0.06-0.32% caffeine, theobromine and theophylline, and tannins. Caffeine is strongly stimulant. Theophylline is stimulant and relaxes smooth muscle. Adenine, xanthine, hypo xanthine, guanine, alk. trigonellene mercaptans.
Medicinal actions and uses: Although it is not often recognised as a medicinal herb, coffee is highly effective when taken as a general stimulant, having a particular effect on the central nervous system temporarily improved perception and physical performance. Coffee increases heart output, stimulates digestive juices and is a powerful diuretic. It can help in headaches and migraine. Coffee's active constituent.
Caffeine, is often combined with conventional analgesics in over-the-coun-

ter headache remedies. In the Aurvedic tradition, the unripe beans are used for headaches, and the ripe roasted beans for diarrhoea. Coffee enemas effectively cleanse the large bowel.

COLA ACUMINATA
(Sterculiaceae)

KOLA NUT, COLA NUT

Parts used: Seeds.

Constituents: Kola nut contains up to 2.5% caffeine (generally higher than coffee), theobromine, tannins, phlobaphene and an anthocyanin.

Medicinal actions and uses: Kola nut stimulates the central nervous system and the body as a whole. It increases alertness and muscular strength, counters lethargy, and has been used extensively both in western African and Anglo-American herbal medicine as an antidepressant, particularly during recovery from chronic illness. Like coffee (*Coffea arabica*) kola is used to treat headaches and migraine. It is diuretic and astringent and may be taken for diarrhoea and dysentery.

COLCHICUM AUTUMNALE
Liliaceae (Meadow-Saffron)

The corms and seeds are used medicinally. Besides the poisonous alkaloid colchicine, which is found in all parts of the plant, the seeds contain other alkaloids, large amounts of fat, tannins and sugars. Colchicine is extracted and used in preparations prescribed by medical practitioners mainly for acute attacks of gout and rheumatism. Tinctures of Meadow Saffron are used in homeopathy for the same complaints. Demecolcine, a derivative of colchicine, was for a time

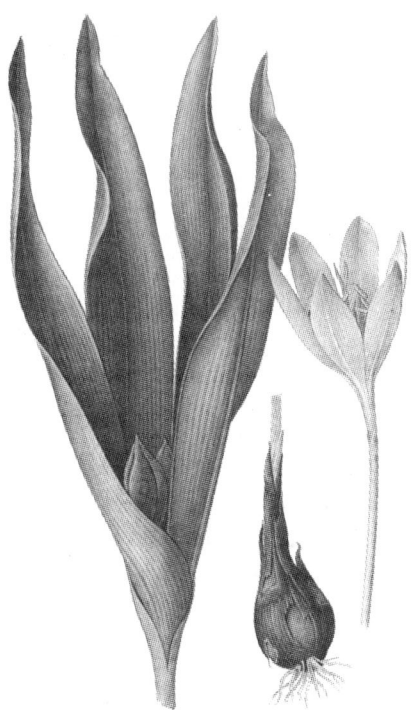

Colchicum autumnale

used in the treatment of chronic leukaemia. **Meadow Saffron is a dangerous plant and it should never be collected and used for self-medication**. If the seeds or flowers are eaten, the outcome is often fatal. Symptoms of poisoning include salivation, vomiting, diarrhoea and abdominal cramps. Convulsions, general paralysis and respiratory failure may follow. The plant is also toxic to animals, particularly when they are fed dry fodder. The alkaloids even pass into milk and can accumulate to reach a toxic level. Colchicine inhibits the division of certain cells and this property has been used to produce new plant varieties.

Best remedy for acute gout pain leukaemia has been successfully treated and also to treat Beheet's syndrome, a chronic disease marked by recurring ulcers and leukaemia.

COLDENIA PROCUMBENS
(Boraginaceae)

Fresh leaves – ground and applied to rheum. swellings.
Dried plant – with equal part of Fenugreek seeds is rubbed to a fine powder and applied warm for causing suppuration of boils.

COLEBROOKEA OPPOSITIFOLIA
(Lamiaceae)

Root – a preparation of it is used in epilepsy.
Leaves – applied to wounds and bruises.

COLEUS AMBOINICUS
(Lamiaceae)

Leaves – in urinary diseases, vaginal discharges.
Juice of leaves – mixed with sugar acts as a powerful arom. carm. given in colic and dyspep.
Essential oil containing carvacrol.

COLEUS FORSKOHLIL
(Labiatae)

COLEUS

An aromatic perennial, with tuber-like roots and an erect stem, reaching 60 cm.
Key Constituents:
Volatile oil
Diterpenes (forskolin)
Key Actions:
Lowers blood pressure
Antispasmodic
Dilates the bronchioles (small airways of the lungs)

Dilates the blood vessels
Heart tonic

COLLINSONIA CANADENSIS
(Labiatae)

STONE ROOT

Parts used: Root, leaves.
Constituents: Stone root contains a volatile oil, tannins and saponins.
Medicinal actions and uses: Stone root has diuretic and tonic properties, and is chiefly employed in the treatment of kidney stones. It is also prescribed to counteract fluid retention. It has been used to reduce back pressure in the veins, which in turn helps prevent the formation or worsening of haemorrhoids and varicose veins. As an astringent, stone root contracts the inner lining of the intestines, and can be helpful in treating disorders of the digestive system such as irritable bowel syndrome and mucous colitis. The fresh leaves or roots of stone root are applied as a poultice to bruises and sores.

COLUTEA NEPALENSIS
(Fabaceae)

Leaves – purg.
Seeds – emetic.

COMMELINA BENGHALENSIS
(Commelinaceae)

Plant – bitter, emol. demulc. refrig. laxat. beneficial in leprosy.

COMMIPHORA MOLOMOL
(Burseraceae)

MYRRH

A spiny, deciduous tree growing to 5 m, with yellow red flowers and pointed fruit.

Key Constituents:
Gum (30-60%), acidic poly-saccharides
Resin (25-40%)
Volatile oil (308%), including heera-boline, eugenol and many furano-sesquiterpenes.

Key Actions:
Stimulant
Antiseptic
Anti-inflammatory
Astringent
Expectorant
Antispasmodic
Carminative

CONIUM MACULATUM
Apiaceae (Hemlock)

The fruits are the most medicinally active parts. The main constituents are poisonous alkaloids, principally coniine, and essential oils and organic acids alkaloids d-coniine, γ-coniceine, conhydrin, n-methyl connine, hesperidin. A tincture of Hemlock is still used by qualified practitioners in homeopathy. The isolated pure coniine is also contained in a few pro-prietary ointments and suppositories to relieve severe pain. **Hemlock is a highly dangerous plant and it should never be collected and used for self-medication.** Even a small dose can be fatal.

It paralysis the body from bottom upwards leading to death. **Socrates** was given Hemlock poison in jail when he was sentenced to death.

Alkaloids poisonous produce paraly-sis of motor nerve treminations and stimulation followed by depression of central nervous system, cause nausea and vomiting.

COIX LACHRYMA-JOBI
(Poaceae)

Seed – tonic, diur.
Root – used in menstrual disorders.
Grains – blood purified and diur. leucine, tyrosine, histidine, lysine, arginine, coicin, glutamic acid.

COMMIPHORA MUKUL
(Burseraceae)

Gum resin – antis. astrin. expect. aphrodis. enriches the blood, demulc. aper. carm. alter. emmen. in snake-bite and scorpion sting. Commercial product contains essential oil besides gum and resin.

CONNARUS MONOCARPUS
(Connaraceae)

Pulp of fruit – used in eye diseases.
Decoct. of root – given in syphilis

Conium maculatum

Bark and wood – used in treatment of ulcers.

Oil from roots – used as an application for swellings.

CONVALLARIA MAJALIS
Liliaceae (Lily-of-the-valley)

The flowering stems, but more often the leaves, are used medicinally. They contain poisonous cardiac glycosides, among them convallotoxin, convallatoxol and convalloside, and also saponins, essential oils and organic acids. The individual glycosides are isolated and included in proprietary medicines prescribed for various heart conditions. The glycosides are considered as effective and safer than digitalin from foxgloves (*Digitalis* spp.) or regulating heart action. Tinctures of Lily-of-the-valley are also used in homeopathy. Contrary to the advice in some popular herbal books, **Lily-of-the-valley should never be collected and used for self-medica-**

tion. It is a dangerous plant and should be used only under strict medical supervision. Some components have purgative and emetic actions. The berries may cause paralysis and respiratory failure and a doctor's help should be sought immediately if a child eats them.

The aromatic extracts from the flowers are used in the cosmetic and perfumery industries.

CONYZA CANADENSIS
Asteraceae (Canadian Fleabane)

The flowering stems are used medicinally. Their constituents include as essential oil with limonene and terpineol as its main components, tannins and choline. These substances give Canadian Fleabane astringent, diuretic and haemostatic actions. Herbalists prescribe an infusion or decoction for severe diarrhoea, gravel and kidney disorders, and for throat

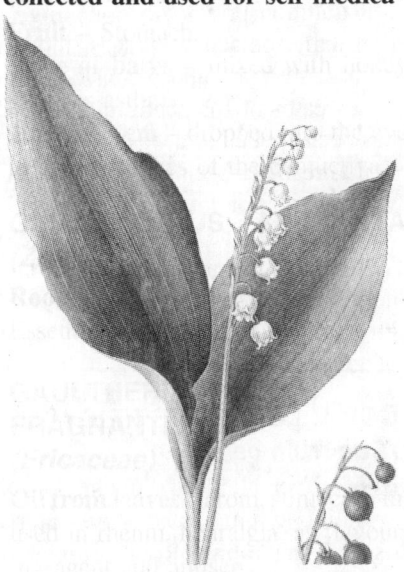

Convallaria majalis

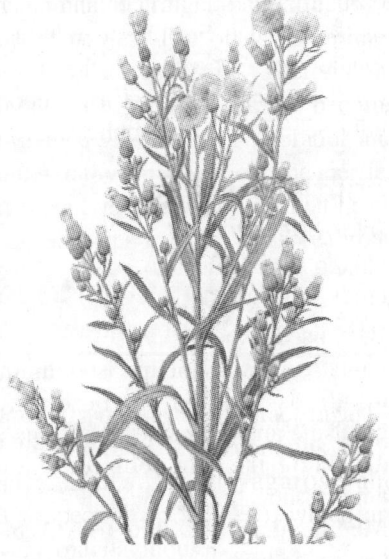

Conyza canadensis

infections. A tincture is used in homeopathy for these complaints and for haemoorhoids and painful menstruation. The essential oil (oil of erigeron) obtained by distillation of fresh plant material has been used to expel intestinal parasites.

COPAIFERA SPP.
(Leguminosae)

COPAIBA

Part used: Oleo-resin.

Constituents: The oleo-resin contains a volatile oil (30-90%), which in turn contains alpha-and beta-caryophyllene, sesquiterpenes, resins and terpenic acids.

Medicinal actions and uses: Antiseptic, diuretic and stimulant, copaiba is still taken extensively in Brazil. Its primary use is to counter mucus in the chest and genito-urinary system. It also irritates the mucuous membranes and promotes the coughing up of mucus. A solution or tincture of copaiba is taken for bronchitis, chronic cystitis, diarrhoea and haemorrhoids. It has been frequently used in the past to treat gonorrhoea. Eczema and other skin diseases reportedly benefit from its application.

COPTIS CHINENSIS
(Ranunculaceae)

HUANG LIAN (CHINESE),
CHINESE GOLDTHREAD

Part used: Root.

Constituents: Contains isoquiniline alkaloids, including berberine, coptisine and worenine. Berberine is antibacterial, amoebicidal and antidiarrhoeal.

Medicinal actions and uses: A bitter-tasting herb, *luang lian* is given in the Chinese herbal tradition as a decoction to "clear heat" and "dry dampness", relieving fever, red and sore eyes, and sore throats. The herb is particularly helpful for diarrhoea and dysentery, and has been used to quell vomiting. Skin problems such as acne, boils, abscesses and burns are also treated with *luang lian*. Like the root of goldthread (*C. trifolia*) *luang lian* is taken as a gargle for mouth and tongue ulcers, and for swollen gums and toothache. Both herbs are also used as an eyewash to treat acute conjunctivitis.

COPTIS TRIFOLIA
(Ranunculaceae)

GOLDTHREAD

Part used: Rhizome.

Constituents: Goldthread contains isoquiniline alkaloids (including berberine and coptisine).

Medicinal actions and uses: A strongly bitter tonic, goldthread has been prescribed in the North American tradition principally for indigestion and stomach weakness, though it has also come under consideration as a treatment for peptic ulcers, and has been applied as a wash for thrush. Goldthread has been used as a mouthwash, gargle or lotion for mouth ulcers, sore lips and throats. It can help to tighten mucous membranes. The herb's constituents (and to some degree its actions) are similar to those of goldenseal.

CORALLOCARPUS EPIGAEUS
(Cucurbitaceae)

Root – aper. alter. used in chr. dysen. syphilitic rheum. chr. mucous enteritis and snake-bite. Root contains a bitter principle like Bryonin.

CORCHORUS CAPSULARIS
(Tiliaceae)

Infusion of leaves – demulc. stomach laxt. carm. stim. to increase appetite, bitter tonic, in desen. fever, dyspep. liver disorders.
Decoction of root and unripe fruit – in diarh.
Leaves contain glucd capsularin, corchorin, cardiac aghyeon corchortoxin having heart action similar to digitalis.

CORCHORUS OLITORIUS
(Tiliaceae)

Leaves – demulc. tonic. diur. useful in chr. cystitis, gonor. dysuria.
Infusion of leaves – tonic, febge.
Seeds – purgative.
Fruit – Vitamin D.

CORCHORUS TRILOCULARIS
(Tiliaceae)

Seeds – in fever and obstruction of the abdominal viscera.
Mucilage – demulc.

CORDIA DICHOTOMA
(Boraginaceae)

Fruit – astrin. anthelm. diur. demulc. expect. used in affections of urinary passages, diseases of lungs and spleen.
Juice of bark – in gripes.
Decoct. of bark – used in dyspep. and fevers.
Kernels – remedy in ringworm
Leaves – application to ulcers and in headache.
Plant – in snake bite
Bark contains tannin.

CORIANDRUM SATIVUM
Apiaceae (Coriander)

The fruits contain up to 1 per cent of an essential oil with a linalool (coriandrol) as its main component, plus fatty oil, proteins, tannins, pectin, sugars and vitamin C. The dried fruits are used by themselves or in tea mixtures, primarily as an aperitif, has a digestive tonic and a carminative. Coriander also has a sedative effect. The essential oil (coriander oil), obtained by distillation from the fruits, is used to make the water solution for windy colic with much the same effect as that made from Caraway (*Carum cavi*). It is also used in many compound preparations and to disguise the taste of other medicines. The fruits – or the oil on its own – are included in ointments for painful rheumatic joints and muscles.
Coriander is an important culinary herb. The fruits and the fresh leaves

Coriandrum sativum

are widely used for flavouring food and the root can be cooked and eaten as a vegetable. An infusion of the herb is a gentle remedy for flatulence, bloating and griping. It settles spasms within the gut and counters the effects of nervous tension.

CORNUS OFFICINALIS
(Cornaceae)

Sʜᴀɴ Zʜᴜ Yᴜ

Part used: Fruit.

Constituents: *Shan zhu yu* contains an iridoid glycoside (verbenalin), saponins and tannins. Verbenalin is known to produce a mild effect on the involuntary nervous system, especially that governin the digestive system.

Medicinal actions and uses: As a herb that "stalilises and binds", *shan zhu yu* is used principally to reduce heavy menstrual bleeding and unusually active secretions, including copious sweating, excessive urine, spermatorrhoea (involuntary discharge of semen) and premature ejaculation. *Shan zhu yu* is astringent, and like all herbs that suppress bodily fluids (even excessive ones), it will simply prolong or lead to a worsening of symptoms if used without tonic or detoxifying herbs. When used in combination with other herbs, for example rehmannia (*Rehmannia glutinosa*) *Shan zhu yu* treats a range of problems, including frequent urination, dizziness and tinnitus (ringing in the ears).

CORONILLA VARIA
Fabaceae (Crown Vetch)

The flowering stems have been used medicinally but nowadays the plant

Coronilla varia

is generally regarded as too toxic for use as a herbal remedy. The constituents include the poisonous water-soluble glycoside coronillin, tannins, bitter compounds, organic salts and vitamin C. Coronillin has an action very similar to that of digitalin from foxgloves (*Digitalis* spp.) in that it regulates heart action. **Crown Vetch is a dangerous plant and it should never be collected and used for self-medication.** Symptoms of poisoning are pallor, diarrhoea, rectching, muscular spasms and eventually coma and death.

CORYDALIS CAVA
Fumariaceae (Bulbous Corydalis)

The tubers are used medicinally. When dried they have a strong aroma and bitter taste. They contain alkaloids, the most important being corydaline and bulbocapnine. Bulbl-

Corydalis cava

capnine has antispasmodic, sedative and hallucinogenic properties. It lowers the blood pressure and inhibits the contractions of striated muscles. In some countries it is used in preparations to treat Parkinson's disease and other serious neurological disorders, vertigo and muscular tremors. Bulbocapnine is also beneficial before and after treatment with anaesthetics. Large doses of the drug can cause severe headache and other side effects. **Bulbous Corydalis is a dangerous plant and it should never be collected and used for self-medication.**

CORYDALIS YANHUSUO (Papaveraceae)

CORYDALIS, YAN HU SUO (CHINESE)

A small herbaceous plants growing to 30 cm with narrow leaves and pink flowers.

Key Constituents:
Alkaloids (including corydalist, corydaline, Tetrahydro-palmatine (THP, protopine).
Protoberine-type alkaloid (leonticine)
Key Actions:
Analgesik
Antispasmodic
Sedative

CORYLUS AVELLANA
Betulaceae (Hazel)

The leaves, bark and fruits have medicinal actions. The main constituents of the leaves are essential oils, glycosides and sugars; those of the bark chiefly tannins and organic acids. These substances give the leaves diuretic properties and they have been used in tea mixtures. They have also been used for treating varicose veins and circulatory disorders. Externally they have been used in bath prepara-

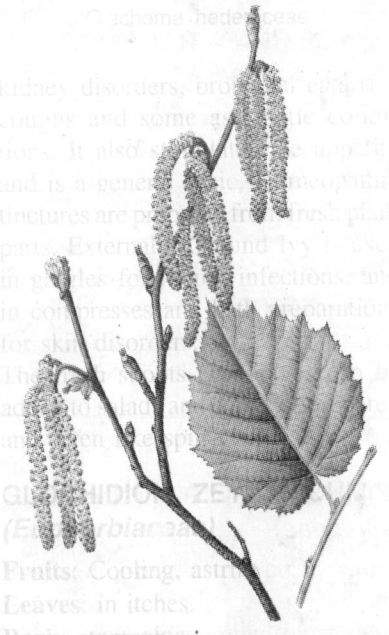

Corylus avellana

tions to treat haemorrhoids and slow-healing wounds. The nuts contain up to 60 per cent fatty oils, plus proteins, sugars and vitamins. They are very nourishing and tasty and are widely used in confectionary and bakery goods. The expressed oil from them is a valuable salad and vegetable oil and it is also used to make soap and cosmetics and as a machine lubricant.

COSCINIUM FENESTRATUM
(Menispermaceae)

Root – bitter, tonic, stomach, in dysen. antisep. used for dressing wounds and ulcers.
Wood – bitter tonic.
Decoct. of bark – in intermittent fevers.
Decoct. of stem – in snake-bite.
Berberine, saponin, caryl alcohol, hentria contane, sitosterol, palmitic acid, oleic acid, sitosterol glued saponin and some resinous material.

COSMOSTIGMA RACEMOSA
(Asclepiadaceae)

Root Bark – chlog. useful in dyspep. accompanied by fever.
Leaves – used to cure ulcerous sores.
Alkaloid, Glucoside.

COSTUS SPECIOSUS
(Zingiberaceae)

Root – bitter, astrin. purg., depurative stim, tonic, anthelm, used in snake-bite.
Root rich in starch.

COTONEASTER BACILLARIS
(Rosaceae)

Stolons – Considered astrin. in indochina.
Leaves yield hydrocyanic acid.

COTULA ANTHEMOIDES
(Asteraceae)

Plant – heated with oil applied externally in rheum.
Infusion – used as eye wash.
Decoct. of plant – said to be beneficial for colds in the heads and chest.

CRATAEGUS LAEVIGATA
Rosaceae (Midland Hawthorn)

The flowers (white forms only) have the strongest medicinal action. They are used on their own, or with the leaves; or the leaves are used by themselves. The main constituents of the flowers are the glycoside quercitrin with the sugar component quercetin, also flavones and traces of an essential oil. The leaves contain the flavonoid glycoside vitexin 4-rhamnosid, sterols and catechins. These substances give Midland Hawthorn hypotensive, vasodilating, antisclerotic and

Crataegus laevigata

sedative properties and the plant is used for various heart and circulatory disorders such as high blood pressure, abnormal heart rate, arteriosclerosis and angina pectoris. Because of its effects on the heart **Midland Hawthorn should be used only under the supervision of a qualified medical or herbal practitioner;** no part is suitable for self-medication. The fruits have the same constituents as those of Hawthorn (*C. monognyna*).

Crataegus monogyna

CRATAEGUS OXYACANTHA & C. MONOGYNA
(Rosaceae)
HAWTHORM

A decidious, thorny tree with small leaves, white flowers and red berries, growing to 8 m.

Key Constituents:
Bioflavonoids (rutin, quercitin)
Triterpenoids
Cyanogenic glycosides
Amines (trimenthylamine in flowers only)
Polyphenols
Coumarins
Tannins

Key Actions:
Cardiotionic
Dilates blood vessels
Relaxant
Antioxidant
Dissolves arterial blocks
Antisclerotic

CRATAEGUS MONOGYNA
Rosaceae (Hawthorn)

The flowers, leaves and fruits are used medicinally. They have basically the same action as parts of other hawthorn species, particularly Midland Hawthorn. The fruits contain flavonoid glycosides, organic acids, tannins, an essential oil, vitamin C, vitamin B complex and pectins. Like the flowers and the leaves the fruits have hypotensive, antisclerotic, vasodilating and sedative properties and they are used in the form of an infusion or tea mixtures in herbal medicine or in propriatary medicines to treat various heart and circulatory disorders, migraine, menopausal conditions and insomnia. Because of its effects on the heart, **Hawthorn should be used only under the supervision of of a qualified medical or herbal practitioner;** no part is suitable for self-medication.

CRATAEVA NURVULA
(Capparaceae)
VARUNA, BARUN (HINDI), THREE-LEAFED CAPER

A deciduous tree growing to 15 m, with pale yellow flowers.

Key Constituents:
Saponins
Flavonoids
Plant sterols
Glucosilinates

Key Actions:
Diuretic
Inhibits the formation of stones

CREPIS ACAULIS
(Asteraceae)
Baked leaves or the root – ground and mixed with goat's milk taken to activate the secretion of milk in women.

Root – eaten raw in urinary complaints.

CRESCENTIA CUJETE
(Bignoniaceae)
Fruit pulp – aper. cooling. diur. and febge. poisonous to birds and small mammals.

Decoct. of bark – used for cleaning wounds.

Leaves – pounded to a poultice applied for headaches.

Fruit pulp contains-crescentic, tartaric, citric and tannic acids chlorogenic acid, fixed oil from seeds.

CRESSA CRETICA
(Convolvulaceae)
Plant – Tonic, aphrodis, expect. and antibil.

Leaves – tonic, aphrodis. Alkaloid.

CRINUM ASIATICUM
(Amaryllidaceae)
Bulb – bitter, tonic, laxt. expect. used in biliousness and in strangway and other urinary troubles.

Fresh root – emetic, nauseant, diaphor.

Seeds – purg., diur., emmen. tonic.

Leaves – expect. applied to skin diseases and to reduce inflam.

Lycorin, root contains alka. narcissine, crinamine.

CRITHMUM MARITIMUM
(Umbelliferae)
SAMPHIRE, SEA FENNEL

Parts used: Aerial parts.

Constituents: Samphire contains a volatile oil, pectin, vitamins (especially vitamin C) and minerals.

Medicinal actions and uses: Though it is currently little used in herbal medicine, samphire is a good diuretic, and it holds out potential as a treatment for obesity. Samphire has a high vitamin C and mineral content, and is thought to relieve flatulence and to act as a digestive remedy. In this, the plant resembles its inland namesake, fennel (*Foeniculum vulgare*).

CROCUS SATIVUS
Iridaceae (Saffron)

The stigmas are the medicinal parts. Among their constituents are a series

Crocus sativus

of crocine glycosides, a bitter glycoside crocetin, (picrocrocine) and an essential oil. These give Saffron stomachic, antispasmodic and emmenagogic properties. **It is a powerful medicine and not suitable for self-medication:** large doses can cause haemorrhage, vomiting, diarrhoea and vertigo. It was once widely used as an abortifacient used in snakebite used in enlargement of liver.

CROTALARIA JUNCEA
(Fabaceae)

Seeds – used to purify blood, in impetigo, psoriasis, emmen. poisonous to live stock.

CROTON OBLONGIFOLIUS
(Euphorbiaceae)

Bark, root, fruits and seeds – purg. in snake bite.
Bark and root – alter.
Bark – used in external applications for sprains, useful in liver diseases.
Fatty oil in seeds.

CROTON TIGLIUM
(Euphorbiaceae)

Seeds and oil – drastic purg. irrit. rubft. cath. fish poison, in snake-bite.
Wood – diaphor. in small doses and purg. and emetic in large doses.
Seed kernels contain croton oil, oil contains toxic resin, contains 2 toxic proteins croton globulin and croton albumin, sucrose and a glycoside, crotonoside.

CRYPTOCARYA WIGHTIANA
(Lauraceae)

Leaves – ground and boiled in oil applied in elephantiasis.

Powdered bark and leaves – taken with sugar as cure for rheum. and swellings.

CUCUMIS MELO
(Cucurbitaceae)

Seeds – cooling, nutri. diuret.
Pulp of fruit – diur. useful in chr. eczema.
Rich in oil and fatty acids.

CUCUMIS SATIVUS

Fruit – nutri. demulc.
Seeds – cooling, tonic, diur.
Fruits contain an enzyme erepsin, vitamin B_1 and C, proteolyte enzymes, ascorbic acid oxidase, succinic and malic dehydrogenases.

CUCURBITA MAXIMA
(Cucurbitaceae)

Seeds – anthel. used as taenicide diur. and tonic
Oil – nerve tonic
Fruit pulp – used as poultice, applied to burns, inflam. and boils.
Saponin.

CUCURBITA PEPO
Cucurbitaceae (Vegetable Marrow)

The seeds contain up to 50 per cent fatty oil, proteins, a glycoside (cucurbitin), resin and substances yet to be identified. They are still used as an anthelmintic, dried, or eaten fresh (chewed or pounded) with the seed coat removed. They have no irritating side effects. A decoction from the dried seeds combined with castor oil can be used for the same purpose. The seeds of the related Cucumber (*Cucumis sativa*) are also anthel-

Cucurbita pepo

mintic. The fruits of Vegetable Marrow contain, besides water, sugars, proteins, a fatty oil, vitamins and minerals. They can be made into compotes and marmalade. The raw fruit juice is slightly diuretic and is sometimes recommended for urinary complaints.

CUMINUM CYMINUM
(Umbelliferae)

CUMIN

Parts used: Seeds.
Constituents: Cumin seeds contain 2-5% volatile oil, which consists of 25-35% aldehydes, pinene and alpha-terpineol. The seeds also contain flavonoids (including apigenin).

Medicinal actions and uses: Cumin, like its close relatives caraway (*Carum carvi*) and anise (*Pimpinella anisum*) relieves flatulence and bloating, and stimulates the entire digestive process. It reduces abdominal gases and distension and relaxes the gut as a whole. In Indian herbal medicine, cumin is used for insomnia, colds and fevers, and mixed into a paste with onion juice, has been applied to scorpion stings. The seeds are also taken to improve breast-milk production – a role it shares with fennel seeds (*Foeniculum vulgare*).

CUPRESSUS SEMPERVIRENS
(Cupressaceae)

CYPRESS

Parts used: Cones, branches, essential oil.
Constituents: Cypress contains a volatile oil (with pinene, camphene and cedrol) and tannins.
Medicinal actions and uses: Applied externally as a lotion or as a diluted essential oil, cypress astringes varicose veins and haemorrhoids, tightening up the blood vessels. A footbath of the cones is used to cleanse the feet and counter excessive sweating. Taken internally, cypress acts as an antispasmodic and general tonic, and is prescribed for whooping cough, the spitting up of blood, and spasmodic coughs. Colds, flu and sore throats, and rheumatic aches and pains, also benefit from this remedy.

CURCUMA AMADA
(Zingiberaceae)

MANGO-GINGER

Constituents: Mango-ginger contains a volatile oil and pungent principles.

Part used: Rhizome.
Medicinal actions and uses:A close relative of turmeric (*C. longa*), mango-ginger is used in traditional Indian herbal medicine to treat flatulence, stomach pain, bad breath, loss of appetite, hiccups, indigestion, colic and constipation. It is also given for coughs and other chest conditions such as bronchitis. The mashed or grated rhizome is applied externally to the skin to treat ulcers, bruises, wounds and sprains.

CURCUMA LONGA
(Zingiberaceae)

TURMERIC, HALDI (HINDI), JIANG HUANG (CHINESE)
A perennial reaching 90 cm, with a short stem, lanceshaped leaves and knobbly rhizome.
Key Constituents:
Volatile oil (3-5%), including zingiberin and turmerone.
Curcumin
Bitter principles
Resin
Key Actions:
Stimulates secretion of bile
Anti-inflammatory
Eases stomach pain
Antioxidant
Antibacterial

CURCUMA ZEDOARIA
(Zingiberaceae)

ZEDOARY
Part used: Rhizome.
Constituents: Zedoary contains a volatile oil, sesquiterpenes, curcumemone, curcumol and curdione. Curcumol and curdione have anti-cancer properties.

Medicinal actions and uses: An aromatic, bitter digestive stimulant, zedoary is used in much the same way as ginger (*Zingiber officinale*) – to relieve indigestion, nausea, flatulence and bloating, and generally to improve the digestion. The rhizome is used in China to treat certain types of tumours.

CUSCUTA EPITHYMUM
(Convolvulaceae)

DODDER, HEELWEED. DEVIL'S GUTS.
Parts used: Aerial parts.
Constituents: Dodder contains flavonoids (including kaempferol and quercitin), and hydroxycinnamic acid.
Medicinal actions and uses: In line with its traditional use to purge black bile, dodder is still considered a valuable, though little-used, herb for problems affecting the liver and gallbladder. It is thought to support liver function and is taken for jaundice. Dodder has a mildly laxative effect, and is also taken for urinary problems.

CUSCUTA REFLEXA
(Convolvulaceae)

Seeds – carmin. anthel. alter.
Plant – purg. used externally against itch, internally in protracted fevers.
Infusion of plant – used as a wash for sores.
Stems – useful in bilious disorders.
Plant contains cuscutalin and cuscutin, very potent drug, contains pigments amarbelin, and cuscutin.
Wax, semi-drying oil.
Parasitic climber.

CYANOPSIS TETRAGONOLOBA
(Leguminosae)
GUAR GUM

Parts used: Pods, seeds.

Constituents: Guar gum contains about 86% water-soluble mucilage, comprising mainly galactomannan.

Medicinal actions and uses: Guar gum is an effective bulk laxative, similar in action to psyllium (*Plantago ovata*). It delays the emptying of the stomach and thus slows down absorption of carbohydrates. As this appears to help stabilise blood sugar levels, guar gum may prove useful in pre-diabetic conditions and in the early stages of late-onset diabetes. Guar gum also lowers cholesterol levels. In Indian medicine, guar seed is a laxative and a digestive tonic.

CYCAS CIRCINALIS
(Cycadaceae)

Pollus – narcotic.

Bark and Seeds – ground to a paste with coconut oil, used as poultice for sores and swellings.

Juice of tender leaves – useful for flatulence and vomiting.

Seeds contain starch, a toxic glucoside, pakocin, phytisterin and a reducing sugar.

CYDONIA OBLONGA
Rosaceae (Quince)

The fruits and the seeds are used medicinally. The fruit pulp contains sugars, pectins, essential oils, tannins and organic acids, including vitamin C. In herbalism an infusion of the dried fruit is used to treat digestive disorders, sore throat, diarrhoea and

Cydonia oblonga

haemorrhage of the bowel. The seeds contain up to 22 per cent mucilage, a fatty oil, a glycoside (amygdalin) and tannins. They have emollient, expectorant, anti-inflammatory and astringent properties and are used dried, **uncrushed,** in an infusion or decoction to treat cough, gastritis and enteritis. Externally mucilaginous compresses made from soaked crushed seeds can be applied to wounds, ulcers, inflamed joints, chapped skin and eye inflammations. The mucilage from the seeds also makes a gargle for sore throats.

The fresh pulp of Quince fruits makes excellent preserves, jellies and syrups. The fruit is only edible when cooked; the **seeds must not be eaten.**

CYMBOPOGON CITRATUS
(Graminaeae)

LEMON GRASS

Parts used: Leaves, essential oil.
Constituents: Lemon grass contains a volatile oil with citral (about 70%) and citronellal as its main constituents. Both are markedly sedative.
Medicinal actions and uses: Lemon grass is principally taken as a tea to remedy digestive problems. It relaxes the muscles of the stomach and gut, relieves cramping pains and flatulence and is particularly suitable for children. In the Caribbean, lemon grass is primarily regarded as a fever-reducing herb (especially where there is significant catarrh). It is applied externally as a poultice or as diluted essential oil to ease pain and arthritis. In India, a paste of the leaves is smeared on patches of ringworm.

CYNARA CARDUNCULUS
Asteraceae (Cardoon)

The leaves are used medicinally. The constituents include a bitter compound (cynarine), mucilage, tannins, organic acids and vitamin A. These substances give Cardoon strong choleretic and diuretic properties and it is used with success in the treatment of gall bladder and liver disorders, including hepatitis. Cardoon is also hypoglycaemic and antisclerotic. On the Continent extracts of Cardoon are included in proprietary digestive tonics in aperitifs and in liqueurs. The fresh juice from the leaves can be used externally to treat some skin disorders. The inner leaf stalks and midribs can be blanched and eaten as a vegetable.

Cynara cardunculus

Lowers blood sugar, cholesterol in diabetics. Diuretic and rheumatic treatment.

CYPERUS ESCULENTUS
(Cyperaceae)

CHUFA, TIGER NUT

Parts used: Tubers.
Constituents: Chufa contains 20-36% fixed oil, known as chufa or tiger nut oil.
Medicinal actions and uses: Chufa is regarded as a digestive tonic, having a heating and drying effect on the digestive system and alleviating flatulence. It also promotes urine production and menstruation. The juice is taken to heal ulcers of the mouth and gums. Ayurvedic medicine classifies the nuts as digestive, tonic, effective

against flatulence and aphrodisiac. In this tradition, they are taken for flatulence, indigestion, colic, diarrhoea, dysentery, debility and excessive thirst.

CYPRIPEDIUM PUBESCENS
(Orchidaceae)

LADY'S SLIPPER, AMERICAN VALERIAN

Parts used: Rootstock.

Constituents: Lady's slipper is poorly researched, but it is known to contain a volatile oil, resins, glucosides and tannins.

Medicinal actions and uses: Due to its scarcity and cost, lady's slipper is now used on a small scale. A sedative and relaxing herb, it treats anxiety, stress-related disorders such as palpitations, headaches, muscular tension, panic attacks, and neurotic conditions generally. Like valerian (*Valeriana officinalis*) lady's slipper is an effective tranquilliser. It reduces emotional tension and often calms the mind sufficiently to allow sleep. Indeed, its restorative effect appears to be more positive than that of valerian.

CYTISUS SCOPARIUS
Fabaceae (Broom)

All parts of the plant have a medicinal value – the flowers, flowering stems, seeds and roots, the flowering stems being used the most. The most important active constituent is the alkaloid sparteine. Other ingredients include a glycoside (scoparin), tannins, essential oils and bitter compounds. The amounts of these substances vary a great deal and Broom is now regarded as too unreliable –

Cytisus scoparius

and therefore potentially too toxic – for general use. Some of the constituents are, however, isolated and included in proprietary preparations. For example, medicines containing sparteine are sometimes prescribed for heart and circulatory disorders because this substance dilates the blood vessels and raises blood pressure (it is hypertensive). Sparteine also stimulates the smooth muscles of the intestines and the uterus and it is utilised in obstetrics. **Dosage and treatment with any preparation containing Broom must be prescribed by a medical practitioner.**

DAEMONOROPS DRACO
(Arecaceae)

Resin – astrin. used in diarh. dysen. eye troubles and in dentrifices. Gum dragon.

DALBERGIA LANCEOLARIA
(Fabaceae)

Bark – used in intermittent fever, its infusion given internally in dyspep.
Seed oil – in rheum. in dyspep.
Tannin in Bark.

DALBERGIA LATIFOLIA
(Fabaceae)

Plant – bitter tonic, stomach-used in dyspep. diarh. leprosy, obesity and worms.
Bark contains tannins.

DALBERGIA SISSOO
(Fabaceae)

Leaves – bitter, stim.
Decoct. of leaves – useful in gonor.
Roots – astrin.
Wood – alter. useful in leprosy, boils, eruptions, and to allay vomiting.
Pods contain tannin.

DAPHNE OLEOIDES
(Thymelaeaceae)

Plant – poisonous
Roots – purg.
Bark and leaves – in cutaneous affections.
Leaves or an infusion – given for gonor. applied to abscesses.

DAPHNE MEZEURUM
Thymeleaceae (Mezereon)

The bark is the medicinal part. Its constituents include the acrid poisonous alkaloid mezereine and the glycoside daphnine. These substances have an irritant, rubefacient effect and cause blistering if used for a long period. Daphnine is also hallucinogenic. A tincture is used in homeopathy for various skin disorders. All

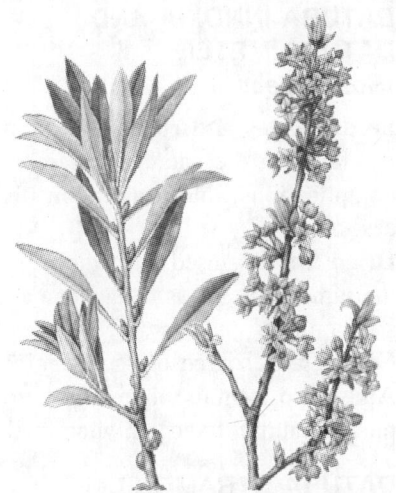

Daphne mezeurum

Mezereon preparations, including ointments and liniments, should be applied only under strict medical supervision. The herb should not be taken internally. Mezereon is dangerous and it should never be collected and used for self-medication. It is used occasionally as an external counterirritant and is effective on rheumatic joints, increasing blood flow to the affected area. It has been considered effective for rheumatic joints, causing inflammation and occasional blistering.

Daphnetoxin and mezerein are highly toxic but they have anti-leukaemic properties, and have been used in a number of countries in the treatment of cancer.

DATISCA CANNABINA
(Datiscaceae)

Herb – bitter, purg. febge, diur.
Root – sedative in rheum. applied to carious teeth.
Glucoside dirtiscin.

DATURA INNOXIA AND DATURA METEL
(Solanaceae)

Seeds, leaves and roots – in insanity, fever in the catar. and cerebral complications, diarh. and skin diseases, antisp.

Dried leaves – used in medicine for the same purposes as belladonna and stramonium.

Commercial source of scopolamine; Alkls. hyoscyamine, hyoscine, atropine, allantoin, fixed oil, vitamin C.

DATURA STRAMONIUM
Solanaceae (Thorn Apple)

The leaves and seeds are used medicinally. Their constituents include tropane alkaloids (0.4 per cent), principally hyoscyamine, atropine and scopolamine, and traces of an essential oil. These substances give Thornapple antispasmodic and hallucinogenic actions, they inhibit glandular secretions and dilate the airways. In medicine today Thornapple is mostly used in tinctures and proprietary preparations to treat asthma, bronchitis and Parkinson's disease. It is occasionally prescribed in the form of cigarettes. **Thornapple is a dangerous plant; it should never be collected and used for self-medication.** It induces symptoms of poisoning similar to those caused by Deadly Nightshade.

DAUCUS CAROTA
Apiaceae (Carrot)

The fully grown fresh root is used medicinally, finely grated or the juice is strained off and used on its own. The constituents include carotenes (provitamin A) and vitamins B complex and C. The alkaloid daucine, which has a nicotine-like odour, and sugars and pectins. Fresh carrot, particularly the carotene constituent, affects the keenness of sight and the ability to see in dimlight. It has anthelmintic, diuretic and stomachic properties. For children it is effective for digestive ailments and tonsillitis. A carrot diet is said to alleviate pain in cancer patients. Carrot juice can however, be toxic if taken in excessive amounts as it induces hypervitaminosis A. An infusion of

Datura Stramonium

Daucus Carota

the fruits is sometimes used in tea mixtures as an anthelmintic.

Carotene in carrot is converted to vitamin A by the liver which improves vision in general.

DECALEPIS HAMILTONII
(Aclepiadaceae)

Root – considered an appetiser and blood purifier.

Volatile principle 4-0-methyl resorcylaldehyde, inositol, saponins, tannins, crystelline resin acid, amorphous acid, a ketonic substance.

DEERINGIA AMARANTHOIDES
(Amaranthaceae)

Leaves – applied to sores
Roots – used as sternutatory
Alkaloid substances.

DELPHINIUM AJACIS
(Ranunculaceae)

Seeds – insecticidal, in form of a tincture applied externally for the destruction of lice in hair.

Alkaloids, ajacin, ajaconine, ajacinine, ajacinoidine, and a base resembling lycoctonine.
Fatty oil.

DELPHINIUM CONSOLIDA
Ranunculaceae (Forking Larkspur)

The flowering stems and seeds contain alkaloids, the glycoside delphinine, aconitic acid (not in the flowers) and constituents yet to be identified. These substances give Forking Larkspur diuretic, anthelmintic, insecticidal and purgative properties. It is

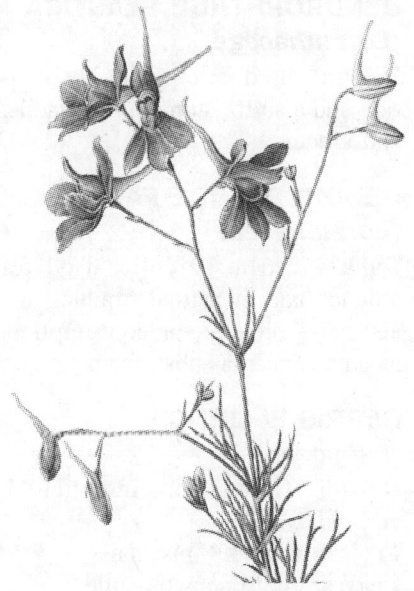

Delphinium consolida

dangerous to take any preparation from it internally. **Forking Larkspur should never be collected and used for self-medication**. The alkaloids act on the nervous system like those of Monkshood causing general weakness & eventually respiratory failure.

DENDROBIUM CRUMENTATUM
(Orchidaceae)

Used by Malays for affections of the brain and nerves and a conserve of the flowers and leaves used for cholera and pounded leaves for poulticing boils and pimples.

Alks. in bulbs and leaves.

DENDROCALAMUS STRICTUS
(Poaceae)

Silicious matter – tonic, astrin.
Leaves – ecbolics to animals.

DENDROPHTHOE ELASTICA
(Loranthaceae)
Leaves – used for checking for abortion and also for stone in the bladder and kidney.

DENDROPHTHOE FALCATA
(Loranthaceae)
Bark – astrin. narcotic, used for wounds and menstrual troubles and also as a remedy for consumption, asthma & mania subst. for betel nut.

DERRIS ELLIPTICA
(Fabaceae)
Roots – fish poison, insecticidal, larvicide.
Fruit and bark – piscidal.
Leaves – poisonous to cattle.
Glucd – derrid, anhydro derrid, tubotoxin, derrin, deguelin, tephrosin, toxicarol, resin, rotenone, l-elliptone.

DERRIS FERRUGINEA
(Fabaceae)
Roots – insecticidal
Rotenone.

DERRIS SCANDENS
(Fabaceae)
Plant – used as fish poison, has no insecticidal value.
Roots contain scandenin, nallanin and chandanin.

DERRIS TRIFOLIATA
Bark – fish poison, useful in rheum. and dysen.
Alk. glucd rotenone.

DESCURIANA SOPHIA
(Brassicaceae)
Flowers and leaves – astrin. antiscor. useful in fevers, broncht. and dysen. given for worms and calculus complaints, mixed in syrup swallowed as cure for fever.
Plant pungent odour when rubbed and an acrid biting taste due to volatile alk.

DESMODIUM GANGETICUM
(Leguminosae)
SALPAN
Part Used: Root.
Constituents: Salpan contains a volatile oil and an alkaloid.
Medicinal Actions and Uses: Salpan root is bitter and tonic, and is used in Ayurveda to improve poor appetite and digestion, and to treat dysentery and haemorrhoids. The plant is also given for feverish and catarrhal conditions such as bronchitis and asthma.

DESMODIUM TRIQUERTUM
(Fabaceae)
Leaves – extract or pills used in piles subst. for tea.
Dried leaves contain tannin.

DIANTHUS ANATOLICUS
(Caryophyllaceae)
Antiper. in intermittent fevers.

DIANTHUS CARYOPHYLLUS
(Caryophyllaceae)
In Spain and North America the flowers considered cardiotonic, diaphor. alexitric, nervine and antisp.
In China plant used as vermifuge.

DIANTHUS SUPERBUS
(Caryophyllaceae)
QU MAI, FRINGED PINK
Parts Used: Aerial parts.
Constituents: *Qu Mai* contains a volatile oil, including eugenol,. benzyl benzoate an methyl salicylate.

Medicinal Actions and Uses: In tradition Chinese medicine, the bitter tasting *Qu Mai* clears "damp-heat", a has been used principally to treat hot, pain conditions of the kidneys and urinary tube such as kidney stones, urinary tract infections, and blood in the urine. Little used on its own, *qu mai* is taken with *dan shenm* (*Salvia militorrhiza*) to induce menstruation. The herb is also used for constipation and some types of eczema.

DICHROA FEBRIFUGA
(Saxifragaceae)

Roots and leafy tops – emetic, febge. antimalaria.
Roots contains two neutral compounds dichrin A and B and two alkaloids, dichroine A and B. Roots and leaves alk, febrifugine and isofebrifugive approximately 100 times as active as QUININE against *plasmodium lophurae* in ducks, possesses greatest antimalarial activity.

DICHROCEPHALA LATIFOLIA
(Asteraceae)

Tender shoots – employed in Cambodia as a poultice for blenorrhagia and for insect bites and stings.
Decoct. of flower buds – considered sudorific in diur. in Java.

DICHROSTACHYS CINEREA
(Fabaceae)

Bruised young shoots – useful in ophthalmia.
Root – astrin. used in rheumatian, urinary calcute and renal troubles.

DICOMA TOMENTOSA
(Asteraceae)

Herb – bitter, febge, especially in febrile attacks after child-birth. Local application to putrescent wounds.

DICTAMNUS ALBUS
Rutaceae (Burning Brush)

The rhizome is the medicinal part. Its constituents include an essential oil, bitter compounds and the alkaloid dictamine. Dictamine in particular causes contractions of the smooth uterine muscles. Burning Bush also has diuretic, laxative, carminative and anthelmintic actions. In herbalism it is prescribed in the form of an infusion. In homeopathy a tincture prepared from the fresh leaves is pre scribed for gynaecological disorders and constipation. **Dictamine is toxic in strong doses.** Burning Bush can also cause allergic skin reactions in hypersensitive individuals.
It has effect on gastro-intestinal tract as antispasmodic. Stimulates muscles

Dictamnus albus

of uterus inducing menstruation and sometimes causing abortion.

Roots contain crystalline toxic alk. dicatmine, trigonelline, choline, obaculactone, crystalline saponin, dictamnolactone, essential oil.

DIDYMOCARPUS PEDICILLATA
(Gesneriaceae)

Leaves – used as a cure for stones in kidney and bladder.

Crystalline colouring matters, including pedicin, pedicellin, pedicinin; and methyl pedicinin isolated from leaves; essential oil-didymocarpene.

DIEFFENBACHIA SEGUINE
(Araceae)

Juice of plant – poisonous

Leaves – used in Malaya for the treatment of rheum. and swellings; they are either powdered and applied as poultice or boiled in oil and used as embrocation –

Rhizome contains calcium oxalate.

DIGITALIS GRANDIFLORA
(Large Yellow Fox Glove)

SCROPHULARIACEAE

The leaves are used medicinally. They provide a basic raw material for making cardenolides (purpurea-glycosides and lanatosides), which are important medicines for strengthening and regulating the contractions of the heart muscles. They are highly toxic and can be prescribed only by qualified medical practitioners. **This and other foxgloves species should never be collected and used for self-medication.**

Digitalis grandiflora

DIGITALIS LUTEA
(Scrophulariaceae)

YELLOW FOXGLOVE

Parts Used: Leaves

Constituents: Yellow foxglove contains cardiac glycosides (including the cardenolides alpha-acetyldigitoxin, acetyldigitoxin and lanatoside). All act to strengthen the beating of a weakened heart.

Medicinal Actions and Uses: Yellow foxglove is little employed in herbal medicine, but in fact it is a less toxic alternative to purple foxglove and woolly foxglove (*D. lanata*). It has similar medicinal actions but its alkaloids are more readily metabolised and flushed out by the body. Like other foxgloves, this plant supports a weakened or failing heart, increasing the strength of contraction, slowing and steadying the heart rate, and lowering blood pressure by strongly stimulating the production of urine, which reduces overall blood volume.

DIGITALIS LANATA
(Woolly Foxglove)

SCROPHULARIACEAE

The principal constituents of the leaves are the cardiac glycosides lanatosides A, B and C, digitoxin and digoxin. Therapeutically lanatosides are four times as potent as the purpurea glycosides of Floxglove. They, but especially digoxin, are used in medical practice in the form of tinctures, tablets, injections, suppositories and other preparations as cardioactive medicines that stimulate and regulate heart action in cases of arrhythmia, tachycardia (an abnormal increase in the heartbeat) and heart failure. These drugs can be prescribed only by qualified medical practitioners.

DIGITALIS PURPUREA
(Foxglove)

SCROPHULARIACEAE

The leaves of Foxglove (known in medicine as digitalis) contain several cardiac glycosides (purpurea-glycosides A and B), saponins, organic acids, tannin and mucilage. The glycosides, among them digitoxin and gitoxin, are isolated and used to make important cardiac medicines which are used, like those of other foxgloves, to strengthen and regulate the heart. The leaves and the isolated glycosides are also strong diuretics. They do not act directly on the kidneys but by improving heart function they improve blood circulation to the kidneys, which then function more effectively in flushing out excess fluid from the

Digitalis lanata   Digitalis purpurea

tissues. The glycosides are dangerous in high doses and they tend to accumulate in the body. Symptoms of poisoning include abdominal pain, irritation of the stomach and bowel, nausea, abnormal heart action and perhaps tremors and convulsions. Death can follow., Digitalis preparations are available only on prescription.

DIOSCOREA OPPOSITA (Dioscoreaceae)

SHAN YAO (CHINESE)

CHINESE YAM

Part Used: Root.
Constituents: Contains steroidal saponins.
Medicinal Actions and Uses: A gentle tonic, *shan yao* is prescribed for tiredness, weight loss and lack of appetite. The root strengthens a weak digestion, improves appetite and may help bind watery stools. It helps to counter excessive sweating, frequent urination and chronic thirst, and it is also given for chronic coughs and wheezing. *Shan Zao* is also taken to treat vaginal discharge and spermatorrhoea (involuntary emission of semen). Like other yam species its traditional use indicates a hormonal effect.

DIOSCOREA VILLOSA (Dioscoreaceae)

WILD YAM

A deciduous perennial vine, climbing to 6 m, with heartshaped leaves and tiny green flowers.

Key Constituents:
Steroidal saponins (mainly dioscin)
Phytosterols
Alkaloids
Tannins
Starch

Key Actions:
Antispasmodic
Anti-inflammatory
Antirheumatic
Increases sweating
Diuretic

DIOSPYROS MELANOXYLON (Ebebaceae)

Bark – astrin.
Decoct. of bark – in diarh. dyspep. tonic, a dilute extract used as astrin. lotion for the eyes.
Leaves – diur. carm. laxat. and styptic.
Dried flowers – useful in urinary, skin and blood diseases.
Tannins

DIOSPYROS MONTANA (Ebenaceae)

Fruits – poisonous, applied externally to boils.
Crushed leaves and fruits – used for stupefying fish.

DIOSPYROS PANICULATA (Ebenaceae)

Leaves – fish poison.
Dried and powdered fruits – applied to heal burns.
Decoct. of fruit – given in gonor. to purify blood and biliousness.
Powdered bark – in rheum. and ulcers.

DIOSPYROS PEREGRINA (Ebenaceae)

Fruit and stem bark – astrin.
Oil of seeds – given in diarh. and dysent.

Unripe fruit – acrid, bitter, oleoginous.

Infusion of fruits – used as gargle in aphthae and sore throats; juice used as application for wounds and ulcers.

Bark – used in dysent. and intermittent fevers.

Bark – tannin.

Fruit juice – antibacterial.

DIPSACUS FULLONUM
(Dipsacaceae)

TEASEL

Part Used: Root

Constituents: Teasel contains inulin, bitter substances and a scabioside.

Medicinal Actions and Uses: Teasel root is little used medicinally today, and its therapeutic applications are disputed. It is thought to have diuretic, sweat-inducing and stomach-soothing properties, cleansing the system and improving digestion. Due to its apparent astringency, teasel is considered helpful in diarrhoea. It is also thought to increase appetite, to tone the stomach, and to act on the liver, helping with jaundice and gallbladder problems. There is no clear picture of teasel's actions, but its closeness to the thirst family means it might well reward careful investigation.

DIPCADI ERYTHRAEUM
(Liliaceae)

Bulbs – used as a subst. or adulterant for Indian squill.

(*Urginea indica*), resembles digitalis in actions and is used mainly as an expect. in treatment of cough.

DIPLOCLISIA GLAUCESCENS
(Menispermaceae)

Leaves – powdered and taken in milk to cure syphilis, biliousness, and gonor.

DIPTERACANTHUS SUFFRUTICOSA
(Acanthaceae)

DIPTERACANTHUS SUFFRUTICOSA

Roots – used in gonor. syphilis, and renal affections, dried and ground root taken in a dose of 2 oz. Causes abortion, also used as medicine for sore eyes.

DIPTEROCARPUS ALATUS
(Dipterocarpaceae)

Balsam – subst. for copaiba, externally used in gonor.

Bark – considered tonic and depurative and prescribed in rheum.

Oil – applied to ulcerated wounds Essential oil, resin, crystalline acrid.

DIPTEROCARPUS TURBINATUS
(Dipterocarpaceae)

Oleoresin – applied to ulcers, ring worm and cutaneous affections, diur., gonor.

DODONAEA VISCOSA
(Sapindaceae)

Leaves – febge, sudorific, in gout and rheum. for wounds, swellings and burns.

Bark – employed in astrin. baths and fermentation.

Plant – fish poison.

Saponin, dodonin.

DOLICHOS BIFLORUS
(Fabaceae)

Seeds – astrin. diur. tonic
Decoct. – used in leucor. and menstrual disorders.
Seeds rich source of urease.

DOLICHOS FALCATUS
(Fabaceae)

Roots – used in piles
Decoct. of seeds – specific for rheum.

DOREMA AMMONIACUM
(Umbelliferae)

AMMONIACUM

Part Used – Oleo-gum-resin.
Constituents – Ammoniacum contains a resin (60-70%), gum, volatile oil (including ferulene and linalyl acetate), free salicylic acid and couramins.
Medicinal Actions and Uses: Used in both Western and Indian medicine, ammoniacum is still listed in the *British Pharmacopoeia* as an antispasmodic and as an expectorant that stimulates the coughing up of thick mucus. It is a specific treatment for chronic bronchitis, asthma and persistent coughs. Ammoniacum is also occasionally used to induce sweating or menstruation in enlargement of liver and spleen.

DORSTENIA CONTRAYERVA
(Urticaceae)

CONTRAYERVA

Part Used: Rhizome.
Contrayerva rhizome is considered aromatic, stimulant and sweat-inducing. Occasionally used in the early stages of serious fevers such as ty-

phoid, it is also given for gastro-intestinal problems such as diarrhoea and dysentery. There is no scientific substantiation of its reputation as an antidote.

DROSERA ROTUNDIFOLIA
(Droseraceae)

SUNDEW

Parts Used: Aerial parts.
Constituents: Sundew contains naphthaquinones, enzymes, flavonoids and volatile oil. The naphthaquinones are antimicrobial, antispasmodic and also cough-suppressing.
Medicinal Actions and Uses: Sundew is of greatest value in the treatment of spasmodic chest conditions such as whooping cough, bronchial asthma and asthma. In relaxing the muscles of the respiratory tract, the plant eases breathing, relieves wheezing and lessens the spasms of whooping cough. Commonly mixed with thyme in a syrup, sundew is a helpful remedy for coughs in children. The herb is also prescribed for gastric problems.

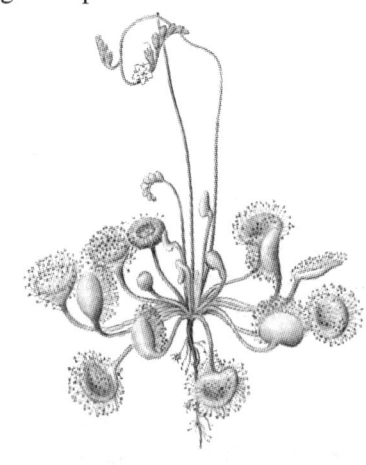

Drosera rotundifolia

DRYOPTERIS FILIX-MASS
Aspidiaceae (Male Fern)

The rhizome along with the frond bases are used medicinally. The main constituents are an oleoresin, filicin, phloro-glucinol compounds (for example, flavasoidic acid), starch and tannins. An ethereal extract is used in both human and veterinary medicine as an anthelminthic, especially against tapeworms. Male Fern preparations are strictly controlled and cannot be obtained from herbal practitioners. It is a dangerous plant; strong doses are very toxic and may lead to permanent blindness, even death. **Male Fern should never be collected and used for self-medication.**

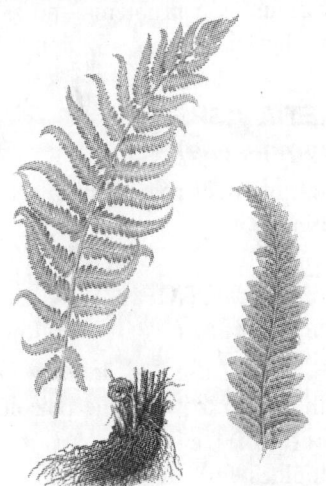

Dryopteris Filix-Mass

DURANTA PLUMIERI
(Verbenaceae)

Plant – poisonous

Leaves contain saponin and fruits narcotine; macerated fruits yield a juice which can kill mosquito larvae even in 1:100 dilution. Larvicide in ponds and swamps.

DURIO ZIBETHINUS
(Bombacaceae)

Decoct. of roots – used for fever.

Fruits – as tonic

Leaves and roots – in a compound fever.

Leaves – used in medicinal bath for fever.

ECHINACEA ANGUSTIFOLIA & E. PURPUREA
(Asteraceae)

ECHINACEA, PURPLE CONEFLOWER

Echinacea: A perennial growing to 50 cm, with daisy-like purple flowers, and leaves covered in coarse hair.

Key Constituents:

Alkamaides (mostly isobutylamides with olefinic and acetylenic bonds)

Caffeic acid esters (mainly echinacoside and cynarin)

Polysaccharides

Volatile oil (humulene)

Echinolene

Betanine

Key Actions:

Immune stimulant

Anti-inflammatory

Antibiotic

Detoxifying

Increases sweating

Heals

Anti-allergenic

ECHINOPS ECHINATUS
(Asteraceae)

Plant – alter, diur., nerve tonic, used in hoarse cough, hysteria, dyspep., scrofula and ophthalmia.

Powdered roots – applied to wounds in cattle to destroy maggots; mixed with acacia gum, applied to hair to destroy lice.

ECHIUM VULGARE
(Boraginaceae)

Viper's Bugloss

Parts Used: Flowering tops.
Constituents: Viper's bugloss contains pyrrolizidine alkaloids, allantoin, alkannins and mucilage. In isolation, pyrrolizidine alkaloids are toxic to the liver. The alkannins are antimicrobial and allantoin helps wounds to heal.
Medicinal Actions and Uses: In many respects, viper's bugloss is similar to borage (*Borago officinalis*) in that both herbs have a sweat-inducing and diuretic effect if taken internally. Viper's bugloss has also been taken to treat chest conditions, as its mucilage soothes dry coughs and encourages expectoration. The significant mucilage content in viper's bugloss has also proved helpful in treating skin conditions. Prepared in a poultice or plaster, it is an effective balm for boils and carbuncles. In recent times, this herb has fallen out of use, due partly to lack of interest in its medicinal potential, and partly to its pyrrolizidine alkaloids, which in isolation are toxic. Viper's bugloss may be safely used externally.

ECLIPTA PROSTRATA
(Compositae)

Trailing Eclipta

Parts Used: Aerial parts.
Constituents: Trailing eclipta contains saponins, including ecliptine and alpha-terthienylmethanol.

Medicinal Actions and Uses: Trailing eclipta has remarkably similar uses in Ayurvedic and Chinese herbal medicine. In both of these traditions, a decoction is used to invigorate the liver, to prevent premature greying of the hair, and to staunch bleeding, especially from the uterus. In the Chinese tradition, the herb is considered a *yin* tonic; in Ayurvedic medicine it is thought to prevent ageing. In the Caribbean, the juice is sometimes taken for asthma and bronchitis. Trailing eclipta is also used there as a treatment for enlarged glands, as well as for dizziness, vertigo and blurred vision. The plant is employed externally for various skin problems and as a wound healer.

EHRETIA ASPERA
(Boraginaceae)

Decoct. of fresh roots – in venereal diseases.

EHRETIA MICROPHYLLA
(Boraginaceae)

Root – alter. used in debility and syphilis, antid to vegetable poison.
Decoct. of leaves – used in phillippines for cough and stomach troubles.

Glucd contains chlorogenic acid.

ELEAGNUS ANGUSTIFOLIA
(Elaeagnaceae)

Oil from seeds – with syrup as an electuary used in catar. and bronchial affections.

Bark contains alk. deagnine and another viscous oil alk.

ELEAGNUS LATIFOLIA
(Eleagnaceae)
Flowers – cardiac. Astrin.
Fruits – astrin.
Eleagnus Umbellata
Flowers – stim., cardiac, astrin.
Seeds – used as stim. in coughs.
Expressed oil – used in pulmonary affections.

ELAEIS GUINEENSIS
(Arecaceae)
African oil palm
In Guinea oil applied to wounds as vulnery, used as liniment for rheumatism and courbature. Oil rich source of carotene, can be used in place of cod-liver oil for correcting vitamin A deficiency.

ELAEOCARPUS FLORIBUNDUS
(Elaeocarpaceae)
Infusion of bark and leaves – used as a mouth wash for inflammed gums.

ELAEOCARPUS GANITRUS
(Elaeocarpaceae)
Fruit – used in diseases of the head and epileptic fits.

ELAEOCARPUS OBLONGUS
(Elaeocarpaceae)
Fruit – used as an emetic, also in rheum., pneumonia, ulcers, leprosy, dropsy and piles.

ELAEOCARPUS SERRATUS
(Elaeocarpaceae)
Leaves – used in rheum. antid. to poison.
Fruits – used in dysen. and diarr. Fruits contain citric acid, leaves contain vitamin C seeds contain fixed oil.

ELAEOCARPUS TUBERCULATUS
(Elaeocarpaceae)
Decoct. of bark – used in haemetamesis, indigestion and biliousness.
Nuts – used as remedy for rheum. typhoid fever and epilepsy.

ELAEODENDRON GLAUCUM
(Celastraceae)
Powdered leaves – Strenulatory, used as fumigatory to rouse women from hysterical syncope, and as snuff to relieve headache.
Fresh root bark – rubbed into paste with water applied to swellings.
Cold water extract of crushed roots – used as emetic.
Bark, leaves contains tannin.

ELEPHANTOPHUS SCABER
(Asteraceae)
Plant – astrin. cardiac tonic, alter. febge, in snake-bite.
Decoct. of roots and leaves – emol. given in dysmia, diarh. dysent. and swellings or pains in stomach.
Root – given to arrest vomiting, powdered with pepper applied to toothache.
Bruised leaves – boiled in coconut oil applied to ulcers and eczema antibiotic.

ELETTARIA CARDAMOMUM
(Zingiberaceae)
Cardamom, Elachi (Hindi)
A perennial growing to 5 m, with mauve-marked, white flowers and very long lance-shaped leaves.
Key Constituents:
Volatile oil (borneol, camphor, pinene, humulene, caryophyllene, carvone, eucalyptole, terpinene, sabinene).

Key Actions:
Eases stomach pain
Carminative
Aromatic
Warming digestive stimulant
Antispasmodic

ELEUSINE CORACANA
(Poaceae)

Grains – tonic, cooling, useful in biliousness, astrin.

ELEUTHEROCOCCUS
SENTICOSUS
(Araliaceae)

SIBERIAN GINSENG

A deciduous, hardly shrub, growing to 3 m. It has 3-7 toothed leaflets on each stem.

Key Constituents:
Eleutherosides, 0.6-0.9%
Phenylpropanoids
Lignans
Sugars
Polysaccharides
Triterpenoid saponins
Glycans

Key Actions:
Adaptogenic
Tonic
Stimulant
Protects the immune system

EMBELIA RIBES
(Myrsinaceae)

EMBELIA

Part Used: Fruit
Constituents: Embelia contains naphthaquinones, including embelin. Embelin stimulates the production of oestrogen and progesterone, and it may has a contraceptive effect.

(anthelm. tonic, coughs, embelic acid, embelin, Quercitol, alk. christembine, resinoid, volatile oil.

Medicinal Actions and Uses: Embelia has been used in Asia as a home remedy for expelling worms. The herb is also diuretic and relieves flatulence, and is used for indigestion, colic, constipation and debility.

EMBLICA OFFICINALIS
(Euphorbiaceae)

INDIAN GOOSEBERRY

Part Used: Fruit.
Constituents: Indian gooseberry contains a fixed oil, a volatile oil and tannins, vitamin C and fixed oil.
Medicinal Actions and Uses: The astringent Indian gooseberry is given to allay the effects of ageing and to restore the organs. In Ayurvedic medicine, the fruit juice is given to strengthen the pancreas of diabetics. The juice is also given to treat eye problems, joint pain, and diarrhoea and dysentery.

Dried fruit – useful in haemorh. diar. dysent., in combination with iron used for anaemia, jaundice and dyspep. cough.

Sherbet of amla and lemon juice for arresting acute baciliary dysent.

EMILIA SONCHIFOLIA
(Asteraceae)

Decoct. of plant – used as febge. in infantile lympanites and in bowel complaints.
Juice of leaves – in eye inflamm. night blindness, sore ears.
Root – used for diarh.

ENHYDRA FLUCTUANS
(Asteraceae)
Leaves – laxat. useful in skin and nervous affections, antibili. demulc. Dry plant yields essential oil, stigmasterol, and a bitter substance.

ENICOSTEMMA LITTORALE
(Gentianaceae)
Plant – bitter, stomach, tonic, laxat. dried and powdered is given with honey as a blood purifier, in DIABETES, and in dropsy, rheum., abdominal ulcers, hernia, swellings, aches, and insect poisoning.
Bitter principle, bitter glycoside, and ophelic acid.

ENTADA PHASEOLOIDES
(Mimosaceae)
MATCHBOX BEAN
Parts Used: Seeds.
Constituents: Matchbox bean contains significant amounts of saponins.
Medicinal Actions and Uses: Australian Aboriginals use the seeds to treat female sterility and indigestion and as a painkiller.
Seeds, stem and bark – poisonous
Seeds – tonic, emetic, antiper. anthelm.
Juice of wood and bark – for ulcers. Two amorphous saponins from seed, have strong haemplytic action on RBCs and have a depressing effect on respiratory system, and has action on muscles of intestine and uterus.

EPHEDRA SINICA
(Ephedraceae)
EPHEDRA, MA HUANG (CHINESE), DESERT TEA
An evergreen shrub growing to 60 cm, with long narrow sprawing stems and tiny leaves.

Key Constituents:
Protoalkaloids (ephedrine, pseudoephedrine)
Tannins
Saponin
Flavone
Volatile oil
Key Actions:
Western herbal medicine
Increases sweating
Dilates the bronchioles (small airways in the lungs)
Diuretic
Stimulant
Raises blood pressure
Chinese herbal medicine
Disperses cold
Helps problems caused by "external cold"
Aids movement of lung

EPILOBIUM ANGUSTIFOLIUM
(Rosebay Willowherb)
ONAGRACEAE

Epilobium angustifolium

Mainly the leaves are used medicinally. Their constituents include tannins (up to 20 per cent), mucilage, sugars, pectin and vitamin C. These substances give Rosebay Willowherb demulcent, astringent and tranquillizing properties. A decoction or infusion is used to treat headache and migraine. Being rich in vitamin C the tea is recommended as a spring tonic. The rhizomes, which contain fewer tannins and no mucilage are used in a decoction or chewed fresh for stomach disorders, including diarrhoea. The fresh rhizome, and also the fresh leaves and young shoots, can be eaten as a vegetable.

EQUISETUM ARVENSE
(Equisetaceae)

HORSETAIL, BOTTLEBRUSH

Parts Used: Aerial parts.

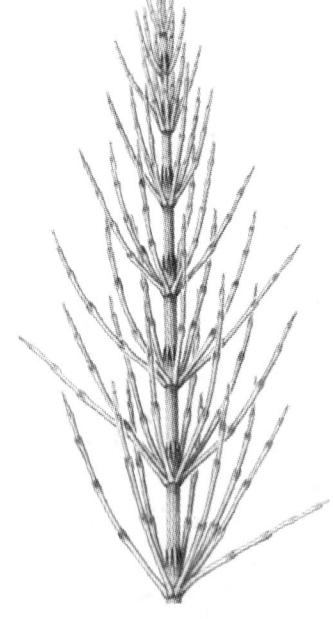

Equisetum arvense

Constituents: Horsetail contains large amounts of silicic acid and silicates (about 15%), flavonoids, phenolic acids, alkaloids (including nicotine) and sterols. Much of the therapeutic effectiveness of this herb is due to its high silica content, a large proportion of which is soluble and can be absorbed. Silica supports the regeneration of connective tissue.

Medicinal Actions and Uses: As its traditional usage indicates, horsetail is an excellent clotting agent. It staunches wounds, stops nosebleeds and reduce the coughing up of blood. In addition, horsetail has an astringent effect on the genito-urinary system, proving especially valuable where there is bleeding within the urinary tract, and in cases of cystitis, urethritis and prostrate disease. Horsetail helps to speed the repair of damaged connective tissue, improving its strength and elasticity. The herb is also prescribed to treat rheumatic and arthritic problems, for chest ailments such as emphysema, for chronic swelling of the legs and for various other conditions. A decoction of the herb added to a bath benefits slow-healing sprains and fractures, as well as certain irritable skin conditions such as eczema.

ERIGERON CANADENSIS
(Compositae)

CANADIAN FLEABANE

Parts Used: Aerial parts.

Constituents: Canadian fleabane contains a volatile oil (including limonene, terpineol and linolool), flavonoids, terpenes, plant acids and tannins, erigeron oil, terpeniol.

Medicinal Actions and Uses: An astringent herb, Canadian fleabane is taken for gastro-intestinal problems such as diarrhoea and dysentery. A decoction of Canadian fleabane is reportedly a very effective treatment for bleeding haemorrhoids. The herb is occasionally used as a diuretic for bladder problems, to clear toxins in rheumatic conditions and to treat gonorrhoea and other urinogenital diseases.

ERIOBOTRYA JAPONICA
(Rosaceae)

Fruit – considered sedative, used in allaying vomiting and thirst.
Flowers – expect.
Infusion of leaves – in diar.
Prinepool constituents of ripe fruit – laevulose, sucrose and malic acid; amygdalin in pericarp of unripe fruit; leaves contain d-sorbitol, ascorbic acid oxidase and vitamin B.

ERIOGLOSSUM RUBIGINOSUM
(Sapindaceae)

Leaves and Roots – used in Malay for poulticing.
Decoct. of roots – given in fevers.
Decoct. of seeds – used for whooping cough.

ERUCA SATIVA
(Brassicaceae)

Seeds – vesic., acrid and used like mustard.
Tender leaves of plant – considered stim., stomach, diur. and antiscorbic.
Essential oil.

ERIODICTYON CALIFORNICUM
(Hydrophyllaceae)
YERBA SANTA
Parts Used: Leaves.
Constituents: Yerba santa contains a volatile oil, flavonoids (including eriodictyol and resin.
Medicinal Actions and Uses: An aromatic herb with a pleasant sweet taste, yerba santa is a valuable expectorant that can be used to treat tracheitis, bronchitis and asthma and similar respiratory tract ailments.

ERVATAMIA CORONARIA
(Apocynaceae)
GRAPE JASMINE, EAST INDIAN ROSEBAY
Parts Used: Root, leaves, latex, wood.
Constituents: Grape jasmine contains alkaloids and resins.
Medicinal Actions and Uses: In Ayurvedic medicine, the root and the latex are used to expel worms. The root is also chewed to relive toothache. The latex is used to treat cataracts (especially in the early stages), eye inflammations and poor eyesight. The leaf juice makes a soothing treatment for skin irritations and wounds. The wood reduces fevers. In Indonesia, a decoction of the root is taken for diarrhoea.
Tabernemontainine and coronarine alkaloids.

ERYNGIUM CAERULEUM
(Apiaceae)
Root – nerve tonic, aphrodis.
Ashes of plant – for haemorrhoids.

ERYNGIUM MARITIMUM
(Umbelliferae)

SEA HOLLY, ERYNGO

Part Used: Root.

Constituents: Sea holly contains saponins, coumarins, flavonoids and plant acids.

Medicinal Actions and Uses: In contemporary European herbal medicine, sea holly is used as a diuretic. It is prescribed as a treatment for cystitis and urethritis, and taken as a means to alleviate kidney stones. It is unlikely that the herb actually dissolves established stones, but it probably helps retard their formation. Sea holly is also used to treat enlargement or inflammation of the prostrate gland, and may be of benefit in treating chest problems.

ERYSIMUM ASPERUM
(Brassicaceae)

Seeds contain cheiroline, resembling quinine in pharmacological action.

ERYTHRAEA CENTAURIUM
(Gentianaceae)

CENTAURY

Parts Used: Aerial parts.

Constituents: Centaury contains many bitter constituents, including secoiridoids, also found in gentian (*Gentiana lutea*).

Medicinal Actions and Uses: One of the most useful bitter herbs, centaury strengthens digestive function, especially within the stomach. By increasing stomach secretions, it hastens the breakdown of food. It also stimulates the appetite and increasing bile production. Centaury needs to be taken over some weeks. The preparation should be slowly sipped so that the components (detectable at a dilution of up to 1:3, 500) can stimulate reflex activity throughout the upper digestive tract.

ERYTHRINA STRICTA
(Fabaceae)

Bark – used in powder form for biliousness, rheum., itch, burning sensation, fever, fainting, asthma, leprosy, epilepsy.
Flowers – antidot. to poison.

ERYTHRINA VARIEGATA VAR. ORIENTALIS
(Fabaceae)

Bark – astrin. febge. anthelm. galact. emmen. as a collyrium in ophthalmia, antid. to snake-poison.
Leaves – laxat. diur. anthelm. galact. emmen. applied externally for dispersing venereal bubocs and for relieving pain in joints.
Juice of leaves – vermfg. cath. poisonous alkaloids.
Leaves – Hypaphorine.
Seeds – Hypaphorine, saponin.
Bark – betaine, choline, along with hypaphorine.

ERYTHRINA VARIEGATA
(Fabaceae)

INDIAN CORAL TREE, DADAP (HINDI)

Parts Used: Bark, leaves.

Constituents: This plant's constituents are unknown.

Medicinal Actions and Uses: In Ayurvedic medicine, Indian coral tree is used to treat inflammatory conditions, period pain and problems related to eating and digestion, including anorexia, flatulence, colic and worms. The bark is used for skin problems, fever and leprosy. A paste made from the leaves is traditionally applied to heal wounds.

ERYTHRONIUM AMERICANUM (Liliaceae)

ADDER'S TONGUE

Parts Used: Leaves.

Constituents: Very little is known about the constituents of this plant. It contains alpha-methylene-butyro-lactone.

Medicinal Actions and Uses: An infusion of the leaves is taken for skin problems such as ulcers and tumours, and for enlarged glands. Adder's tongue is often used to treat scrofulous skin arising from tubercular infection. The leaves (or the whole plant) may also be applied as a poultice for skin conditions. Although the fresh leaves are strongly emetic, they are rarely used to stimulate vomiting.

ERYTHROXYLUM COCA (Erythroxylaceae)

COCA

Parts Used: Leaves.

Constituents: Coca contains cocaine and various other alkaloids, a volatile oil, flavonoids, vitamins A and B^2, and minerals. The plant's stimulant and anaesthetic action is due largely to cocaine.

Medicinal Actions and Uses: In Bolivia and Peru, coca leaves play an important part in the culture and herbal medicine of the indigenous Aymara and Quechua peoples. High altitures, cold and an impoverished diet place great physical demands on the population. Coca leaves, chewed with lime or ashes, release small amounts of the active constituents which act as a tonic and help block the effects of cold, exhaustion and poor nutrition. Coca leaves are also used in South American herbal medicine to treat nausea, vomiting and asthma and have been used to speed convalescence. Cocaine extracted from coca leaves is used legally in conventional medicine as a local anaesthetic. It is also taken illegally as a narcotic, stimulant drug. As an isolated chemical, cocaine is extremely addictive.

ESCHSCHOLZIA CALIFORNICA (Papaveraceae)

CALIFORNIAN POPPY

Parts Used: Aerial parts.

Constituents: Californian poppy contains alkaloids (including protopine, cryptopine and chelidonine) and flavene glycosides.

Medicinal Actions and Uses: Though Californian poppy is closely related to the opium poppy (*Papaver somniferum*), it has a markedly different effect on the central nervous system. Californian poppy is not a narcotic. In fact, rather than disorientating the user, it tends to normalise psychological function. Californian poppy's gently antispasmodic, sedative and analgesic effects make it a valuable herbal medicine for treating physical and psychological problems in children. Californian poppy may also prove beneficial in attempts to overcome bedwetting, difficulty in sleeping and nervous tension and anxiety.

EUCALYPTUS GLOBULUS
(Myrtaceae)

EUCALYPTUS, BLUE GUM

An evergreen tree growing to 100 m, with a blue-grey trunk and green leaves.

Key Constituents:
Volatile oil (cineole; up to 80%)
Flavonoids
Tannins
Resin
Key Actions:
Antiseptic
Expectorant
Stimulates local blood flow
Oil from leaves – antisep. used in infections of the upper respiratory tract, and certain skin infections, mixed with equal amount of olive oil used as rubft. for rheum; also used in ointments for burns; mosquito repellent, internally used as a stimulating expect. in chr. broncht. and asthma.
Leaf Decoct. – insect and vermin repellent.
Root – purg.

EUCALYPTUS SMITHII
(Myrtaceae)

EUCALYPTUS

Part Used: Essential oil.
Constituents: The volatile oil contains about 70% eucalyptol (1, 8-cineole), as well as pinene, limonene, alpha-terpineol and linalool. While it is similar to the oils of related species, this oil appears to be better tolerated by the skin.
Medicinal Actions and Uses: Eucalyptus essential oil is used in aromatherapy, and also as a disinfectant and antiseptic for the treatment of viral conditions, skin and other infections, and as a decongestant.

EUCOMMIA ULMOIDES
(Eucommiaceae)

DU ZHONG (CHINESE), GUTTA PERCHA

Part Used: Bark.
Constituents: *Du zhong* contains gutta percha, alkaloids, iridoids and other glycosides, and potassium.
Medicinal Actions and Uses: *Du zhong* is considered an excellent tonic for the liver and kidneys. It is thought to act specifically to help lower back pain and weakness, knee weakness and frequent urination. *Du zhong* is said to "tonify the *yang*", to improve the circulation, and also to prevent miscarriage in women who are weak or suffering from back pain.

EUGENIA CARYOPHYLLATA
(Myrtaceae)

CLOVE

An evergreen, pyramid-shaped tree growing to 15 m. The tree is strongly aromatic.
Key Constituents:
Volatile oil containing eugenol (up to 85%), acetyl eugenol methyl salicylate, pinene, vanillin.
Gum
Tannins
Key Actions:
Antiseptic
Carminative
Stimulant
Analgesic
Prevents vomiting
Antispasmodic
Eliminates parasites

EUONYMUS ATROPURPUREUS
(Celastraceae)

WAHOO BARK

Parts Used: Stem bark, root bark.
Constituents: Wahoo bark contains cardenolides (cardiac glycosides) similar to digitoxin, asparagine, sterols and tannins.
Medicinal Actions and Uses: Wahoo bark is considered a gallbladder remedy with laxative and diuretic properties. It is prescribed for biliousness and liver problems, as well as for skin conditions such as eczema (which may result from poor liver and gallbladder function) and for constipation. In the past, it was often used in combination with herbs such as gentian (*Gentiana lutea*) as a fever remedy, especially if the liver was under stress. Following the discovery that it contains cardiac glycosides, wahoo bark has been given for heart conditions.

EUPATORIUM CANNABINUM
(Compositae)

HEMP AGRIMONY

Parts Used: Aerial parts, roots.
Constituents: Hemp agrimony contains a volatile oil (with alpha-terpinene, p-cymene, thymol and an azulene), sesquiterpene lactones (especially eupatoriopicrin), flavonoids, pyrrolizidine alkaloids, and poly-saccharides. P-enzyme is antiviral, while cupatoriopicrin has anti-cancer properties and inhibits cellular growth. The polysaccharides stimulate the immune system. In isolation, the pyrrolizidine alkaloids are toxic to the liver.

Medicinal Actions and Uses: Hemp agrimony has been employed chiefly as a detoxifying herb for fever, colds, flu and other acute viral conditions. It also stimulates the removal of waste products via the kidneys. The root is laxative, and the whole plant is considered to be tonic. Recently, hemp agrimony has found use as an immune-stimulant, helping to maintain resistance to acute viral and other infections.

EUPATORIUM PERFOLIATUM
(Compositae)

BONESET

Parts Used: Aerial parts.
Constituents: Boneset contains sesquiterpene lactones (including eupafolin), polysaccharides, flavonoids, diterpenes, sterols and volatile oil. The sesquiterpene lactones and polysaccharides are significant immune-stimulants.
Medicinal Actions and Uses: A hot infusion of boneset will bring relief to symptoms of the common cold. The plant stimulates resistance to viral and bacterial infections, and reduces fever by encouraging sweating. Boneset also loosens phlegm and promotes its removal through coughing, and it has a tonic and laxative effect. It has been taken for rheumatic illness, skin conditions and worms.

EUPATORIUM PURPUREUM
(Compositae)

GRAVEL ROOT, JOE PYE WEED

Parts Used: Root.
Constituents: Gravel root contains a volatile oil, flavonoids and resin.

Medicinal Actions and Uses: As its common name indicates, gravel root is a valuable herb for urinary tract problems. It helps to prevent the formation of kidney and bladder stones and may diminish existing stones. Gravel root is also useful for cystitis, urethritis, prostrate enlargement (and other forms of obstruction), and for rheumatism and gout. The root is thought to help the latter two conditions by increasing the removal of waste by the kidneys.

EUPHORBIA HIRTA
(Euphorbiaceae)
PILL-BEARING SPURGE, ASTHMA PLANT

Parts Used: Aerial parts.

Constituents: Pill-bearing spurge contains flavonoids, terpenoids, alkanes, phenolic acids, shikimic acid and choline. The latter two constituents may be partly responsible for the antispasmodic action of this plant.

Medicinal Actions and Uses: A specific treatment for bronchial asthma, pill-bearing spurge relaxes the bronchial tubes and eases breathing. Mildly sedative and expectorant, it is also taken for bronchitis and other respiratory tract conditions. It is most often used along with other anti-asthmatic herbs, notably gumplant (*Grindelia camporum*) and lobelia (*Lobelia inflata*). In the Anglo-American tradition, pill-bearing spurge is taken to treat intestinal amoebiasis.

EUPHORBIA LATHYRUS
(Euphorbiaceae)
CAPER SPURGE

Parts Used Seeds, latex.

Constituents: The seeds contain a fixed oil and resin; the latex contains euphorbone and resin.

Medicinal Actions and Uses: Caper spurge is so violent a purgative that it is rarely if ever used in contemporary herbal medicine. This indicates the extent to which medicine as a whole has changed in modern times. Purging was the first resort of many traditional medical systems, and never more enthusiastically so than in Western medicine in the 18th century. Caper spurge seeds were commonly employed, but an oil extracted from them was also used in very small doses (the oil is highly toxic). In the past, the milky latex of caper spurge was used as a depilatory and to remove corns, but is too irritant to be used safely.

EUPHORBIA PEKINENSIS
(Euphorbiaceae)
DA JI

Part Used: Root.

Constituents: *Da ji* contains euphorbon.

Medicinal Actions and Uses: *Da ji* is classified as a toxic herb in Chinese herbal medicine, and therefore it is prescribed only for relatively serious illnesses. It is taken as a cathartic to purge excess fluid in conditions such as pleurisy and ascites (excess fluid in the abdomen), and for the treatment of kidney problems, especially nephritis. *Da ji* is applied externally to inflamed sores to reduce swelling. The herb is incompatible with liquorice species (*Glycyrrhiza glabra* and *G. uralensis*), as it neutralises their medicinal effects.

EUPHORBIA NERIIFOLIA
(Euphorbiaceae)

Milky juice – used as purg. rubft. and expect. to remove warts and cutaneous eruptions.

Root – in scorpion-sting and snake-bite, antisp. fish poison.

EUPHORBIA THYMIFOLIA
(Euphorbiaceae)

Dried leaves and Seeds – arom., astrin, stim, laxt., given to children in bowel complaints.

Juice of plant – for ringworm, in snake-bite and skin diseases.

Root – used for amenor.

Essential oil; leaves and stems contain 5, 7, 4-trihydroxy flavone-7-glycoside.

EUPHORBIA TIRUCALLI
(Euphorbiaceae)

Milky juice – vesic, rubft. purg, counter-irrit, application for warts, rheum, neuralagia, toothache, in cough, asthma and ear-ache, fish poison.

Euphorbon, from fresh latex isoeuphorol isolaterol, dried latex contains a ketone euphorone.

EUPHARASIA OFFICINALIS
Scrophulariaceae (Eyebright)

The flowering stems are used medicinally. They contain the iridoid glycoside aucubin, tannins, an essential oil, bitter compounds and pigments. These substances give Eyebright anti-inflammatory properties and it is primarily used as an eyebath and in compresses applied to the eyes. Washing out the eyes with an infusion helps treat inflammation of the conjunctiva, photophobia (an ab-

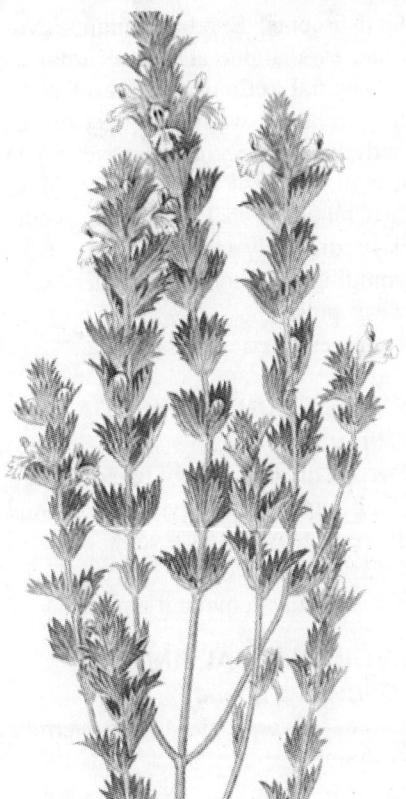

Eupharasia officinalis

normal sensitivity to light) and general tiredness of the eyes. A tincture prepared from fresh plant material is used in homeopathy for the same complaints. Internally Eyebright is used as a stomachic in jaundice. In the past Eyebright was recommended for headache, hysteria and insomnia as well. Poultices of Eyebright are sometimes used to treat stubborn wounds.

EVODIA RUTAECARPA
(Rutaceae)

EVODIA, WU ZHU YU (CHINESE)

Part Used: Fruit

Constituents: Evodia contains evodine, evodiamine and rutaecarpine.
Medicinal Actions and Uses: Evodia has a marked warming effect on the body, helping to relieve headaches and a wide range of digestive problems. In Chinese herbal medicine, evodia is used mainly for abdominal pains, vomiting, diarrhoea, headaches and a weak pulse.
Increases arterial pressure.

EVODIA LUNU-ANKENDA
(Rutaceae)
Decoct. of root, or root bark – boiled in oil given to improve complexion.
Juice of leaves – in fever
Infusion of flowers and leaves – used in Malaya as tonic and emmen.

EVOLVULUS ALSINOIDES
(Convolvulaceae)
Plant – bitter, tonic, febge., vermif. in dysen.
Leaves – made into cigarettes smoked in chr. broncht. and asthma.
Alkaloid.

EXOECARIA AGALLOCHA
(Euphorbiaceae)
Juice – boiled in oil applied in rheum. leprosy and paralysis.
Decoct. of leaves – in epilepsy, application to ulcers.
Bark – emetic, purg.
Root – an ingredient of embrocations used for swellings of hand and feet.
Latex – purg. abortif. fish poison. Acrid latex, tannin.

EXOGONUM PURGA
(Convolvulaceae)
Dried tubercles – hydrogne cath. Purgative action due to resinous constituents – jalap resin, trease.

F

FAGONIA CRETICA
(Zygophyllaceae)
Plant – bitter, astrin, tonic, febge, prophylactic against small pox, in dropxy, dilirium and any disorder which arises from poisoning.
Leaves and twigs – Cooling.

FAGOPYRUM ESCULENTUM
(Polygonaceae)
BUCKWHEAT
Parts Used Leaves, flowers.
Constituents Buckwheat contains bioflavonoids, especially rutin, which is strongly antioxidant. Rutin strengthens the inner lining of blood vessels, roots-oxymethyl anthraquinone.
Medicinal Actions & Uses Used for a wide range of circulatory problems, buckwheat is best taken as a tea or tablet, accompanied by vitamin C or lemon juice (*Citrus limon*) to aid absorption. Buckwheat is used particularly to treat fragile capillaries (seen as small bruises with no apparent cause), but also helps strengthen varicose veins and heal chilblains. Often combined with lime flowers (*Tilia spp.,*) buckwheat is a specific treatment.

FAGOPYRUM TATARICUM
Polygonaceae
(Green Buckwheat)
The discovery of rutin, a substance that affects the strength and permeability of the capillary walls, was of primary importance for the pharma-

Fagopyrum tataricum

ceutical use of Green Buckwheat because the flowering stems contain up to 1 per cent of rutin, a flavonoid glycoside. The substance is used in proprietary medicines for treating circulatory disorders. Buckwheat has, however, been supplanted as a major source of rutin of Japanese Pagoda Tree (*Sophora japonica*).

Feeding buckwheats to cattle is suitable only if the livestock are not out in the sun. If exposed to sun the animals are subject to fagopyrism, a disease characterized by skin rash, swelling of the head and neck, and spasms.

FARSETIA JACQUEMONTII
(Brassicaceae)

Plant – Considered specific for rheum and taken as a cooling medicine after pounding.

FERONIA LIMONIA
(Rutaceae)

WOOD APPLE

Parts Used Fruit, leaves.

Constituents The fruit contains fruit acids, vitamins and minerals. The leaves contain tannins and a volatile oil.

Medicinal Actions & Uses Wood apple fruit is used mainly to stimulate the digestive system. In India, the fruit forms part of a paste applied to tone the breasts. The astringent leaves are used to treat indigestion flatulence, diarrhoea, dysentery (particularly in children) and haemorrhoids.

Pulp – applied externally as a remedy for bites of venomous insects and reptiles.

FERULA ALLIACEA
(Apiaceae)

Gum resin – in scorpion-sting, intestinal antisep, carmin, in hysteria and epilepsy.

Essential oil.

FERULA ASA-FOETIDA
(Umbelliferae)

ASAFOETIDA, DEVIL'S DUNG

Part Used Oleo-gum-resin.

Constituents Asafoetida exudate contains 6-17% volatile oil, as well as resin and gum. The volatile oil contains disulphides, which have an expectorant action. The oil also settles the digestion. Asafoetida resin contains sesquiterpenoid coumarins, including foetidin, pinene, Organic disulphide, unbelloferone.

Medicinal Actions & Uses In Middle Eastern and Indian herbal medicine, asafoetida is used for simple digestive problems such as wind, bloating, indigestion and constipation. Asafoetida's volatile oil, like that of garlic, has components that leave the body via the respiratory system and aid the coughing up of congested mucus. Asafoetida is taken (usually in tablet form) for bronchitis, bronchial asthma, whooping cough and other chest problems. Asafoetida also lowers blood pressure and thins the blood. The herb has a reputation for helping in neurotic states. Improvement may be a psychosomatic response, since the herb's unpleasant smell suggests potency.

FERULA GUMMOSA
(Apiaceae)

GALBANUM

Part Used Oleo-gum-resin.

Constituents Galbanum exudate contains a volatile oil, resins, gums and a coumarin (umbelliferone).

Medicinal Actions & Uses Galbanum is a digestive stimulant and antispasmodic, reducing flatulence, griping pains and colic. It is also expectorant. Applied as an ointment, the gum may help heal wounds.

FERULA JAESCHKEANA
(Apiaceae)

Gum resin – applied wounds and bruises.

Latup contains resin, essential oil – of fruit contains d-α-pinene, cumaldehyde, azulene, sulphur compounds, and an aldehyde.

FICUS ASPERRIMA
(Moraceae)

Juice and Bark – in enlargement of liver and spleen.

FICUS BENGHALENSIS
(Moraceae)

BANYAN TREE

Parts Used: Fruit, bark, leaves, latex, aerial roots.

Medicinal Actions & Uses: The astringent leaves and bark of the tree are employed to relieve diarrhoea and dysentery and to reduce bleeding. As with other *Ficus* species, the latex is applied to haemorrhoids, warts and aching joints. The fruit is laxative and the roots are chewed to prevent gum disease. The bark is used in Ayurvedic medicine for diabetes.

FICUS CARICA
(Moraceae)

FIG

Parts Used: Fruit, latex.

Constituents: Figs contain around 50% fruit sugars (mainly glucose), flavonoids, vitamins and enzymes.

Medicinal Actions & Uses: The fruit sugars within the fig (especially the dried fruit) have a pronounced but gentle laxative effect; syrup of figs is still a remedy for mild constipation. The fruit's emollient pulp helps relieve pain and inflammation, and it has been used to treat tumours, swellings and gum abscesses – the fruit often being roasted before application. Figs are also mildly expectorant and, when used with herbs such as elecampane (*Inula helenium*), are helpful in treating dry and irritable coughs and bronchitis. The milky latex from leaves and stems is reputed to be analgesic, and has long been used to treat warts, insect bites and stings.

FICUS GIBBOSA
(Moraceae)

Root bark – stomach, aper,
Decoct. of root – aper,

FICUS HETEROPHYLLA
(Moraceae)

Juice of root – used internally in colic.

Juice of leaves – mixed with milk in dysent.

Root bark – pulverized and mixed with coriander seed used as remedy in cough and asthma.

FICUS PALMATA
(Moraceae)

Fruit – dermule, laxt, used as diet in cases of constipation and in diseases of lungs and bladder.

FICUS RACEMOSA
(Moraceae)

Bark – astrin. given to cattle when suffering from rinder-pest.

Root – in dysent.

Sap of root – in diabetes

Leaves – powdered and mixed with honey given in billions affections.

Fruit – astrin, stomach, carmin, given in menor and harmoptysis.

Milky juice – in piles and diarh.

FICUS RELIGIOSA
(Moraceae)

PEEPAL

Parts Used Fruit, leaves, bark, latex.
Constituents The fruit contains fruit sugars, flavonoids and enzymes.
Medicinal Actions & Uses Peepal's uses are similar to those of the banyan (*F. benghalensis*). Its astringent bark and leaves are taken for diarrhoea and dysentery, whereas the leaves alone are used for constipation. The leaves are applied with *ghee* (clarified butter) as a poultice to boils and to swollen salivary glands in mumps. The powdered fruit may be taken for asthma and the latex is used to treat warts.

FICUS RETUSA
(Moraceae)

Juice of bark – in liver disease
Powdered leaves and bark – in rheum, headache.
Root bark and leaves – boiled in oil application for wounds and bruises.

FICUS RUMPHII
(Moraceae)

Juice – used to kill worms, and given internally with turmeric, pepper and ghee for the relief of **asthma**.

FILIPENDULA ULMARIA
Rosaceae (Meadow Sweet)

The flowers, and sometimes the young leaves and rhizomes, are used medicinally. All parts contain the glycosides gaulatherin and spiraein, traces of an alkaloid (helio-tropine), tannins, a yellow pigment, vanillin and free salicylic acid, produced by the splitting of gaulthering and citric acid. These substances give the plant anti-pyretic, weak antispasmodic, astringent and antirheumatic properties. The flowers are used in an infusion to treat influenza, and to alleviate headache and rheumatic and arthritic pain. Meadowsweet is gentler on the stomach than aspirin and it is one of the most effective herbal remedies for gastritis and peptic ulcers. Both the leaves and flowers are also strongly diuretic and are used to treat certain bladder and kidney disorders. The

Filipendula ulmaria

fresh root is used in homeopathic preparations.

FIMBRISTYLIS JUNCIFORMIS
(Cyperaceae)

Root – given in dysent.

FLACOURTIA JANGOMAS
(Flacourtiaceae)

Fruit – for biliousness, in liver complaints.
Leaves – in diarh. diaphoret.
Decoction of bark – for biliousness.

FLACOURTIA INDICA
(Flacourtiaceae)

Fruit – in jaundice and enlarged spleen.

Gum – given with other ingredients for cholera.

FLACOURTIA SEPIARIA
(Flacourtiaceae)

Infusion of leaves and roots – in snake-bite
Bark – tritiated in sesamum oil is used as a liniment in rheum.

FLEMINGIA MACROPHYLLA
(Fabaceae)

Roots – Used as external application to ulcers and swellings, mainly of the neck.

FLEMINGIA STROBILIFERA
(Fabaceae)

Roots – used in spiliphy, hysteria.

FLUEGGEA LEUCOPYRUS
(Euphorbiaceae)

Leaves – made into paste with tobacco used to destroy worms in sores.
Plant – fish poison.
Tannin

FOENICULUM VULGARE
(Umbelliferae)
FENNEL

Parts Used Seeds, essential oil.
Constituents Fennel seeds contain about 8% volatile oil (about 80% anethole, plus fenchone and methylehavicol), flavonoids, coumarins (including bergapten) and sterols. The volatile oil relieves wind and is antispasmodic, fatty oil, proteins, sugars and mucilage.
Medicinal Actions & Uses – The primary use of fennel seeds is to relieve bloating, but they also settle stomach pain, stimulate the appetite

Foeniculum vulgare

and are diuretic and anti-inflammatory. Like anise (*Pimpinella anisum*) and caraway (*Carrum carri*), the seeds make an excellent infusion for settling the digestion and reducing abdominal distension. The seeds help in the treatment of kidney stones, and combined with urinary antiseptics such as uva-ursi (*Arctostaphylos uva-ursi*) make an effective treatment for cystitis. An infusion of the seeds may be taken as a gargle for sore throats and as a mild expectorant. Fennel is safe for children and, as an infusion or syrup, can be given for colic and painful teething in babies. Fennel increases breast-milk production and the herb is still used as an eyewash for sore eyes and conjunctivitis. The seeds have a long-standing reputation as an aid to weight loss and to longevity. Essential oil from the sweet variety is used for its digestive and relaxing properties.

FORSYTHIA SUSPENSA (*Oleaceae*)

LIAN QIAO (CHINESE), WEEPING FORSYTHIA

Part Used: Fruit.
Constituents: The fruit contains forsythin.
History & Forlklore: *Lian qiao* was first
Medicinal Actions & Uses: A bitter-tasting, pungent herb with an antiseptic effect, *lian qiao* is chiefly used to treat boils, carbuncles, mumps and infected neck glands, It is also a remedy for colds, flu, sore throats and tonsillitis, and for the early stages of fevers. It is given in combination with other berbs for dysentery and skin infections. *Lian qiao* has been used for "cold" swellings of the neck (as in tuberculosis of the lymph glands). In Chinese folk medicine, it is a treatment for breast cancer. The herb is sometimes taken to induce menstruation.

FRAGARIA VESCA (*Rosaceae*)

WILD STRAWBERRY

Parts Used: Leaves, fruit.
Constituents: The leaves contain flavonoids, tannins and a volatile oil. The fruit contains fruit acids and a volatile oil. The fruit contains fruit acids and a volatile oil with methyl salicylate and borncol.
Medicinal Actions & Uses: Wild strawberry leaves are mildly astringent and diuretic. The plant is little

Fragaria vesca

Fraxinus excelsior

used medicinally today, but it can be taken to treat diarrhoea and dysentery. The leaves where used as a gargle for sore throats, and in a lotion for minor burns and grazes. In Europe, the fruit is considered to have cooling and diuretic properties, and has been prescribed as part of a diet in cases of tuberculosis, gout, arthritis and rheumatism.

FRAXINUS EXCELSIOR
(Ash)

OLEACEAE

The bark and leaves are used medicinally. The bark contains the coumarin glycoside fraxin, tannins and bitter compounds. The leaves also contain fraxin and tannins and also the sugar alcohol mannite and organic acids. Both parts are mildly laxative and diuretic. An infusion is used in herbalism to regulate bowel movements, to expel uroliths and to alleviate rheumatic and gouty pains. Externally the leaves are used in compresses or in bath preparations to treat suppurating wounds.

FRITILLARIA ROYLEI
(Liliaceae)

Bulbs – powdered and boiled with orange skin used for Tuberculosis and asthma, alkaloid primine; roots contain alks, peimine and propeimene.
Also alks, peimisine, peimiphine, peimidine, peimitidine.

FRITILLARIA THUNBERGII
(Liliaceae)

ZHE BEI MU
Part Used: Bulb.

Constituents: The bulb contains alkaloids, including peimine, which affects the parasympathetic nervous system.

Medicinal Actions & Uses: Zhe bei mu increases the coughing up of mucus and relieves irritability in the respiratory tract. It is given for bronchitis and tonsillitis, and for fever and respiratory symptoms accompanying other acute infections such as flu. *Zhe bei mu* is thought to act specifically on tumours and swellings of the throat, neck and chest, and is taken for thyroid gland nodules, scrofula (tuberculosis of the lymph glands of the neck), abscesses and boils, and breast cancer. It has also been used to treat dysentery, and to increase breast milk production.

FUCUS VESICULOSUS
(Fucaceae)

BLADDERWRACK, KELP

Part Used: Whole plant.

Constituents: Bladderwrack contains phenols, polysaccharides and minerals, especially iodine (up to 0.1%). The polysaccharides are immune-stimulant.

Medicinal Actions & Uses: Due to its iodin content, bladderwrack is taken as an anti-goitre remedy. The plant appears to raise the metabolic rate by increasing hormone production by the thyroid gland, but this increase may be limited to poorly functioning thyroids. Bladderwrack is reputedly helpful in rheumatic conditions.

FUMARIA OFFICINALIS
Papaveraceae
(Common Funitory)

The flowering stems are used medicinally. Their constituents include alkaloids (mostly fumarine); tannins and mucilage. They stimulate the appetite and are laxative by increasing the peristaltic action of the smooth muscles of the intestines. They are also diuretic and choleretic. Externally Common Fumitory is a good skin cleanser and it is used in the treatment of certain skin disorders such as eczema. In a mixture with the leaves of Walnut (*Juglans regia*) it heals haemorhoids. Common Fumitory is toxic and large doses can cause severe diarrhoea, muscular spasms, and even respiratory failure. **It should therefore be taken internally only under the supervision of a qualified medical or herbal practitioner.**

Fumaria officinalis

GALEGA OFFICINALIS
(Leguminosae)

GOAT'S RUE

Parts Used: Aerial parts.

Constituents: Goat's rue contains alkaloids (including galegine), saponins, flavonoids and tannins. Galegine strongly reduces blood sugar levels.

Medicinal Actions & Uses: Today, goat's rue is chiefly used as an antidiabetic herb, having the ability to reduce blood sugar levels. It is not a substitute for conventional treatments but can be valuable in the early stages of late-onset diabetes and is best used as an infusion. The herb has the effect of increasing breast-milk production. It is also a useful diuretic.

GALEOPSIS SEGETUM
Lamiaceae
(Dewny Hempnettle)

The flowering stems are the medicinal parts. Their constituents include tannins, saponins, a glycoside, traces of an essential oil and silicic acid. These substances give Downy Hempnettle diuretic, astringent, stromachic and expectorant properties. It is used in tea mixtures to treat chest colds, cough, whooping cough, bronchitis, to stimulate the appetite, improve digestion and to treat disorders of the spleen. A tincture is used in homeopathy for the same purposes. Other species of hempnettle have similar medicinal uses.

Galeopsis segetum

GALEOPSIS TETRAHIT
(Lamiaceae)

Plant – Expect, in physical complaints, antisp, resolv. detergent.

Galega officinalis

Infusion of plant – in pulmonary troubles.

GALIPEA OFFICINALIS
(Rutaceae)

ANGOSTURA

Part Used: Bark.

Constituents: Angostura bark contains bitter principles, alkaloids, including cusparine, and 1-2% volatile oil.

Medicinal Actions & Uses A strong bitter with tonic properties, angostura stimulates the stomach and digestive tract as a whole. It is antispasmodic and is reported to act on the spinal nerves, helping in paralytic conditions. Angostura is typically given for weak digestion, and is considered valuable as a remedy for diarrhoea and dysentery. In South America, it is sometimes used as a substitute for cinchona (*Cinchona* spp.,) to control fevers.

GALIUM APARINE
(Rubiaceae)

CLEAVERS, GOOSE GRASS

Parts Used: Aerial parts.

Constituents: Cleavers contains irridoids (including asperuloside), polyphenolic acids, anthraquinones (only in the root), alkanes, flavonoids and tannins. Asperuloside is a mild laxative.

Medicinal Actions & Uses: A valuable diuretic, cleavers is often taken for skin diseases such as seborrhoea, eczema and psoriasis; for swollen lymph glands; and as a general detoxifying agent in serious illnesses such as cancer. The plant is commonly prepared in the form of an infusion, but for conditions such as cancer, it is best taken as a juice, which is strongly diuretic. The juice and the infusion are also taken for kidney stones and other urinary problems.

GALIUM ODORATUM
Rubiaceae (woodruff)

The flowering stems are used medicinally. Their constituents include a coumarin glycoside (responsible for the hay-like aroma), tannins and bitter compounds. These substances give Woodruff sedative, diuretic, vulnerary and antispasmodic properties. In herbalism it is used in the form of an infusion or decoction to treat nervous irritability, overwork, muscular spasms of various kinds, heart palpitations and insomnia. **Strong doses may cause vertio, vomiting and headache.** Externally compresses of

Galium odoratum

Woodruff are used to treat slow-healing wounds, skin rashes and ulcerous conditions. The herb is also occasionally used in homeopathy. It is an effective moth repellent.

GALIUM VERUM
(Rubiaceae)

LADY'S BEDSTRAW

Parts Used: Aerial parts.
Constituents: Lady's bedstraw contains iridoids (including asperuloside), flavonoids, anthraquinones and alkanes.
Medicinal Actions & Uses: A slightly bitter-tasting remedy, lady;s bedwtraw is used mainly as a diuretic

and for skin problems. Like its close relative, cleavers (*G. aparine*), the herb is given for kidney stones, bladder stones and other urinary conditions, including cystitis. It is occasionally used as a means to relieve chronic skin problems such as psoriasis, but in general, cleavers is preferred as treatment for this condition. Lady's bedstra has had a long-standing reputation, especial in France, of being a valuable remedy for epilepsy, though it is rarely used for this purpose today.

GARCINIA INDICA & G. CAMBOGIA
(Guttifferae)

Fruit – antiscor, cooling, cholag, emol. dimule, astrin.
Oil – sooting, used in skin diseases. Milky juice contains arabin, essent. oil, resin, **rind of fruit**–acids tartaric, citric and phosphoric.

GARCENIA MANGOSTANA
(Guttiferae)

Rind of fruit – astrin, useful in chr. diar, and dysin.
Bitter substance mangostin. α-β-mangostin.

GARCINIA MORELLA
(Guttiferae)

Gum resin –purg. anthelm. used in dropsical affections, amenor, obstinate constipation, and as vermifuge.
Seeds contain morethin.

GARDENIA JASMINOIDES
(Rubiaceae)

ZHI ZI (CHINESE), GARDENIA
Part Used: Fruit.
Constituents: Zhi zi fruit contains a

Galium verum

volatile oil, gardenin, crocin and geniposide.

Medicinal Actions & Uses: In the Chinese herbal tradition (pp.) zhi zi is a "bitter, cold" herb used mostly to relieve symptoms associated with heat. These include fever, irritability and restlessness, insomnia, painful urination and jaundice. The herb also treats cystitis, headaches and difficulty is breathing. It staunches bleeding, and is taken for nosebleeds and for urinary and rectal bleeding. *Zhi zi* is mixed with egg white and applied as a powder to bruises.

GARDENIA GUMMIFERA
(Rubiaceae)

Gum – antisep. carmin, antisp. stim, in dyspep, anthelm. used in veterinary medicine to keep off flies from sores. Gum dikenate

GARUGA PINNATA
(Burseraceae)

Fruit – Stomach.
Juice of bares – mixed with honey given is asthma.
Juice of stem – dropped into the eye to cure opacities of the conjuctiva.

GASTROCHILUS PANDURATA
(Zingiberaceae)

Roots – used in dysent.
Essential oil.

GAULTHERIA
FRAGRANTISSIMA
(Ericaceae)

Oil from leaves – arom, stim, carmin, used in rhenm, neuralgia, as flavouring agent and antisep.
Essential oil.

GAULTHERIA PROCUMBENS
(Ericaceae)

WINTERGREEN

Parts Used: Leaves, fruit, essential oil.

Constituents: Wintergreen contains phenols (including gaultherin and salicylic acid), 0.8% volatile oil (up to 985 methyl salicylate), Mucilage, resin and tannins.

Medicinal Actions & Uses: Wintergreen is strongly anti-inflammatory, antiseptic, and soothing to the digestive system. It is an effective remedy for rheumatic and arthritic problems, and, taken as a tea, it relieves flatulence and colic. The essential oil, in the form of a liniment or ointment, brings relief to inflamed, swollen or sore muscles, ligaments and joints, and can also prove valuable in treating neurological conditions such as sciatica (pain resulting from pressure on a nerve in the lower spine) and trigeminal neuralgia (pain affecting a facial nerve). The oil is sometimes used to treat cellulitis, a bacterial infection causing the skin to become inflamed. The Inuit of Labrador and other native peoples eat the berries raw, and use the leaves to treat headaches, aching muscles and sore throat.

GELIDIUM AMANSII
(Rhodophyceae)

AGAR

Part Used Seaweed extract (agar).
Constituents: Agar contains polysaccharides, mainly agarose and agaropectin (up to 90%), which are very mucilaginous.
Medicinal Actions & Uses: Like

most seaweeds and their derivatives, agar is nutritious and contains large amounts of mucilage. Its chief medicinal use is as a bulk laxative. In the intestines, agar absorbs water and swells, stimulating bowel activity and the subsequent elimination of faeces.

GELSEMIUM SEMPERVIRENS
(Loganiaceae)

YELLOW JASMINE, GELSEMIUM

Part Used: Rootstock.

Constituents: Yellow jasmine contains indole alkaloids (including gelsemine and gelsedine), iridoids, coumarins and tannins. The alkaloids are toxic and act as a depressant to the central nervous system.

Medicinal Actions & Uses: A potent medicinal herb, yellow jasmine is prescribed in small doses as a sedative and antispasmodic, most commonly for neuralgia (pain caused by nerve irritation or damage). Yellow jasmine is often given for nerve pain affecting the face. The herb is also applied externally to treat intercostal neuralgia (nerve pain between the ribs) and sciatica (pain resulting from pressure on a nerve in the lower spine). Yellow jasmine's antispasmodic property is used in treating whooping cough and asthma. The herb is occasionally taken for migraine, insomnia, and bowel problems, and also to reduce blood pressure. Yellow jasmine is also used in homeopathic medicine.

GENISTA TINCTORIA
Fabaceae
(Dyer's Greenweed)

The flowering stems are the medicinal parts. Their constituents include

Genista tinctoria

the alkaloids cytisine and sparteine, the flavonoid genistein and a yellow glycoside (luteolin). These substances give Dyer's Greenweed strong diuretic, weak cardioactive and laxative properties. Besides being a remedy for kidney and urinary disorders, it has also been used to strengthen heart action, to raise blood pressure and to alleviate rheumatic and arthritic pain. It is not a suitable remedy for individuals with hypertension or for pregnant women.

GENTIANA LUTEA
Gentianaceae
(Great Yellow Gentian)

The roots of old plants are used.

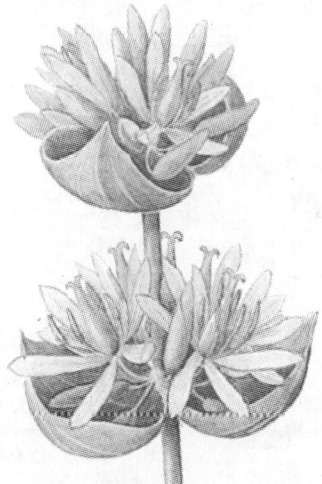

Gentiana lutea

Among the constituents are the bitter glycosides (for example, gentiopicrosides), a flavonoid derivative (gentisin), alkaloids, sugars and pectin. These substances give Great Yellow Gentian a markedly bitter taste and it acts as a tonic on the whole digestive system. Taken at least half and hour before meals it stimulates the appetite, and promotes the flow of digestive juices and bile. It is used in conventional and homeopathic medicine as well as in herbalism.

The fermented root is used in the manufacture of bitter foodstuffs and liqueurs. Before fermentation the root is dried slowly until it turns a reddish-brown colour (for medicinal uses the root is dried quickly so that it retains its yellow colour).

GENTIANA MACROPHYLLA
(Gentianiaceae)
Qin Jiao

Part Used: Root.

Constituents: *Qin jiao* contains alkaloids such as gentainine and gentianindine, and bitter principles.

Medicinal Actions & Uses: Like its European cousin, gentain (*Gentia lutea*), *qin jiao* is a strongly bitter herb. It is commonly taken in the form of a tincture in order to stimulate the digestion and "cool" the body in general. Unlike gentian, however, *qin jiao* is also mildly pungent, and is therefore appropriate for a somewhat different range of illnesses. In Chinese herbal medicine, it is prescribed for the treatment of "wind-damp" conditions such as fever, jaundice and "dry" constipation, and is used generally to help support the function of the liver and digestive system. As it is anti-inflammatory and mildly sedative, *qin jiao* is also prescribed as a treatment for various rheumatic and arthritic conditions.

GERANIUM MACULATUM
(Geraniaceae)
American Cranesbill

Parts Used: Root, aerial parts.

Constituents: American cranesbill contains up to 30% tannins.

Medicinal Actions & Uses: An astringent and clotting agents, American cranesbill is used today much as in earlier times. The herb is often prescribed for irritable bowel syndrome and haemorrhoids, and it is used to staunch wounds. It may also be used to treat heavy menstrual bleeding and excessive vaginal discharge.

GERANIUM ROBERTIANUM
Geraniaceae (Herb Robert)

The flowering stems are used medicinally. Their constituents include an essential oil, tannins and the bitter compound geraniin. These substances

Geranium robertianum

Geum urbanum

give Herb Robert astringent and diuretic properties and it is used in an infusion to check pulmonary haemorrhage and nosebleeding, and to treat severe diarrhoea, kidney and bladder disorders. Externally compresses and ointments are used to treat various skin disorders, boils and septic cuts. A decoction can be used as a gargle for tonsillitis.

GERBERA PILLOSELLOIDES
(Asteraceae)

The Zulus apply infusion of the root, made with human urine, to the ear for earache, the sutos use a milk decoct. or infusion for chest complaints.

GEUM URBANUM
Rosaceae
(wood Avens)

The rhizomes are used medicinally. Their constituents include tannins,

bitter compounds, an essential oil with gein and eugenol, and organic pigments (leucoanthocyanins). These substances give Wood Avens astringent, antiseptic, tonic, anti-inflammatory and antidiarrhoeal properties. It is used as a powder or decoction to treat gastritis and enteritis, intestinal colic, liver disorders and to check internal bleeding. Externally Wood Avens is used in gargles for stomatitis, gingivitis and other mouth inflammations and bad breath (halitosis), and in compresses and bath preparations for skin disorders and haemorrhoids.

GINKGO BILOBA
(Ginkgoaceae)

GINKGO, MAIDENHAIR TREE, BAI GUO (CHINESE)
A deciduous tree with one or several main trunks and spreading branches. It grows to 30 m.

Key constituents
Flavonoids
Ginkgolides
Bilobalides
Key Actions
Circulatory stimulant & tonic
Anti-asthmatic
Antispasmodic
Anti-allergenic
Anti-inflammatory

GIRARDINIA ZEYLANICA
(Urticaceae)

Leaves – used for headaches and
swollen joints.
Decoct. given in fever.

GIRONNIERA RETICULATA
(Ulmaceeae)

Plant: scraped fine and mixed in the
lemon juice used internally as a blood
purifier in itch and other cutaneous
eruptions, the body being at the same
time anointed with it externally.
Crystalline substance -methyl-indole
or skatole.

GISEKIA PHARNACEOIDES
(Ficoidaceae)

Plant: arom. aper., anthelm.
Gisekia, tannin.

GLECHOMA HEDERACEAE
Laminaceae
(Ground Ivy)
The flowering stems are used medicinally. The constituents include the
bitter compound glechomine, an essential oil, tannins (6-7 per cent),
saponin and potassium salts. These
substances give Ground Ivy astringent, anti-inflammatory and tonic
properties and an infusion is used for
gastritis, enteritis and diarrhoea, for

Glechoma hederaceae

kidney disorders, bronchial catarrhs,
coughs and some asthmatic conditions. It also stimulates the appetite
and is a general tonic. Homeopathic
tinctures are prepared from fresh plant
parts. Externally Ground Ivy is used
in gargles for mouth infections, and
in compresses and bath preparations
for skin disorders.
The fresh shoots and leaves can be
added to salads and soups or prepared
and eaten like spinach.

GLOCHIDION ZEYLANICUM
(Euphorbiaceae)

Fruits: Cooling, astrin.
Leaves: in itches.
Bark: stomach.

GLORIOSA SUPERBA
(Liliaceae)

Root: purg., cholag., anthelm, used in leprosy, parasitical affections of skin, piles, colic, in snake-bite, and scorpion sting.
Starch from root: given internally in gonor.
alkls: superbine, glorisine, colelincine and other alkls.

GLOSSOCARDIA BOSVALLIA
(Asteraceae)

Plant: emmen., used in female complaints.

GLYCINE MAX
Fabaceae
(Soybean)

The seeds (beans) are used medicinally. They contain valuable nutritive substances – proteins (40 per cent), fatty oil (20 per cent), carbohydrates, lecithin (a phospholipid), vitamins and minerals. Soybean is an important constituent of some infant foods and milk substitutes. It is also an ideal food for diabetics because its sugars remain largely unabsorbed. Because foodstuffs made from Soybean flour and Soybean oil are low in cholesterol they are though to help prevent arteriosclerosis and coronary heart disease. The protein-rich beans are increasingly being mixed with meat products or used on their own as meat substitutes.

The expressed oil from the beans is used to make plastics and many other products. It is an important cooking oil and is used as a major raw material of margarine. Fermented beans are used to make oriental sauces and pastes and the well-known Worcester Sauce in Britain. Blanched Soybean seedlings are a popular salad vegetable.

GLYCOSMIS PENTAPHYLLA
(Rutaceae)

Roots: pounded and mixed with sugar given in low fever.
Wood: used in snake-bite.

GLYCYRRHIZA GLABRA
Fabaceae
(Liquorice)

The roots and underground stolons of three-year old plants are used medicinally. The constituents include a sweet substance, glycorrhizin (7 per cent), potassium and calcium salts of glycyrizinic acid, a triterpenoid saponin flavonoid glycosides, traces of essential, starch sugars, a phyto-

Glycine max

Glycyrrhiza glabra

sterol (sitosterol), tannins and enzymes. These substances give Liqorice expectorant, laxative and antispasmodic actions. It is used either cut into pieces (in tea mixtures) or ground into a powder (in medicines). The extract is made into sticks that have pleasant spicy flavour. Liquorice is of due for coughs and bronchitis, peptic and duodenal ulcers, and rheumatoid arthritis. It is also used to sweeten and flavour pharmaceutical preparations and, now rarely, as a binding agent in pills. A mixture of powdered Liquorice, Fennel (*Foeniculum vulgare*) and Senna (*Cassia angustifolia*) leaves is a popular natural laxative. **In large doses Liqourice uses side effects,** notably headache, high blood pressure and water retention.

GMELINA ARBOREA
(Verbenaceae)
Juice of leaves: demuk. used in gonor. cough and to remove foetid discharges and wounds and ulcers.
Plant: Used in snake-bite and scorpion sting.

GNAPHALIUM ULIGINOSUM
(Compositae)
MARSH CUDWEED
Parts Used: Aerial parts.
Constituents: Marsh cudweed contains a volatile oil and tannins.
Medicinal Actions & Uses: While little used medicinally today, marsh cudweed has astringent, antiseptic and anticatarrhal properties. In British herbal medicine, it is occasionally taken for tonsillitis, sore throat and hoarseness, and for catarrh in the throat, nasal passages and sinuses. Marsh cudweed is used in Russia to reduce high blood pressure. It is thought to be antidepressant and aphrodisiac.

GNETUM SCANDENS
(Gnetaceae)
Stems and roots: antiper.
Plant: used as fish poison.

GOSSYPIUM HERBACEUM
(Malvaceae)
COTTON
Parts Used: Root bark, seed oil.
Constituents: Cotton root bark contains gossypol (a sesquiterpene) and flavonoids. Cotton seed contains a fixed oil, which is about 2% gossypol, and flavonoids. Gossypol causes infertility in men.

Medicinal Action & Uses: Cotton root bark is rarely used medicinally today. It was once employed as a substitute for ergot (*Claviceps purpurea*), the widely used labour-inducing herb. Cotton root bark is both milder-acting and safer in effect, stimulating uterine contractions and hastening a difficult labour. It also promotes abortion or the onset of a period, and reduces menstrual flow. Cotton root bark encourages the blood to clot and the secretion of breast-milk. Cotton seed oil is used to treat heavy menstrual bleeding and endometriosis.

GRATOPHYLLUM PICTUM
(Acanthaceae)

Leaves: emol. resolv. used in scorpion sting. inflamed breast.
Alk.

GRATIOLA OFFICINALIS
Scrophulariaceae
(Hedge Hyssop)

Gratiola officinalis

The flowering stems are used medicinally. The constituents include the cardiac glycosides gratiolin, gratiogenin, gratiotoxin and an essential oil (gratiolon). These substances give Hedge Hyssop strong purgative, emetic, cardioactive, diuretic and anthelmintic properties. Tinctures from the fresh plant are used in homeopathy for certain gastrointestinal and liver disorders. Hedge Hyssop is a dangerous plant; **it should never be collected and used for self-medication**. Long-term use may affect the eyesight; in large doses it may be fatal. Milk from cows that have grazed on the plant may also be poisonous.

GREWIA ASIATICA
(Tiliaceae)

Fruit: astrin. cooling, stomach.
Infusion of bark: in rheum.
Leaves: used as applications to pustular eruptions.

GREWIA MICROCOS
(Tiliaceae)

Plant: used in indign. typhoid fever, dysen. and syphititic ulcerations of the mouth and in small pox, eczema and itches.

GREWIA SCLEROPHYLLA
(Tiliaceae)

Root: prescribed in cough and irritable conditions of the intestines and bladder.
Decoct.: used as an emol. enema.

GREWIA TENAX
(Tiliaceae)

Decoct. of wood: given to cure coughs and pains in the side.

GREWIA VILLOSA
(Tiliaceae)

Root: used in diarh.

Juice of fresh bark: used with sugar and water for gonor. and urinary complaints attended in the irritability of the bladder.

GRINDELIA CAMPORUM
(Asteraceae)

GUMPLANT

Parts Used: Leaves, flowering tops.
Constituents: Gumplant contains diterpenes (including grindelic-acid), resins and flavonoids.
Medicinal Actions & Uses: Gumplant is a valuable remedy for bronchial asthma, and for states where phlegm in the airways impedes respiration. Both antispasmodic and expectorant, gumplant helps to relax the muscles of the smaller bronchial passages and to clear congested mucus. Additionally, it is thought to desensitise the nerve endings in the bronchial tree and to slow the heart rate, both leading to easier breathing. Gumplant is also taken for bronchitis and emphysema, and to clear catarrh in the throat and nose. It has been employed in the treatment of whooping cough, hay fever and cystitis, and externally to help speed the healing of skin irritation and burns.

GUAIACUM OFFICINALE
(Zygophyllaeae)

LIGNUM VITAE

Parts Used: Wood, resin.
Constituents: Lignum vitae contains lignans (furoguaiacidin, guaiaicin and others), 18-25% resin, vanillin and terpenes.

Medicinal Actions & Uses: Used in Europe, especially in Britain, as a remedy for arthritic and rheumatic conditions, lignum vitae has anti-inflammatory properties that help to reduce joint pain and swelling. It is also diuretic, laxative and sweat-inducing and speeds the elimination of toxins, which makes it valuable for treating gout. Tincture of lignum vitae is commonly used as a friction rub on rheumatic areas. Cotton wool moistened with the resin may be applied to aching teeth. A decoction of the woodchips acts as a local anaesthetic, and is used to treat rheumatic joints and herpes blisters.

GUAZUMA TOMENTOSA
(Sterculiaceae)

Bark: Sudorific, toxic, demuk. useful in skin diseases and elephantiasis.

GUAREA RUSBYI
(Meliaceae)

CCILLANA

Part Used Bark.
Constituents: Cocillana contains beta-sitosterl, and probably also resins, a fixed oil, tannin, an alkaloid and a glycoside.
Medicinal Actins & Uses: Cocillana is used in cough mixtures, being an even more powerful expectorant than ipecacuanha (*Cephaelis ipecacuanha*). Cocillana is taken as a treatment for coughs, excessive mucus production in the throat and chest, and bronchitis. At a high dosage, the plant induces vomiting.

GUIZOTIA ABYSSYNICA
(Asteraceae)

Oil from seeds: in rheum.

GYMNEMA SYLVESTRE
(Asclepiadaceae)
Leaves: in diabetes, chewed to reduce glycosuria, reduces blood sugar.
Root: emetic, expect.
Gymnemic acid; leaves contain anthraquinone compd.

GYMNOPETALUM COCHINSINENSIS
(Cucurbitaceae)
Root: pounded and mixed with hot water, rubbed on the body in body-ache and atrophy of limbs.
Plant: Used in the composition of a special drug given to women in labour.

GYMNOSPORIA SPINOSA
(Celastraceae)
Bark: ground to paste applied with mustard oil to destroy pediculi.

GYMNOSTACHYUM FEBRIFUGUM
(ACANTHACEAE)
Root: febge.

GYMANDROPSIS GYNANDRA
(Capparidaceae)
Decoct. of root: used in fever.
Leaves: rubft. vesic. in rheumatism.
Juice of leaves: remedy for otalgia.
Seeds: anthelm. rubft.
Plant: in scorpion-sting and snake-bite.
Essential oil, cleomin.

GYMNOCARDIA ODORATA
(Flacourtiaceae)
Oil from Seeds in leprosy and other skin diseases.
Fruit: fish poison.
Glued. gynocardin. HCN.

GYNURA PSEDO-CHINA
(Asteraceae)
Plant: emol, resolv, used as poultice in erysipelas and for tumours on breast.
Juice of leaves: as gargle for inflamm. of throat.

HACKELOCHLOA GRANULARIS
(Poaceae)
Plant: in conjunction with little sweet oil, used internally in enlarged spleen and liver.

HAEMATOXYLON CAMPECHIANUM
(Caesalpiniaceae)
Decoct. and extract of heart wood: astrin. toxic. used in chr. diarh. and atomic dyspep, decoct. a valuable injection in leucor.

HAGENIA ABYSYNICA
(Rosaceae)
Dried flowers and tops: anthelm.
α and β kosin and kosotoxin profokosin.

HAMAMELIS VIRGINIANA
(Hamamelidaceae)
WITCH HAZEL
A small deciduous tree growing to 5 m (15 ft), with coarsely smoothed, broadly oval leaves.
Key Constituents:
Tannins (8-10%)
Flavonoids

Bitter principle
Volatile oil (leaves only)
Key Actions:
Astringent
Anti-inflammatory
Stops external and internal bleeding

HAMILTONIA SUAVEOLENS
(Rubiaceae)

Infusion of roots: given in courbature.

HARDWICKIA PINNATA
(Caesalpiniaceae)

Balsam: used for gonor.
Balsam, essential oil, oleoresin

HARONGA MADAGASCARIENSIS
(Guttiferae)

Haronga

Parts Used: Leaves, bark.
Constituents: Haronga bark contains phenolic pigments, triterpenes, anthraquinones and tannins. The leaves contain phenolic pigments, hypericin, flavonoids and tannins. Hypericin, which is also found in St. John's wort (Hypericum perforatum), has antiviral and antidepressant properties.
Medicinal Actions & Uses: Thought to stimulate bile secretion, haronga is used in European herbal medicine to treat indigestion and poor pancreatic function. In African herbal medicine, haronga is chiefly employed as an astringent and mild laxative, and is also given for digestive system ailments such as diarrhoea and dysentery.

HARPOGOPHYTUM PROCUMBENS
(Pedaliaceae)

Devil's Claw

A trailing perennial, reaching 1.5 m is length, with fleshy lobed leaves and barbed woody fruit.
Key Constituents
Fibroid glycosides (harpago side)
Sugars (stachyose)
Phytosterols
Flavonoids
Harpagoquinone
Key Actions:
Anti-inflammatory
Analgesic
Digestive stimulant

HEDERA HELIX
Araliaceae (Ivy)

The young leaves are used medici-

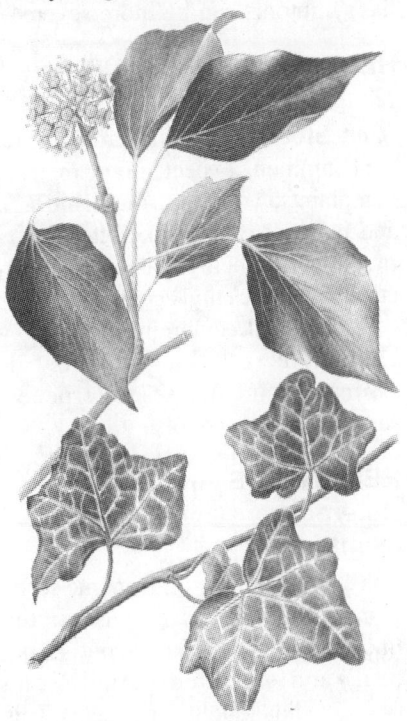

Hedera helix

nally. The constituents include tannins, a saponin (hederine), its aglycone (hederagenin), arsenic oxide, organic acids and iodine. These give Ivy expectorant, antispasmodic and cardiac actions. Small doses cause dilatation of the blood vessels, larger doses cause constriction of the vessels and the slowing of the heart beat. For these reasons Ivy can be dangerous if taken internally. **It should never be collected and used for self-medication**. In the past Ivy has been used to treat respiratory diseases and rheumatic pain. Externally the fresh leaves can be applied as a compress to slow-healing wounds, bruises and arthritic joints but they are an irritant and may cause dermatitis. It eaten the berries have a mainly purgative action, though if many are consumed the symptoms can be more severe.

Helianthus annuus

HEDYCHIUM SPICATUM
(Zingiberaceae)
Root Stock: stomach. carm. toxic, stim, emmen, expect. good in liver complaints, vomiting, diarh, inflam. and pains, used in snake-bite.
Essential oil, methyl paracumarin acetate, cinnamic ethyl acetate rhizomes – essential oil containing p-methoxy cinnamate, ethyl cinnamate, d-sabinene, cineole, sesquiterpenes, sesquiterpene alcohol.

HELIANTHUS ANNUUS
Asteraceae
(Sunflower)
The expressed fatty oil from the seeds contains glycerides of unsaturated linoleic and oleic acids (about 45 per cent) and saturated palmitic and arachic acids (about 4 per cent). It is used in salves, plasters and liniments for rheumatic pain. It is also widely used in foodstuffs as a salad and margerine oil, in soaps and as a lubricant. In homeopathy a tincture from the seeds is used internally to relieve constipation and externally on cuts and bruises. The seeds are also roasted and eaten, used as a coffee substitute and ground into flour.
The dried flowerheads are also used medicinally in some countries. They have diuretic, carminative, anti-inflammatory and antidiarrhoeal properties.

HELICTERES ISORA
(Sterculiaceae)
Fruits: demuk. astrin. useful in the griping of bowels and flatulence of children.
Bark: in dysent. and diur.
Juice of root: in diabetes empeyema, stomach affections and snake-bite.

Root and Bark: expect. demuk. astrin. bowels, antigalactagogne; lessen griping, a cure for scabbies when applied topically.

HELIOTROPIUM BREVIFOLIUM
(Boraginaceae)

Plant: Laxat. diur.
Juice: Applied to sore eyes, gum boils, and sores generally, in sting of nettles and insects.

HELIOTROPIUM INDICUM
(Boraginaceae)

Leaves: applied to boils, ulcers, wounds and in stings of insects and reptiles.
Plant: diur.
Alkaloid.

HELLEBORUS NIGER
(Ranunculaceae)

BLACK HELLEBORE, CHRISTMAS ROSE

Parts Used: Rhizome, root, leaves.
Constituents: Black hellebore con-tains cardiac glycosides (helleborin, helleborein and hellebrin). These substances have an action similar to that of the glycosides found in common foxglove. (*Digitalis purpurea*).

Medicinal Actions & Uses: Toxic when taken in all but the smallest doses, the acrid black hellebore has purgative and cardiotonic properties, expels worms and promotes menstrual flow. In the 20th century, the cardiac glycosides in the leaves came into use as a heart stimulant for the elderly. The herb has also been taken to stimulate delayed menstrual periods. However, black hellebore is now considered too strong to be safely used.

Caution: Black hellebore is extremely toxic. Do not use under any circumstances.

HEPATICA NOBILIS
Ranunculaceae
(Hepatica)

The leaves are used medicinally. Their constituents include saponins, tannins and the glycoside hepatilobin. These

Helleborus niger

Hepatica nobilis

substances give Hepatica strong diuretic, astringent and vulnerary properties. An infusion or decoction is used in herbalism to treat kidney, gall bladder and liver disorders, and coughs and bronchitis. A tincture is also used in homeopathy. **Large doses can be toxic;** for this reason Hepatica should be used only under the supervision of a qualified medical or herbal practitioner. The herb is used in gargles for stomatitis and for chronic irritations of the throat and pharynx.

HEMIDESMUS INDICUS
(Asclepiadaceae)

Roots: Used as subst. for sarsaparika, demuk. alter. diaphor., diur., tonic, in loss of appetite, disinclination for food, fever, skin diseases, as blood purifier, in leucor. skin diseases, syphilis, rheumatism, and in scorpion-sting and snake-bite.
Essential oil, 2-hydroxy-4-methoxy benzaldehyde, sterols and a glued, saponin, resin acid, tannin.

HERACLEUM WALLICHI
(Apiaceae)

Root – aphrodis.

HERNIARIA GLABRA
Caryophyllaceae
(Smooth Ruturewort)

The flowering stems are the medicinal parts. Those of Smooth Rupturewort contain a triterpene saponin that decomposes into quillaiac acid and simple sugars. Hairy Rupturewort contains saponins that decompose into galactonic acid, glucose and other sugars. Both plants also contain the flavonoid glycoside

Herniaria glabra

rutin, the coumarin glycoside herniarin and an essential oil. These substances give the plants mild diuretic, mild antispasmodic and antiseptic properties. They are used in infusions on their own or mixed with other herbs to treat bladder, kidney and gall bladder disorders and they help prevent the formation of kidney stones and gravel.

HETEROPHRAGMA QUADRILOCULARE
(Bignoniaceae)

Root: presented as drink in viper-bite.
Tar from wood: used in skin diseases.

HIBISCUS ABELMOSCHUS
(Malvaceae)

Seeds stim., antisp. stomach, cooling, toxic, carmin, aphrodis, rubbed to paste with milk used to cure itch, in snake-bite.
Essential oil.

HIBISCUS CANNABINUS
(Malvaceae)

Juice of flowers: With sugars and black pepper in biliousness with acidity.
Seeds: aphrodis, fattening, as external applications to pains and bruises.
Leaves: purg.
Seeds contain fatty oil take groundnut oil, radium, thorm, rubidium; flower petals contain glucoside – cannabiscitrin and flavanol cannabiscetin.

HIBISCUS ESCULENTUS
(Malvaceae)

Immature capsules: used in the form of a decoct. as emol. demulc. diur. in catarrhal affections, ardor urinae, dysuria and gonor.
Mucilage from fruits and seeds: emol. demulc. useful in gonor.

HIBISCUS FURCATUS
(Malvaceae)

Infusion of roots in water: Cooling drink for hot weather.

HIBISCUS MICRANTHUS
(Malvaceae)

Plant: febfge.

HIBISCUS MUTABILIS
(Malvaceae)

Flowers: used in Malaya and China for pectoral and pulmonary complaints, and as stim.
Leaves: applied to swellings.
Plant: emol.

HIBISCUS ROSA-SINENSIS
(Malvaceae)

Root: in cough, subst. for **Alhea**
Leaves: emol. aper.

Flowers: emol.
Infusion of petals: given as demulc. and refrig. drink in fevers.

HIBISCUS SABDARIFFA
(Malvaceae)

Leaves: emol. much used in Guinea as diur., sedative and refrig.
Fruit: antiscorb.
Leaves, seeds and ripe calyces: diur. antiscor.
Succulent calyx: boiled in water used as a drink in bilious conditions.
Organic acids from flowers, gossypectin, anthocyanin, and a glucd. hibiscin; infusion of drug contains – citric, tartaric, malic; possesses diur. and chloretic effect, decreases the viscosity of blood, reduces blood pressure and stimulates intestinal peristalsis; dried fruits contain cal. oxalate, gossepectin, anthocyanines and vitamin C; dry petals contain flavanol glucd hibiscitrin.

HIBISCUS SURRATENSIS
(Malvaceae)

Mucilaginous flowers: used as emol. and pectoral in La Reunion.
Stem and leaf: in a lotion used by the Zulus for the treatment of penile irritation of any sort, including venereal sores and urethrites.

HIBISCUS TILIACEUS
(Malvaceae)

Root: Febge, employed in the preparation of embrocations.

HIBISCUS TRIONUM
(Malvaceae)

Infusion of flowers: in China and Malay taken for itching and painful skin diseases and as diur.
Dried leaves: considered stomach.

HIBISCUS VITIFOLIUS
(Malvaceae)
Roots provide a preparation used by Gold Coast women to kill head-lice.

HIERACIUM PILOSELLA
(Compositae)
MOUSE-EAR HAWKWEED

Parts Used: Aerial parts.

Constituents: Mouse-ear hawkweed contains a coumarin (umbelliferone), flavonoids and caffeic acid. It is thought to be mildly antifungal.

Medicinal Actions & Uses: Mouse-ear hawkweed relaxes the muscles of the bronchial tubes, stimulates the cough reflex and reduces the production of catarrh. This combination of actions makes the herb effective against all manner of respiratory problems, including asthma and wheeziness, whooping cough, bronchitis and other chronic and congested coughs. Mouse-ear hawkweed's astringency and its diuretic action also help to counter the production of catarrh, sometimes throughout the respiratory system. The herb is used to control heavy menstrual bleeding, and to ease the coughing up of blood. It may be applied as a poultice to heal wounds.

HIPPOPHAE RHAMNOIDES
Elaneagnaceae
(Sea Buckthorn)
The fruits are used medicinally as a tonic on the Continent. They contain organic acids, tannins, the flavonoid glycoside quercetin, provitamin A, abundant vitamin C and vitamins B complex and E. The freshly pressed juice from the berries is made into syrups and preserves and used both preventively as a protection against possible infection, particularly in late

Hippophae rhamnoides

winter and early spring, and as a general tonic during convalescence. The constituents of the berries also strengthen the eyesight and have an antisclerotic effect.

HIPTAGE BENGHALENSIS
(Malpighiaceae)
Leaves: useful in chr. rheum. skin diseases and asthma.

Juice of leaves: insecticide and application for scabies.

Roots contain glucd hiptagin.

HOLARRHENA
ANTIDYSENTERICA
(Apocynaceae)
Bark: in dysent. dried and ground rubbed over the body in dropsy.

Seeds: astrin. febge. in fever, dysent. diarh. and intestinal worms.

Alkls.: Conessine, kurchine, kurchicine, from **bark** alks. conessine, holarrhimine, conarrhimine, a base, conamine, conessinine, isoconessinine, coninine. Toxic, paralyses central nervous system (kurchicine) raises blood pressure. Conkurchine alk.

HOLOPTELEA INTEGRIFOLIA
(Ulmaceae)

Juice of boiled bark: applied to rheum. and swellings.

HOLOSTEMMA ANNULARE
(Asclepiadaceae)

Roots: alter., used as a remedy for scalding in gonor, beaten into a paste applied to the eyes in ophthalmia; rubbed to a paste given in cold milk in diabetes; dried and powdered with equal quantity of the root of *Ceiba pentandra* given in spermatorrhea.

HOMALOMENA AROMATICA
(Araceae)

Rhizome: arom. stim.

HOMONOIA RIPARIA
(Euphorbiaceae)

Decoct. of root: used in piles, stone in bladder, gonor. and syphilis.
Root: laxt. diur. given for ulcers, stranguary, urinary discharges, vesical calculi.
Milky juice contains toxalbumin, crepetin.

HOPEA ODORATA
(Dipterocarpaceae)

Resin: in powder form used as styptic.
Bark: astrin. resin.

HOPPEA DICHOTOMA
(Gentianaceae)

Plant: used in piles and in snake-bite.

HORDEUM DISTICHON
(Poaceae)

BARLEY

Parts Used: Seeds.
Constituents: Barley contains proteins, sugars, starch, fats and B vitamins. The young seedlings also contain the alkaloids hordenine and gramine.
Medicinal Actions & Uses: An excellent food for convalescence in the form of porridge or barley water, barley is soothing to the throat and provides easily assimilated nutrients. It can also be taken to clear catarrh. Its demulcent quality soothes inflammation of the gut and urinary tract. Barley aids in the digestion of milk and is given to babies to prevent development of curds within the stomach. It is commonly given to children suffering from minor infections or diarrhoea, and it is particularly recommended as a treatment for feverish states. Made into a poultice, barley is an effective remedy for soothing and reducing inflammation in sores and swellings.

HORDEUM VULGARE
(Poaceae)

Grains: demulc. easy of digestion, used in dietary sick, parched and powdered much employed in the form of a gruel in cases of painful and atomic dyspep.

HUGONIA MYSTAX
(Linaceae)

Roots: used externally in reducing

inflammatory swellings, and as antid. to snake-bite, internally in the form of a powder as febg. and anthelm.

HUMULUS LUPULUS
Cannabaceae (Hops)

The cones or strobiles are used medicinally. When dried they have a spicy aroma and a bitter taste. The constituents include a resin with bitter compounds (chiefly humulone and lupulone), polyphenolic tannins, flavonoids, asparagin, oestrogenic substances and an essential oil with humulene. These substances give Hop mild sedative, hypnotic, stomachic, diuretic and weak antiseptic properties. An infusion is used in herbalism for digestive disorders, nervous irritability, to induce sleep and as an antiaphrodisiac (in men). An effective way of using hops for insomnia is in pillows. A tincture of the fresh cones is used in homeopathy.

The distilled essential oil is contained in some perfumes. Young shoots and immature leaves can be added to salads.

HURA CREPITANS
(Euphorbiaceae)

Juice of bark: a Brazilian remedy for leprosy
Seeds: emetic, purg.
Leaves: used for chr. pains
Toxic substance creptin and alk.

HYBANTHUS ENNEASPERMUS
(Violaceae)

Plant: tonic, diur.
Leaves and tender stalks: demuk.
Root: in bowel complaints of children.
Fruit: in scorpion-sting
Alkaloid.

HYDRANGEA ARBORESCENS
(Hydrangeaceae)

WILD HYDRANGEA
Part Used: Root.
Constituent: Hydrangea is thought to contain flavonoids, a cyanogenic glycoside (hydrangein), saponins and a volatile oil.
Medicinal Actions & Uses: Western herbal medicine considers the diuretic hydrangea as being particularly helpful in the treatment of kidney and bladder stones. It is thought both to encourage the expulsion of stones and to help dissolve those that remain. The herb is given for many other troubles

Humulus lupulus

affecting the genito-urinary system, including cystitis, urethritis, enlarged prostrate and prostatitis.

HYDNOCARPUS ANTHELMINTICA
(Flacourtiaceae)

Oil form seed: Used in leprosy and many skin infections.

Kernel contains oil, hydrocarpic acid, chaulmoogric acid, gorlic acid, oleic acid, palmitic acid, lower homologus of hydrocarpic acid.

HYDNOCARPUS KURZII
(Flacourtiaceae)

Seeds: Yield chaulmoogra oil which is used in leprosy and many skin diseases.

Fruit: fish poison.

HCN is fresh seeds, hydrocarpic acid, chaulmoogric acid, garlic acid, oleic acid, Palmitic acid, lower homologous of hydrocarpic acid.

HYDNOCARPUS LAURIFOLIA
(Flacourtiaceae)

Seed and Oil: used in leprosy and skin diseases, fish poison.

Seeds yield fatty oil, fixed oil, hydrocarpic acid, chaulmoogric acid, garlic acid, oleic acid, palmitic acid and lower homologous of hydrocarpic acid.

HYDRANGEA ASPERA
(Saxifragaceae)

Fresh plant contains HCN.

HYDROCOTYLE JAVANICA
(Apiaceae)

Leaves: tonic, blood purifier, for indign, nervousness and dysent.

HYDRASTIS CANADENSIS
(Ranunculaceae)

GOLDENSEAL

A small herbaceaous perennial, with a thick yellow root and an erect stem growing to 30 cm.

Key Constituents:
Isoquinoline alkaloids (hyrastine, berberine, canadine)
Volatile oil
Resin

Key Actions:
Tonic
Mild laxative
Anti-inflammatory
Antibacterial
Bitter
Uterine stimulant Stops internal bleeding
Astringent

HYDROLEA ZEYLANICA
(Hydrophyllaceae)

Leaves: antisep. used as poultice for callous ulcers.

HYGROPHILA SPINOSA
(Acanthaceae)

GOKULAKANTA

Parts Used: Aerial parts, root.
Constituents: Gokulakanta contains mucilage, fixed and volatile oils and an alkaloid.
Medicinal Actions & Uses: Commonly used as a remedy in India, gokulkanta is taken chiefly for its reputed aphrodisiac properties. Both the aerial parts and ash of the burned plant are strongly diuretic, and are used to flush water from the body in cases of excess fluid retention. Gokulkanta root is demulcent, and is used to alleviate the inflammation produced by urinary tract infections. The herb is

also thought to support the liver in conditions such as jaundice and hepatitis.

HYGRORYZA ARISTATA
(Poaceae)

Seeds: Cooling, astrin. to urinary tract, useful in biliousness.

HYMENODICTYON EXCELSUM
(Rubiaceae)

Inner bark: astrin., febge., toxic alk., hymenodictine, bitter substance, aerenlin.

HYOSCYAMUS NIGER
(Solanaceae)
HENBANE

Parts Used: Leaves, flowering tops.
Constituents: Henbane contains 0.045%-0.14% tropane alkaloids,

Hyoscyamus niger

especially hyoscyamine and hyoscine. Hyoscyamine and hyoscine are common to other members of the *Solanaceae* family, but henbane's relatively high hyoscine content gives it a more specifically sedative action than its relatives thornapple (*Datura stramonium*) and deadly nightshade (*Atropa belladonna*)

Medicinal Actions & Uses: Henbane is used extensively in herbal medicine as a sedative and painkiller. It is specifically used for pain affecting the urinary tract, especially pain due to kidney stones, and is also given for abdominal cramping. Its sedative and antispasmodic effect makes it a valuable treatment for the symptoms of Parkinson's disease, relieving tremor and rigidity during the early stages of the illness. Henbane has also been used to treat asthma and bronchitis, usually as a "burning powder" or in the form of a cigarette. Applied externally as an oil, it can relieve painful conditions such as neuralgia, sciatica and rheumatism. Henbane reduces mucus secretions, as well as saliva and other digestive juices. Like its cousin deadly nightshade, it dilates the pupils. One of henbane's active components, hyoscine, is sometimes used as a substitute for opium (from *Papaver somniferum*).

HYPERCOUM PROCUMBENS
(Papveraceae)
Juice: has the same effect as opium.
Leaves: diaph.
Alkaloid protopine

HYPERICUM PERFORATUM
Hypericaceae
(Perforate St. John's Wort)
The flowering stems are used medici-

Hypericum perforatum

nally. Their constituents include tannins (the flowers as much as 16 per cent), a red-pigmented flavonoid glycoside (hypericin), the flavonoid glycosides rutin and hyperin, catechol, an essential oil and resin. These substances give Perforate St. John's Wort mild sedative, cholagogic, anti-inflammatory, diuretic, antiseptic and astringent properties. It is used to treat chronic inflammation of the internal organs and for gynaecological disorders. The oil, prepared by macerating the fresh stems in olive or sunflowerseed oil, is used to heal wounds, burns, bruises and haemorrhoids. Excessive use of St. John's Wort, however, causes a skin allergy in hypersensitive individuals, which becomes aggravated by exposure to the sun.

HYPTIS SUAVEOLENS
(Lamiaceae)

Plant: Pounded and applied to parasitical cutaneous diseases. In Brazil infusions used as carmin. and as su-

dorific in catarrhal conditions. Plant yields essential oil containing menthol.

HYSSOPUS OFFICINALIS
(Labiatae)
HYSSOP

Parts Used: Flowering tops, essential oil.

Constituents: Hyssop contains terpenes (including marubiin, a diterpene), a volatile oil (consisting mainly of camphor, pinocamphone and beta-pinene), flavonoids, hyssopin, tannins and resin. Marubiin is a strong expectorant. Pinocamphone is toxic, and the volatile oil can cause epileptic seizures.

Medicinal Actions & Uses: Currently an undervalued medicinal herb, hys-

Hyssopus officinalis

sop is potentially useful as it is both calming and tonic. It has a positive effect when used to treat bronchitis and respiratory infections, especially where there is excessive mucus production. Hyssop appears to encourage the production of a more liquid mucus, and at the same time gently stimulates expectoration. This combined action clears thick and congested phlegm. Hyssop can irritate the mucous membranes, so it is best given after an infection has peaked, when the herb's tonic action encourages a general recovery. As a sedative, hyssop is a useful remedy against asthma in both children and adults, especially where the condition is exacerbated by mucus congestion. Like many herbs with a strong volatile oil, it soothes the digestive tract and can be an effective remedy against indigestion, wind, bloating and colic.

I

IBERIS AMARA
(Cruciferae)

WILD CANDYTUFT

Parts Used: Aerial parts, seeds.
Constituents: Contain mustard seed oil glycosides and vitamin C.
Medicinal Actions & Uses: Little used in herbal medicine today, wild candytuft is a bitter-tasting tonic, aiding digestion and relieving wind and bloating. It is traditionally taken to treat gout, rheumatism and arthritis. It also has a high vitamin C content.

ICHNOCARPUS FRUTESCENS
(Apocynaceae)

Root: Properties similar to

Hemmidesmus indicus; alter, tonic, subst. for sarsaparilla.
Decoct. of leaves and stalks: in fever.

ILEX AQUIFOLIUM
(Aquifoliaceae)

Holly
Parts Used: Leaves, berries.
Constituents: Holly contains illicin (a bitter principle), ilexanthin, theobromine (only in the leaf) and caffeic acid. Theobromine is a caffeine-type alkaloid, used to treat asthma.
Medicinal Actions & Uses: Holly is little used today. Its leaves are diuretic, fever-reducing and laxative, and they have been employed to treat fevers, jaundice and rheumatism. Holly berries purge the bowels and cause vomiting if taken in large doses.

ILEX PARAGUARIENSIS
(Aquifoliaceae)

Mate
Parts Used: Leaves.
Constituents: Mate contains xanthine derivatives, including about 1.5% caffeine, about 0.2% theobromine, theophylline, and up to 16% tannins. The high tannin conent means that mate should not be consumed with meals, as tannins impair the absorption of nutrients.
Medicinal Actions & Uses: Mate is a traditional South American tea that increases short-term physical and mental energy levels. It is taken as a fortifying beverage in much the same way as tea (*Camellia sinensis*) is consumed throughout Asia and Europe. Mate has properties similar to those

of tea and coffee (*Coffea arabica*). It stimulates the nervous system, is mildly analgesic and diuretic. As a medicinal herb, mate is used to treat headaches, migraine, neuralgic and rheumatic pain, fatigue and mild depression. It has also been used in the treatment of diabetes.

ILLICIUM VERUM
(Illiciaceae)
STAR ANISE, BA JIAO HUI XIAN (CHINESE)
Part Used: Fruit.
Constituents: Star anise has a volatile oil containing about 85% anethole, methyl-chavicol and safrole. An extract has antibacterial properties.
Medicinal Actions & Uses: Used in Chinese herbal medicine as a remedy for rheumatism, back pain and hernias, star anise has stimulant, diuretic and digestive properties. It makes an effective remedy for wind and indigestion – especially colic – and can safely be given to children. To treat hernias of the intense or bladder, star anise is often mixed with fennel (*Foeniculum vulgare*). Both herbs help to relax the organ's muscles and relieve spasm. Star anise is also used for toothache.

IMPATIENS BALSAMINA
(Balsaminaceae)
Flowers: Cooling, tonic, useful when applied to burns and scalds.
Plant: used for pains in the joints internally acts as emetic, cath. and diur.

IMPATIENS CHINENSIS
(Balsaminaceae)
Plant: used externally for burns and internally for gonor.

IMPERATA ARUNDIANACEA
(Poaceae)
Roots: used as emol. in Cambodia, mostly in the fumigation of piles; in China appreciated for their restor; haemostatic and anti-febrile properties.

IMPERATORIA OSTRUTHIUM
(Umbelliferae)
Part Used: Root.
Constituents: Masterwort contains a camphoraceaous volatile oil (including limonene, phellandrene, alpha-pinene and a sesquiterpene), peucadanin, oxipeucadanin and ostrutool.
Medicinal Actions & Uses: Masterwort is little used today, but it may well be a herb that bears further investigation. The root is aromatic, warms central areas of the body and is a bitter tonic. It has a strong action within the stomach and gut, settling indigestion and relieving wind and griping. Masterwort is also beneficial for chest conditions, and is used for colds, asthma and bronchitis. It can also be helpful for menstrual problems.

INDIGOFERA ASPALATHOIDES
(Fabaceae)
Leaves, flowers and tender shoots: Cooling, demulc. employed in decoct. in leprosy and cancerous affections.
Root: chewed as a remedy for toothache.
Plant: rubbed up with butter applied to reduce oedematous tumours.
Leaves: applied to absceses.
Oil from the root: used to anoint the head in erysipelas.

Decoct. of entire plant: given as an
alter. in secondary syphilis, psoriasis
etc.

INDIGOFERA ENNEAPHYLLA
(Fabaceae)

Juice of plant: antiscor. alter. diur.
used in old venereal affections.

INDIGOFEREA LINIFOLIA
(Fabaceae)

Plant: given in febrile eruptions and
used in amenor.
Yields a before tinifolin and a wax.

INDIGOFERA TINCTORIA
(Fabaceae)

Juice of leaves: Prohylactic against
hydrophobia.
Extract of plant: given in epilepsy
and nervous disorders, used in
broncht. and as ointment in sores, old
ulcers, and haemorrhoids.
Glucd indican in plant.

INULA HELENIUM
Asteraceae
(Elecampane)

The roots of second or third-year
plants are used medicinally. When
dried they have a pungent aroma and
a bitter taste. The constituents include
abundant inulin (up to 50 per cent),
essential oils, bitter compounds, resin
and mucilage santonin, alantobetone,
isoalantolactone, dihydroalanto-
betone. The bitter substance helenine
has an anthelmintic action and was
once used for ridding children of
pinworm and round-worm.
Elecampane also has expectorant,
antitussive, diuretic, cholagogic, antiseptic and tonic properties. Nowa-

Inula helenium

days it is mainly used for cough,
bronchitis, lack of appetite and *digestive disorders*. The essential oil distilled from the fresh roots is used as
an insecticide and anthelmintic in
some countries.

INULA JAPONICA CHINENSIS
(Compositae)
XUAN FU HUA

Parts Used: Flowers, aerial parts.
Constituents: Xuan fu hua contains
a volatile oil, flavonoids, phenolic
acids and triterpenes (including
taraxasterol).
Medicinal Actions & Uses: Used in
traditional Chinese medicine, as a
mildly warming expectorant remedy,
xuan fu hua is especially suitable
when phlegm has accumulated in the
chest. The herb is often prescribed for
bronchitis, wheeziness, chronic
coughing and other chest complaints
brought on by "cold conditions".
Xuan fu hua also has a bitter action
and it helps to strengthen digestive

function. It is prescribed to stop vomiting and, on occasions, hiccups. The flowers are normally used in medicinal preparations, but the aerial parts are also taken, generally for less serious conditions.

INULA ROYLEANA
(Asteraceae)

Plant: considered poisonous, disinfectant, insecticidal.
Root contain alkl. produces fall in blood pressure and stimulates tone and peristaltic movements of intestines.

IPOMOEA HISPIDA
(Convoluvulaceae)

Plant: boiled in oil used to cure rheum. headache, epilepsy, leprosy and ulcers.
Phytosterin glucd. euparanol, resin etc.

IPOMOEA PURGA
(Convolvulaceae)

JALAP

Part Used: Root.
Constituents: Jalap contains the resin convolvulin.
Medicinal Actions & Uses: Jalap is such a powerful cathartic that its medicinal value is questionable. Even in moderate doses it stimulates the elimination of profuse watery stools, and in larger doses it causes vomiting.

IPOMOEA PES-CAPRE
(Convolvulaceae)

Herb: astrin. stomach. laxt.
Leaves: externally applied in rheum. and colic.

Juice: given as diur. in dropsy and at the same time bruised leaves applied to the dropsical part.
Alk. in root, resin, essential oil, pentatriacontane, triacontane, sterol, behemic acid, melissic acid, butyric acid and myrstic acid.

IPOMOEA REPETANS
(Convolvulaceae)

Juice: emetic, purg., antid., to opium and arsenal poisoning.
Plant: Considered wholesome for females suffering from nervous and general debility.

IRIS VERSICOLOR
(Iridaceae)

BLUE FLAG, WILD IRIS
Part Used: Rhizome.
Constituents: Blue flag contains triterpenoids, salicylic and isophthalic acids, a very small amount of volatile oil, starch, resin, an oleo-resin and tannins.
Medicinal Actions & Uses: Blue flag is currently used mainly to detoxify the body. It increases urination and bile production, and has a mild laxative effect. This combination of cleansing actions makes it a useful herb for chronic skin diseases such as acne and eczema, especially where gallbladder problems or constipation contribute to the condition. Blue flag is also given for biliousness and indigestion. In small doses, it relieves nausea and vomiting. However, in large doses blue flag will itself cause vomiting. The traditional use of blue flag for gland problems persists. It is also believed by some to aid weight loss.

IRIS GERMANICA
Iridaceae
(Garden Iris)

The rhizomes of second-to-fourth-year plants are used medicinally. During drying they acquire a violet-like scent, caused by the ketone irone; they become yellowish and fragile. The other constituents include an essential oil, glycosides, sugars, resin, starch, mucilage and tannins. These substances give Garden Iris expectorant and diuretic properties but is rarely used nowadays, even in herbal medicine, because it can cause vomiting and nausea. **The fresh root is strongly purgative.** The dried rhizomes are, however, still used in pharmaceutical preparations to disguise the taste and smell of other medicines.

Iris germanica

IXORA COCCINEA
(Rubiaceae)
Flowers: in dysent. and dysmen.

JATERORHIZA PALMATA
(Menispermaceae)
CALUMBA

A twining perennial growing to 15 m, with large palm like leaves and green-white flowers.

Key Constituents:
Isoquinoline alkaloids (palmatine, columbamine, jatrorrhizine)
Bitter principles (furaniditerpenol, palmanin)
Volatile oil (upto 1%–mostly thymol)
Mucilage

Key Actions:
Bitter
Eases stomach pain
Tonic
Reduces fever
Expels worms

JATROPHA CURCAS
(Euphorbiaceae)
Roasted Nuts: purg.
Seeds: purg.
Juice of plant: useful in scabies, eczema, ringworm.
Twigs: used for tooth brushing in swollen gums.
Leaves: used in form of decoct. and cateplasm to the mammal as a betal. rubeft.
Plant: Fish poison.
Seeds contain toxic principle curcin, fatty oil, 2 phytosterols, phytosterolin, glued; sucrose, resinous matter having nauseating, purging and griping effect.

JATROPHA GLANDULIFERA
(Euphorbiaceae)

Fixed oil from seeds: purg., used in chr. ulcerations, foul wounds, ringworm, in rheum. and paralysis.

Juice of the plant: used to remove film from the eyes.

Root: sprayed with water given to children suffering from abdominal enlargements, purg., said to reduce glandular swellings.

JATROPHA GOSSYPIFOLIA
(Euphorbiaceae)

Leaves: applied to boils and carbuncles, eczema and itches.

Decoct. of bark: emmen.

Seeds: cause insanity and act as an emetic.

Leaves and seeds: purg.

JUGLANS CINEREA
(Juglandaceae)

BUTTERNUT

Part Used: Inner bark.

Constituents: Contains naphthaquinones (including juglone, juglandin and juglandic acid), a fixed and a volatile oil, and tannins. The naphthaquinones have an approximately similar laxative effect to the anthraquinones found in plants such as senna (*Cassia senna*), and Chinese rhubarb (*Rheum palmatum*). Juglone is purgative, antimicrobial, antiparasitic and cancer inhibiting.

Medicinal Actions & Uses: Used to this day as a laxative and tonic, butternut is a valuable remedy for chronic constipation, gently encouraging regular bowel movements. It is especially beneficial if combined with a carminative herb such as ginger (*Zingiber officinale*) or angelica (*Angelica archangelica*). Butternut also lowers cholesterol levels, and promotes the clearance of waste products by the liver. It has a positive reputation in treating intestinal worms, and, being antimicrobial and astringent, it has been prescribed as a treatment for dysentery.

JUGLANS NIGRA
(Walnut)
(Juglandaceae)

JUGLANDACEAE

The green outer layer (pericarp) of the its and leaflets are used medicinally. Their constituents include the tannin, glandin, organic acids, an essential oil, glycoside hydrojuglone and a bitter compound (juglone). These give Walnut astringent, haemostatic, anti-inflammatory, antispasmodic and mild sedative properties. In herbal medicine it is used mainly for stomach and intestinal disorders and as a skin treatment. A tincture prepared from the fresh leaves is used in homeopathy for the same ailments.

The ripe nuts contain a large amount of oil and are widely used in the food industry and in confectionery. The expressed oil is used in cooking, in artists paints and in soaps.

JUGLANS REGIA
(Juglandaceae)

Bark: anthel. detergent.

Leaves: astrin. tonic in decoct. considered to be specific in strumous sores, anthelm.

Fruit: alter. in rheum.

Alkaloid, Barium, Arsenic.

Juglans regia

JUNIPERUS COMMUNIS
(Juniper)

CUPRESSACEAE

The constituents of the berries include myrcene, sabinine, α, β-pinene, cineole, tannins, ditepenes, sugars, resin, vit. C a resin (10 per cent), an essential oil (juniper berry oil, 0.5-2 per cent) with pinene and borneol, inositol, a flavonoid glycoside and a bitter compound (juniperin). These give Juniper strong diuretic, tonix, rubefacient, carminative, antiseptic and aromatic properties. The crushed dried berries are used on their own in an infusion or in tea mixtures for dropsy and bladder and kidney disorders, and for rheumatic pain. The wood is used for the same purposes but it is not as potent as the berries. **Juniper must not be taken internally when the kidneys are inflamed or during pregnancy. Long-term**

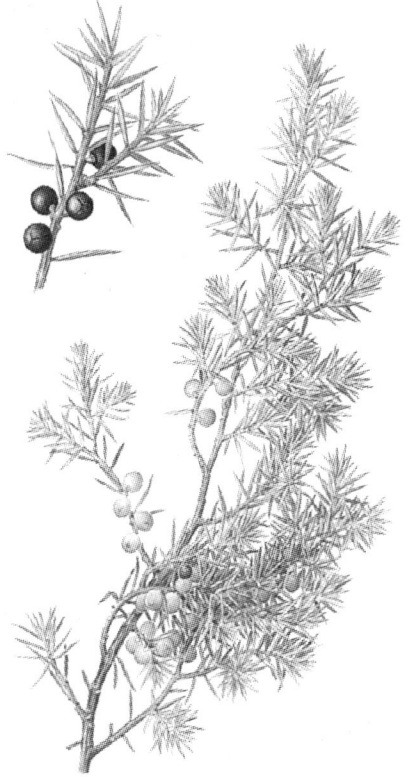

Juniperus communis

use may also damage the kidneys. Juniper is also used externally in compresses and bath preparations to relieve rheumatic and arthritic pain, for wounds and as a tonic.

JUSTICIA GENDARUSSA
(Acanthaceae)

Leaves and Tender shoots: diaphor. given in form of decoct. in chr. rheum.

Infusion of leaves: given internally in cephalogia, hemiplegia and fetal paralysis.

juice of leaf: for earache.

Plant: febge. emetic.

Alkaloid

KAEMPFERIA GALANGA
(Zingiberaceae)

Tubers: stim., expect. diur. carmin, reduced to powdered and mixed with honey given in coughs and pectoral affections. Boiled in oil externally applied to stoppages of nasal organs. Essential oil, alk.

KAEMPFERIA ROTUNDA
(Zingiberaceae)

Root: stomach, applied to reduce swellings, in form of a poultice used to promote suppuration.
Tuber: in powder form local application in mumps.
Plant: reduced to powder used in form of an ointment efficacious in healing fresh wounds; taken internally removes any coagulated blood or purulent matter.
Essential oil.

KALANCHOE LACINIATA
(Crassulaceae)

Juice of leaves: styptic, used on fresh cuts and abrasions, on bruises, burns and superficial ulcers, given in billious diar. and lithiasis.
Succulent leaves: valued as applications to wounds and sores; allay irritation and promote cicatrization.

KALANCHOE PINNATA
(Crassulaceae)

Leaves: bitter slightly toasted used as an application to bruises, wounds, boils and bites of insects.
leaves contain malic, isocitric, and citric acids.

KANDELIA RHEEDII
(Rhizophoraceae)

Bark: mixed with dried given or long pepper and rose water considered a cure for diabetes.
Mangrove.

KEDROSTIS ROSTRATA
(Cucurbitaceae)

Root: prescribed internally in electuary cases of piles, in powder used as demulc. in humoral asthma.

KICKXIA RAMOSISSIMA
(Scrophulariaceae)

Plant: valued as a remedy for diabetes.

KLEINHOVIA HOSPITA
(Sterculiaceae)

Decoct. of leaves: used in the Phillippine islands for scabies, and as lotion for cleaning skin eruptions.

KOCHIA INDICA
(Chenopodiaceae)

Plant: cardiac stimulant, used in cases of weak and irregular heart.

KOKOONA ZEYLANICA
(Celastraceae)

Powdered inner bark: used as snuff in headache.
Oil: used for protection against leeches.

KRAMERIA TRIANDRA
(Krameriaceae)

RHATANY

Part Used: Root.
Constituents: Rhatany contains 10-20% tannins, including phlobaphene, benzofurans, and n-methyltryrosine.

Medicinal Actions & Uses: Rhatany is astringent and antimicrobial. It is a useful remedy taken principally for problems affecting the gastro-intestinal tract. It is most commonly used for diarrhoea and dysentery. In addition, rhatany makes a good mouthwash and gargle for bleeding and infected gums, mouth ulcers and sore throats. The plant's astringency makes it useful in the form of an ointment, suppository or wash for treating haemorrhoids. Rhatany may also be applied to wounds to help staunch blood flow, to varicose veins and over area of capillary fragility that may be prone to easy bruising.

Laburnum anagyriodes

KYDIA CALYCINA
(Malvaceae)

Leaves: pounded and made into a paste applied in rheum. and lumbago.

KYLLINGA TRICEPS
(Cypereaceae)

Decoct. of roots: based to relieve thirst in fevers and diabetes.
Oil boiled in the roots: used to relieve pruritus of the skin.
Roots yield an oil which is used to promote action of liver and relieve pruritus.

LABURNUM ANAGYRIODES
Fabaceae
(Laburnum)

The seeds contain the highly toxic alkaloid cytisine, also proteins, tannins, glycosides and choline. Cytisine is isolated and used in pharmaceutical preparations to treat, for example, hypotension. In homeopathy a tincture prepared from the fresh leaves and flowers is sometimes used to treat various neurological and digestive disorders. Laburnum is a dangerous plant; **it should never be collected and used for self-medication**. Symptoms of poisoning include dilatation of the pupils, stomach cramps, vomiting, giddiness, muscular weakness, convulsions and respiratory failure and death. If poisoning is suspected medical aid should be sought immediately.

LACTUCA SERRIOLA
(Asteraceae)
Decoct. of seeds: used as demulc.
Plant: cooling, sedative, diur. diaphor. antisp. hypnotic, expect. useful in the treatment of the coughs in phthisis, broncht. asthma and pertussis.

Grain contains alk. bitter substance lactucin, oxalic acid, lactucopicrin.

LACTUCA VIROSA
(Compositae)

Wild Lettuce

Parts Used: Leaves, latex.

Constituents: The latex contains sesquiterpene lactones (including lactucopicrin and lactucerin); the leaves also contain flavonoids and coumarins. The sesquiterpene lactones have a sedative effect.

Medicinal Actions & Uses: Wild lettuce is a safe sedative that can be given to adults and children to encourage a sound night's sleep or to calm overactivity or overstimulation. Most commonly, it is recommended for excitability in children. It is also taken to treat coughs, often in combination with herbs such as liquorice (*Glycyrrhiza glabra*). Wild lettuce is thought to lower the libido. It may also be used to relieve pain.

Lactuca virosa

LAGENANDRA TOXICARIA
(Araceae)

Plant: very poisonous, remedy for itch, insecticidal.

LAGENARIA SICERARIA
(Cucurbitaceae)

Pulp: emetic, purg., applied to the soles in burning of the feet.
Decoct. of leaves: mixed with sugar given in jaundice.
Saponin, fatty oil.

LAGERSTROEMIA SPECIOSA
(Lythraceae)

Seeds: narcotic.
Bark and leaves: purg.
Root: astrin., stim., febge.
Fruit: used as local application for aphtahe of the mouth.
All parts of the plant especially old leaves and ripe fruit contain hypoglycemic principle having activity equivalent to 6-7.7 units of INSULIN.

LAGGERA ALATA
(Asteraceae)

Herb: used in Madagascar as a disinfectant.

LALLEMANTA ROYLEANA
(Lamiaceae)

Seeds: cooling, sedative, used in flatulence, constip.

LAMARCKIA AUREA
(Poaceae)

HCN: glucd

LAMINARIA SACCHARINA
(Phaeophyceae)
Used as cure for goitre, scrofule and
syphilis.
Source of Iodine.

LAMINUM ALBUM
Lamiaceae
(White Dead Nettle)
The flowers are used medicinally.
When dry they smell of honey and
have a bitter taste. The constituents
include mucilage, tannins, saponin,
flavonoid glycosides, tyramine,
methylamine and potassium salts.
These substances give White Dead-
nettle mild astringent, anti-inflammatory, expectorant, diuretic, tonic and
hypnotic properties. In herbalism an
infusion is used for catarrh of the
upper respiratory tract, bronchitis,

Laminum album

insomnia and for urinary and gynaecological disorders. In homeopathy a
tincture prepared from fresh material
is used for the same purposes. Externally a strong infusion is used in healing compresses or bath preparations
and as a gargle.
Tea made from the fresh flowers,
sweetened with honey, can be drunk
as a 'spring cure'. The tender young
leaves can be prepared and eaten like
spinach.

LAMPRACHAENIUM
MICROCEPHALUM
(Asteraceae)
Plant: used as arom. bitter.

LANNEA GRANDIS
(Anacardiaceae)
Bark: astrin, used as a lotion in
impetigenous eruptions, leprons
ulcres and obstinate ulcers.
Leaves: boiled and applied for local
swellings, and pains of the body.
Decoct. of bark: used for toothache.

LANSIUM DOMESTICUM
(Meliaceae)
Fruit peels contain lansinic acid
(toxic, heart poison); fruit contains
vitamin C; seeds contain alkl. and
resin;
Resin checks diarh; relieves intestinal spasm.

LANTANA CAMARA
(Verbenaceae)
Plant: in Guiana and La Reunion
considered vulnerary, diaphor., carmin
and antisp.
Decoct: Given in tetanus, rheum. and
malaria, tonic, much used in atoxy of
abdominal viscera.

Essential oil, containing camerene, isocamarene, and micranene bears lantanine, toxic.

LAPORTEA CRENULATA
(Urticaceae)

Seeds: used in the same way as coriander.
Juice of root: used in long-standing fevers.
Plant: poisonous.

LASIA ACULEATA
(Araceae)

Root: remedy for affections of the throat.
Leaves and Roots: remedy for piles.
Petioles: ground and mixed with water given to drink to cattle affected with throat disease.

LASIOSIPHON ERIOCEPHALUS
(Thymelaeaceae)

bark: fish poison
Plant: Vesic.
Leaves: applied to swellings and contusions: Resin from Bark.

LATHYRUS SATIVUS
(Fabaceae)

Oil from Seeds: powerful and dangerous.
Seeds contain poisonous principle.

LAUNEA GLOMERATA
(Asteraceae)

Decoct. of plant: mixed with some wheat meal into a poultice applied to the eyes to cure eyeache.

LAURUS NOBILIS
(Lauraceae)

Berries: emmen. used in diarh. leucor.

and dropsy; used in Europe to promote miscarriages.
Essential oil.

LAVANDULA ANGUSTIFOLIA
Lamiaceae
(Garden Lavender)

The flowering stems or the flowers alone are used medicinally. Among

Lavandula angustifolia

the constituents are an essential oil (up to 3 per cent) with linalyl acetate, linalool, camphor and borneal as the main components, also tannins (12 per cent). These give Garden Lavender mild sedative, carminative, antispasmodic, rubefacient and tonic properties. In herbalism it is still used internally for headache, nervous disorders and insomnia, as a cough suppressant and for flatulence, but mostly it is used externally as a skin freshener. The essential oil, which is obtained from the fresh plants by steam distillation, is a component of various proprietary preparations. Stimulates blood flow. The oil's chief use, however, is in perfumes, colognes and toilet articles. It is also used to mask unpleasant odours in medicines.

LARIX DECIDUA
(Pinaceae)
LARCH

Parts Used: Inner bark, resin.
Constituents: Larch contain lignans, resins and volatile oil (consisting mainly of alpha- and -beta -pinene and limonene).
Medicinal Actions & Uses:: Larch has astringent, diuretic and antiseptic properties. The bark may be used to treat bladder and urinary tubule infections such as cystitis and urethritis, and respiratory problems, including bronchitis. The resin is applied to wounds, where it protects and counters infection. A decoction of the bark is used to soothe eczema and psoriasis.

LARREA TRIDENTATA
(Zygophyllaceae)
CHAPARRAL, CREOSOTE BUSH
Parts Used: Aerial parts.

Constituents: Chaparral contains about 12% resin and nordihydroguaiaretic acid. The latter is reportedly harmful to the lymph glands and the kidneys.
Medicinal Actions & Uses: Until recently, chaparral remained in wide use in the US, with an average of 9.07 tonnes (10 tons), consumed there each year. It was thought to be a beneficial remedy for rheumatic disease, venereal infections, urinary infections and certain types of cancer, especially leukaemia. Chaparral was also taken internally for skin afflictions such as acne and eczema, and applied as a lotion to sores, wounds and rashes. Recently, however, its sale was banned in the US, due to concern over its potential toxic effect on the liver.

LAURUS NOBILIS
(Lauraceae)
BAY LAUREL

Parts Used: Leaves, essential oil.
Constituents: Bay laurel contains up to 3% volatile oil (including 30-50% cineole, linalool, alpha-pinene, alpha-terpineol, acetate, mucilage, tannin and resin).
Medicinal Actions & Uses: Bay laurel is used mainly to treat upper digestive tract disorders and to ease arthritis aches and pains. It is settling to the stomach and has a tonic effect, stimulating the appetite and the secretion of digestive juices. When used as an ingredient in cooking, bay laurel leaves promote the digestion and absorption of food. The leaves have much the same kind of positive effect as spearmint (*Mentha spicata*) and rosemary (*Rosmarinus officinalis*) in assisting meat. Bay laurel has also

been used to promote the onset of menstrual periods. The essential oil is chiefly employed as a friction rub, being well diluted in a carrier oil and massaged into aching muscles and joints. A decoction of the leaves may be added to a bath to ease aching limbs.

LAWSONIA INERMIS
(Lythraceae)

Parts Used: Leaves, bark.
Constituents: Henna contains coumarins, naphthaquinones (including lawsonee), flavonoids, sterols and tannins.
Medicinal Actions & Uses: Used mainly within Ayurvedic and Unani medicine, henna leaves are commonly taken as a gargle for sore throats, and as an infusion or decoction for diarrhoea and dysentery. The leaves are astringent, prevent haemorrhaging and strongly promote menstrual flow. A decoction of the bark is used to treat liver problems. Applied in the form of a plaster, henna treats fungal infections, acne and boils.

LEEA INDICA
(Vitaceae)

Root: used in diar. dysent. and as sudorific.
Decoct. of root: given in colic, cooling and relieves thirst.
Leaves: roasted and applied to the head in vertigo.
Leaves contain amorphous froth forming acid.

LEEA MACROPHYLLA
(Vitaceae)

Root: astrin. used as remedy for ringworm and in cure of guinea worm;

provided and applied to obstinate sores to promote cicatrization; applied externally to allay pain.

LENS CULINARIS
(Fabaceae)

Seeds: mucilaginous, laxt. useful in cases of constipation and other intestinal affections; made into a paste, useful in cleansing application in foul and indolent ulcers.
Arsenic and protein.

LEONOTIS NEPETAEFOLIA
(Lamiaceae)

Ashes of flower-heads: applied to burns and scalds; mixed with curd applied to ringworm and itchy skin diseases.
Root: crushed and rubbed on the breast when it swells and milk does not pass through the nipples.
Bitter principle in leaves.

LEPIDAGATHIS HAMILTONIANA
(Acanthaceae)

The Whorl in flower or in seed: Pounded and fried in Koronje oil rubbed on itch, also on head sore of children.
Root: crushed and rubbed on breast when swollen and milk does not pass through the nipples.

LEPIDIUM SATIVUM
(Brassicaceae)

Plant: administered in cases of asthma, cough with expectoration, and bleeding piles.
Root: used in secondary syphilis and tenesmus.
Seeds: galactag. and administered af-

ter being boiled with milk to cause abortion, applied to pains or hurts as a poultice, used as aper.
Leaves: stim., diur, useful scorbutic diseases.
Essential oil, glucd-glucotropoeolin.

LEPIDIUM VIRGINICUM
(Cruciferae)
Virginia Peppergrass
Parts Used: Leaves, root.
Constituents: Virginia peppergrass contains high levels of vitamin C.
Medicinal Actions & Uses: Virginia peppergrass is nutritious and generally detoxifying. It has been used to treat vitamin C deficiency and diabetes, and to expel intestinal worms. The herb is also diuretic and of benefit in easing rheumatic pain. The root is taken to treat excess catarrh within the respiratory tract.

LEPTADENIA RETICULATA
(Asclepiadaceae)
Plant: stim., tonic

LEPTANDRA VIRGINICA VIRGINICUM
(Scrophulariaceae)
Black Root
Part Used: Dried root.
Constituents: Black root contains a volatile oil, saponins, sugars and tannins.
Medicinal Actions & Uses: Black root is used in small doses today as a laxative and a remedy for liver and gallbladder disorders. The herb also treats flatulence and bloating, and eases the discomfort of haemorrhoids, chronic constipation and rectal prolapse. It may be given for skin problems if poor liver function is a factor.

LEUCAENA GLAUCA
(Mimosaceae)
Bark: eaten for internal pain
Seeds yield mimosine, leaves glucd quercitrin.

LEUCAS ASPERA
(Lamiaceae)
Plant: antipyr. insecticide.
Flowers: used in cold.
Juice of leaves: applied in psoriasis, scabies and chr. skin eruptions.
Leaves: considered useful in chr. rheum.
Leaves contain glucd.

LEUCAS CEPHALOTES
(Lamiaceae)
Plant: stim., diaphor., insecticide.
Fresh juice: external application in scabies.
Flowers: in form of a syrup used as a remedy for cough and colds.
Essential oil, alkl.

LEUCAS LAVANDULAEFOLIA
(Lamiaceae)
Leaves: roasted and eaten with salt as a febge, used for loss of appetice, in snake-bite.
Juice: employed in headaches and colds.

LEUCAS STELLIGERA
(Lamiaceae)
Plant: stim., carmin, emmen.

LEUCONOTIS EUGENIFOLIUS
(Apocynaceae)
Used in ringworm and skin diseases.

LEVISTICUM OFFICINALE
Apiaceae
(Lovage)
The roots of two – or three year plants

Levisticum officinale

or the flowering stems (collected before flowering) are used medicinally. Their constituents include an essential oil with terpineol and butyl phthalidine as its main components, furanocoumarins, sugars, esters of organic acids and resin. These substances give Lovage stomachic, carminative, cholagogic, diuretic, mild expectorant, antidiaphoretic and antirheumatic properties. In herbalism an infusion is used mainly to relieve flatulence, as an appetizer, for dropsy and urinary disorders rheumatism and nervous exhaustion. A hot infusion can be used an inhalent and if added to bath water, it has a cleansing and deodorising effect on the skin. **If taken internally in excess. Lovage may cause nausea and vertigo**. In particular, large doses should never be taken by pregnant women or by persons with kidney diseases.

LILIUM GIGANTEUM
(Liliaceae)
Leaves: employed as external cooling application to alleviate the pains of wounds and bruises.

LIMNOPHILA INDICA
(Scrophulariaceae)
Plant: considered antisep., made into liniment with coconut oil used in elephantiasis.
Juice of plant: rubbed over the body in pestilent fevers, and given internally in dysent., combined with gingere, cumin and other aromatics.
Essential oil.

LIMONIA CRENULATA
(Rutaceae)
Root: purg., sudorific, used for the cure of colic and cardialgia.
Dried fruit: antid. to various poisons, tonic, diminishes intestinal fermentation, resists the contagion of small pox, malignant and pestilent fevers.
Leaves: considered remedy for epilepsy.

LINARIA MINOR
(Scrophulariaceae)
HCN is young branches.

LINDENBERGIA INDICA
(Scrophulariaceae)
Juice: given in chr. bronch. and mixed with that of coriander applied to skin eruptions.

LINARIA VULGARIS
Scrophulariaceae
(Common Toadflax)
The flowering stems contain

Linaria vulgaris

Linum usitatissimum

flavonoid glycosides (linarin and pectolinarin), pectins, phytosterine, antirrhinic acid, tannic acid and vitamin C. These substances give Common Toadflax diaphoretic, diuretic, mild laxative and anti-inflammatory properties. An infusion is used for constipation, dropsy, inflamed kidneys and disorders of the liver and spleen. Externally it is used in compresses or bath preparations for skin rashes, varicose veins and haemorrhoids. A tincture is used in homeopathy.

LINUM USITATISSIMUM
Linaceae
(Flax)

The seeds are widely used medicinally. Their constituents include 30-40 per cent of a fatty acid (60 per cent), linoleic acid (20 per cent), stearic acid (8 per cent) and oleic acid; also mucilage, proteins, a cyanogenic glycoside (linamarin) HCN glucd and enzymes. Whole or crushed, the seeds are a reliable means of relieving constipation. After they have been swallowed the mucilaginous layers of the seed coat swell and, with the oil which acts as a demulcent and laxative, there is a rapid evacuation of the bowels. Externally, crushed seeds mixed to a paste with water are used to make hot poultices to relieve pain and to heal septic wounds, skin rashes and ulcers. The extracted oil is used in the pharmaceutical industry to make liniments for burns and rheumatic pain.

The oil is also important in the manufacture of paints, soap & printer's ink.
bark and leaves: used in gonor.
Flowers: nervine and cardiac tonic.

LIPPIA CITRODORA
(verbenaceae)
LEMON VERBENA
Parts Used: Leaves.
Constituents: Lemon verbena con-

tains a volatile oil (mainly consisting of citral, cineole, limonene and geraniole), mucilage, tannins and flavonoids.

Medicinal Actions & Uses: An undervalued medicinal herb, lemon verebena shares qualities with lemon balm (*Melissa officinalis*). Both herbs contain a strong lemon-scented volatile oil that has calming and digestive properties. Lemon verbena has a gentle sedative action and has a reputation for soothing abdominal discomfort. Its tonic effect on the nervous system is less pronounced than that of lemon balm, but it nonetheless helps to lift the spirits and counter depression.

LIPPIA NODIFLORA
(Verbenaceae)

Plant: febge. diur. made into a poultice used as maturant for boils.
Infusion of leaves and tender stalls: given to children in indig. and to women after delivery.
Bitter substance from plant.

LIQUIDAMBAR ORIENTALIS
(Hamamelidaceae)

LEVANT STORAX

Part Used: Bark extract.
Constituents: Levant storax contains cinnamic acid, cinnamyl cinnamate, phenylpropyl cinnamate, triterpene acids and a volatile oil.
Medicinal Actions & Uses: Levant storax balsam acts as both an irritant and an expectorant within the respiratory tract, and it is one of the ingredients of Friar's Balsam, an expectorant mixture that is inhaled to stimulate a productive cough. In addition, levant storax balsam is applied externally to encourage the healing of skin diseases and problems such as scabies, wounds and ulcers. Mixed with witch hazel (*Hamamelis virginiana*) and rosewater (*Rosa species*), levant storax makes an astringent face lotion. In China, storax balsam is used to clear mucus congestion and to relieve pain and constriction in the chest.

LIRIOSMA OVATA
(Oleaceae)

MUIRA PUAMA

Parts Used: Root, bark, wood.
Constituents: Muira puama contains esters and plant sterols.
Medicinal Actions & Uses: Muira puama has long been used by indigenous Amazonians as a tonic and aphrodisiac. It is still considered a valuable remedy for impotence. The bark is strongly astringent and may be employed as a gargle for sore throats and taken in the form of an infusion in order to treat diarrhoea and dysentery.

LITCHI CHINENSIS
(Sapindaceae)

Fruit: tonic
Leaves: used as a cure for bites of animals.

LITHOSPERMUM ARVENSE
(Boraginaceae)

Infusion of leaves: used as a sedative in Spain.
Alk. cynoglossine.

LITSEA CHINENSIS
(Lauraceae)

Bark: demuk. astrin. used in diar. and dysent. aphrodis. anodyne local antid. to bites of venomous animals; freshly

ground used either dry or triturated in water or milk, as emol. application to bruises and as styptic dressing for wounds.
Oil from berries: used in rheum.
Leaves: mucilaginous considered antisp. and emol.
Alk. laurotetanine

LITSEA POLYANTHA
(Lauraceae)
Bark: astrin. used in diarh. stomach, stim., after being bruished, applied fresh or dry to contusions, powdered and applied to the body for pains arising from blows or bruises or from handwork, also applied to fractures in animals.
Seeds contain fatty oil.

LOBARIA PULMONARIA
(Stictaceae)
TREE LUNGWORT
Part Used: Lichen.
Constituents: Tree lungwort contains a variety of plant acids (including stictic and sticinic acid), fatty acids, mucilage and tannins.
Medicinal Actions & Uses: A beneficial but under-used remedy, tree lungwort has expectorant and tonic properties. It aids in clearing congested mucus, reduces catarrh and helps to increase the appetite. In a decoction sweetened with honey, it is appropriate for all conditions that are marked by chronic respiratory catarrh, especially coughs and bronchitis. The plant also treats asthma, pleurisy and emphysema. Being astringent and demulcent, tree lungwort makes a useful treatment for pulmonary ulcers as well as for a variety of gastro-intestinal problems. It is highly suitable for treating ailments in children.

LOBELIA INFLATA
(Campanulaceae)
LOBELIA, INDIAN TOBACCO
An annual growing to 50 cm, with lance-shaped leaves and pale blue, pink-tinged flowers.
Key Constituents:
Piperidine alkaloids (principally lobeline, but many others present)
Carboxylic acids
Key Actions:
Respiratory stimulant
Antispasmodic
Expectorant
Induces vomiting
Increases sweating

LODOICEA MALDIVICA
(Arecaceae)
DOUBLE COCONUT
Fruit: used as tonic, preservative, alixipharmic, febfge., used to check diar. and vomiting especially in cholera, mixed with the root of Nux Vomica given to children for colic; the water or soft kernel considered antibil. and antacid when taken after meals.

LOLIUM TEMULENTUM
(Poaceae)
Darnel meal: considered a sedative poultice and believed to cure freckles, cattle poison.
Toxic alk. temuline.

LONICERA GLAUCA
(Caprifoliaceae)
HONEYSUCKLE & JIN YIN HUA (CHINESE)
Parts Used: Flowers, leaves, bark.
Constituents: Honeysuckle's constituents include a volatile oil, tannins and salicylic acid. *Jin yin hua* contains a volatile oil (which includes

linalool and jasmine), tannins, luteolin and inositol.

Medicinal Actions & Uses: Honeysuckle is rarely used in contemporary Western herbal medicine. Traditional usage indicates that different parts of the plant have significantly different therapeutic benefits. The bark is diuretic and may be taken to relieve gout, kidney stones and liver problems. The leaves are astringent and make a good gargle and mouthwash for sore throats and mouth ulcers. The flowers, which relieve coughs and are antispasmodic, are traditionally taken as a treatment for asthma. *Jin yin hua* is prescribed for an entirely different range of diseases in Chinese herbal medicine. It is principally employed to counter "hot" infectious disorders such as abscesses, sores, inflammation of the breasts and dysentery. *Jin yin hua* is also taken to bring down fever.

LOPHOPHORA WILLIAMSII
(Cactaceae)

PEYOTE, MESCAL

Part Used: Whole plant.

Constituents: Peyote contains alkaloids, principally mescaline, which is a powerful hallucinogen.

Medicinal Actions & Uses: Peyote is a shamanistic plant, taken in Native American rituals to deepen spiritual understanding. It plays an important part in the emotional and mental state of the community. It is also used to treat fevers, as a painkiller for rheumatism and to treat paralysis. It is applied as a poultice for fractures, wounds and snake bite. Peyote is also used to induce vomiting.

LOPHATHERUM GRACILE
(Poaceae)

Leaves: considered as antifebrile and diur. in China.

LOTUS CORNICULATUS
(Fabaceae)

HCN - glued.

LUFFA ACUTANGULA
(Cucurbitaceae)

Plant: laxat. purg. useful in skin diseases and asthma, bitter tonic, diur. used in spleenic enlargements.

Fruits and Seed: emetic, cath.

Kernel of Seeds: used in dysent. Seeds contain fixed oil.

LUFFA CYLINDRICA
(Cucurbitaceae)

LOOFAH, SI GUA LUO (CHINESE)

Part Used: Fruit.

Constituents: Loofah contains xylan, xylose and galactan, bitter substance Luffin, saponin glycoside, fixed oil, enzyme.

Medicinal Actions & Uses: In Chinese medicine, the inner skeleton of the dried fruit is used to treat pain in the muscles, joints, chest and abdomen. It is prescribed for chest infections accompanied by fever and pain, and is used to clear congested mucus. Loofah is also given to treat painful or swollen breasts.

LUFFA ECHINATA
(Cucurbitaceae)

Plant: emetic, anthelm, in jaundice, phthis, hiccough.

Fruit: used in dropsy, purg., in form of infusion given in colic and in cholera after each stool.

Amorph. bitter substance.

LUPINUS ALBUS
(Fabaceae)
Used as anthelm. diur. pectoral and tonic.

Alkls. lupinine, lupinidine, lupamine, decoct. of seeds increases sugar tolerance in diabetic patients; reduces blood pressure.

LYCHNIS CORONARIA
(Caryophyllaceae)
Extract or decoct. of the root: used in Spain for infraction of the lymph glands of the misentery and for diseases of the lungs and liver.

LYCIUM CHINESE
(Solanaceae)
LYCIUM, CHINESE WOLFBERY

A deciduous shrub growing to 4 m, with bright green leaves and scarlet berries.

Key Constituents:
Berries only.
Physalien
Carotene
Vitamins B1, B$_{12}$ and C
Root only
Cinnamic acid
Psyllic acid
Key Actions:
Berries
Tonic
Protect the liver
Root
Reduces fever
Lowers blood pressure

LYCOPERSICON ESCULENTUM
(Solanaceae)
TOMATO

Fruit contains: oxalic acid, narcotine, rutin, citric and malic acids, vitamin A and C.

LYCOPODIUM CLAVATUM
(Lycopodiaceae)
CLUB MOSS

Parts Used: Moss, spores.
Constituents: Club moss contains about 0.1%-0.2% alkaloids (including lycopodine), polyphenols, flavonoids and triterpenes.

Medicinal Actions & Uses: Club moss is diuretic, sedative and antispasmodic, and it is particularly useful for treating chronic urinary complaints. The herb may also be taken for indigestion and gastritis. The spores may be applied to the skin to relieve and protect itchy or irritated areas. Clavatine, clavatoxine alks.

LYCOPUS EUROPAEUS
Lamiaceae
(Gipsywort)
The flowering stems are used medicinally. Their constituents include an essential oil, tannins, a bitter com-

Lycopus europaeus

pound (lycopin) and organic acids.
These substances give Gipsywort
sedative, astringent and cardioactive
properties; they also reduce the activity of iodine in the thyroid gland. It
was once prescribed for hyper-
thyroidism and related disorders such
as Basedow's disease. In homeopathy tinctures of the fresh plant are used
to treat anxiety and various heart disorders such as angina and spasms.
Gipsywort preparations take a long
time to take effect and need careful
medical supervision. **It is not a suitable herb for self-medication**.

LYCOPUS VIRGINICUS
(Labiatae)

BUGLEWEED

Parts Used: Aerial parts.
Constituents: Bugleweed contains
phenolic acids (including derivatives
of caffeic, chlorogenic and ellagic
acids).
Medicinal Actions & Uses:
Bugleweed has sedative properties
and today the herb is principally prescribed to treat an overactive thyroid
gland and the racing heartbeat that
often accompanies this condition.
Bugleweed is also considered an aromatic and tonic astringent that reduces
the production of catarrh.

LYSIMACHIA NUMMULARIA
Primulaceae
(Creeping Jenny)

The flowering stems are used medicinally. When dry they are pale green
and have a bitter, astringent taste.
Their constituents include saponins,
tannins and silicic acid. These sub-

Lysimachia nummularia

stances give creeping Jenny astringent
and antiseptic properties. It is sued in
an infusion for gastroenteritis and
severe diarrhoea. The fresh, macerated flowering stems are used in compresses applied externally to slow-
healing wounds, ulcers and skin
rashes and they also relieve rheumatic
pain. The decoction and the fresh juice
have a soothing emollient action on
the skin.

LYSIMACHIA VULGARIS
(Primulaceae)

YELLOW LOOSESTRIFE

Parts Used: Aerial parts.
Constituents: Yellow loosestrife

contains a benzoquinone, saponins, flavonoids and tannins.

Medicinal Actions & Uses: An astringent herb, yellow loosestrife is principally used to treat gastro-intestinal conditions such as diarrhoea and dysentery, to stop internal and external bleeding, and to cleanse wounds. It makes a serviceable mouthwash for sore gums and mouth ulcers, and may be used to treat nosebleeds. Yellow loosestrife has also been taken as an expectorant.

LYTHRUM SALICARIA
(Lythraceae)

PURPLE LOOSESTRIFE

Parts Used: Aerial parts.

Constituents: Purple loosestrife contains salicarin, a glycoside (vitexin), tannins, a volatile oil, mucilage and plant sterols.

Medicinal Actions & Uses: The astringent purple loosestrife is mainly employed as a treatment for diarrhoea and dysentery. It can be safely taken by people of all ages; some herbalists recommend it to help arrest diarrhoea in breast-feeding babies. The herb may also be used to treat heavy periods and for inter-menstrual bleeding. Externally, it is applied as a poultice or lotion to wounds, leg ulcers and eczema, and used to treat excess vaginal discharge and vaginal itching. Purple loosestrife is now little used to treat eye problems, but, as Culpeper's experience suggests, the herb could be worth further investigation as a remedy for disorders of the eyes and vision.

MACHILUS MACRANTHA
(Lauraceae)

Bark: Asthma, rheum.
Leaves: applied to ulcers.

MACROTOMIA BENTHAMI
(Boraginaceae)

Plant: considered useful in diseases of tongue and throat.

MADHUCA BUTYRACEA
(Sapotaceae)

Fat: used as ointment in rheumatism, emol. for chapped hands, etc. in winter.
Saponin, Saponogenin and basic acid.

MADHUCA SPP.
(Sapotaceae)

BUTTER TREE

Parts Used: Flowers, seed oil.
Constituents: The leaves contain an alkaloid and a saponin; the seeds a saponion and fixed oil.

Medicinal Actions & Uses: The expectorant flowers are used to treat chest problems such as bronchitis. They are also taken to increase breast-milk production. The leaves are applied as a poultice to eczema. In Indian folk medicine, the leaf ash is mixed with *ghee* (clarified butter) to make a dressing for wounds and burns. The seed oil is laxative and is taken for constipation and to loosen the stool of haemorrhoid sufferers. The oil is also applied to itchy skin.

MADHUCA INDICA
(Sapotaceae)

Flowers: yield a distilled spirit which is astrin., tonic, appetizing, regarded as cooling, tonic, nutritive, used in coughs in form of decoction; dried ones used as a fermentation, in cases of orchitis for their sedative effect; fried in *ghee* eaten by persons suffering from piles.
Bark: used in decoct. as atrin., and tonic, fish poison.
Alk, glucosidic saponin sapogenin and basic acid.

MADHUCA LONGIFOLIA
(Sapotaceae)

Bark: in decoct. used as an astrin., and emol. also as remedy in itch.
Flowers: laxat. stim. anthelm. used in snake-bite, as fish poison.
Oil from seeds: good for skin diseases.
Gummy juice: rheum.
A poisonous saponin, mowrin, bitter substance, fruits contain essential containing ethyl cinnamate. Mowrin has a digitalis-like action upon the heart and a saponin-like haemolytic effect.

MADHUCA MALABARICA
(Sapotaceae)

Fruits: given in rheum. biliousness, consumption, asthma and worm.
Oil from seeds: rheum., and for improvement of hair.
Flowers: soaked in water used in kidney complaints.

MAERUA AREWARIA
(Capparidaceae)

Root: alter, tonic and stim.

MAESA INDICA
(Myrsinaceae)

Berries: anthelm.
Root: used in syphilis.
Leaves: used as fish poison.

MAGNOLIA OFFICINALIS
(Magnoliaceae)

HOU PO (CHINESE), MAGNOLIA
Part Used: Bark.
Constituents: *Hou Po* contains a volatile oil and magnocurarine. Extracts of the plant have a slight muscle-relaxing effect when injected.
Medicinal Actions & Uses: *Hou po* bark is aromatic, warming and pungent. It relieves griping pain and flatulence, and is taken for abdominal distension, indigestion, loss of appetite, vomiting and diarrhoea.

MAHONIA NAPAULENSIS
(Berberidaceae)

Berries: considered diur. and demuk. in dysent.
Roots of old plants yield alks. umbellatine and neprotine.

MAJORINA HORTENSIS
(Lamiaceae)

Plant: carmin., expect., tonic to the liver.
Leaves and Seeds: astrin. remedy for colic.
Essential oil from leaves – used for hot formentations in acute diarrh.
Essential oil, bitter substance.

MALLOTUS PHILLIPPINENSIS
(Euphorbiaceae)

Glands and Hairs on the fruits: bitter, anthelm. cath., styptic.

Rotterin and Isorotheltrin. Both are resins and wax.

MALUS SYLVESTRIS
(Rosaceae)

Fruit: eaten to obviate constipation.
Infusion of bark: given in intermittent, remittent and billious fevers.
Root: anthelm. refrig. hyptonic
Leaves yield glucoside, seeds amygdalin, phlorizin glucds.

MALVA ROTUNDIFOLIA
Malvaceae

Seeds: demuk., used in broncht. coughs, ulceration of bladder and haemorrhoids, applied externally in skin diseases.
Leaves: emol. used in piles.
Plant: used as a cooling drug.

MALVA SYLVESTRIS
Malvaceae
(Common Mallow)

The flowers and leaves, free of mallow rust, are used medicinally. When properly dried the flowers are blue. The constituents include abundant mucilage, tannins, an essential oil, organic pigments (anthocyanins) in the flowers, and pro-vitamin A and vitamins B and C (in the leaves). These substances give Common Mallow emollient, expectorant, anti-inflammatory and astringent properties and it is used in much the same way as Marshmallow – for bronchitis, laryngitis and other respiratory disorders, gastritis and enteritis, and as a mild laxative. It also promotes the healing of internal injuries, damaged mucosa and gastric ulcers. Externally Common Mallow

Malva sylvestris

is used in compresses and bath preparations for skin rashes, boils and ulcerous conditions, and in gargles and mouth washes.
The fresh leaves and young shoots can be added to salads and soups and cooked as a vegetable.

MALVA SYLVESTRIS
MAURITANICA
Malvaceae
(Common Mallow)

MALVACEAE

Musk Mallow is a perennial with erect stems, which are often purple-spotted. It is distinguished from Common Mallow in having reinform basal leaves, deeply and narrowly divided stem leaves, usually solitary rose-pink flowers in the leaf axils, and hairy, blackish mericarps.

Malva sylvestris Mauritanica

MALVA VETICILLATA
(Malvaceae)

Root: used in Indochina to produce vomiting in whooping cough.
Leaves and stems: considered digestive, given to women in advanced stage of pregnancy.
Ash and dried leaves: employed in the preparation of a drink given in scabies.

MALVASTRUM COROMANDELIANUM
(Malvaceae)

Plant: considered emol. resolv. and bechic in West Indies.
Leaves: applied to inflamed sores and wounds as a cooling and healing solve.
Flowers: given as pectoral and diaphor.

MANDRAGORA OFFICINARUM
(Solanaceae)

MANDRAKE
Part Used: Root.
Constituents: Mandrake contains 0.4% tropane alkaloids (hyoscine and hyoscyamine).
Medicinal Actions & Uses: Mandrake has now largely fallen out of use. The herb is sometimes applied as a poultice or plaster for rheumatic and arthritic pains, or, as a decoction, for ulcers and similar kinds of skin disorders. Narcootic, anaesthetic Mandragorine a mixture of Hyoscyanine and scopolamine alkaloids.

MANGIFERA INDICA
(Anacardiaceae)

MANGO
Leaves: in scorpion sting.
Ripe fruit: Laxt., diur., astrin., useful in haemor. from uterus, lungs and intestines.
Unripe fruit: useful in ophthalmia and eruptions.
Rind of fruit: astrin. stim. tonic in debility of stomach.
Seeds: used in asthma.
Kernel: astrin., used in haemor. in diar., anthelm., its juice of snuffed can stop nasal bleeding.
Bark: astrin., used in uterine haemor. haemoptysis, and melaena, diar., and other discharges.
Fruit contains vitamins A, C, B & D.

MANIHOT ESCULENTA
(Euphorbiaceae)

MANIOC, CASSAVA
Part Used: Root
Constituents: Manioc contains

cyanogenic glycosides (0.02-0.03% in the bitter varieties; 0.007% in the sweet) and starch. Essential oils, saponins.

Medicinal Actions & Uses: Manioc root is easily digestible and makes a suitable, if low-protein, food for convalescence. The bitter variety may be used to treat scabies, diarrhoea and dysentery. Manioc flour may be used to help dry weeping skin. In China, a poultice is made of manioc, wheat flour and ginger (*Zingiber officinale*), to draw out pus when infection is present.

Caution: Raw bitter manioc is toxic and has caused many deaths. The root must be carefully soaked and cooked before eating.

MANILKARA KAUKI
(Sapotaceae)

Seeds: tonic, febge, anthelm., made into powder used in opthalmia, prescribed in leprosy, thirst, delirium and disorders of many secretions.

Root and bark: astrin, given in infantile diarh., after being ground with water and mixed with honey.

Leaves: boiled in gingerly oil and added to pulverised bark used as a remedy for beri-beri.

MARANTA ARUNDINACEA
(Marantaceae)

ARROWROOT

Part Used: Rhizome.

Constituents: Arrowroot contains 25-27% neutral starch.

Medicinal Actions & Uses: Arrowroot is used in herbal medicine in much the same manner as slippery elm (**Ulmus rubra**), as a soothing demulcent and a nutrient of benefit in

convalescence and for those with weak digestions. It helps to relieve acidity, indigestion and colic, and is mildly laxative. It may be applied as an ointment or poultice mixed with antiseptic herbs such as myrrh (*Commiphora mohnol*).

MARRUBIUM VULGARE
(Labiatae)

WHITE HOREHOUND

Parts Used: Leaves.

Constituents: White horehound contains the diterpenes marrubiin (0.3-1.0%) and marrubenol, flavonoids, alkaloids (including betonicine and stachydrine) and 0.6% volatile oil. Marrubiin is strongly expectorant and bitter.

Medicinal Actions & Uses: White horehound is helpful for wheeziness, bronchitis, bronchiectasis (a damaged air passage within the lung), bronchial asthma, non-productive coughs and

Marrubium vulgare

whooping cough. The herb apparently causes the secretion of a more fluid mucus, readily cleared by coughing. As a bitter tonic, white horehound increases the appetite and supports the function of the stomach. The herb may also act to normalise heart rhythm, improving its regularity. It is less commonly used as a decoction for skin conditions.

It has also proved beneficial for menstrual pain and menstrual irregularities. Extremely it is used in compresses to treat painful and inflamed wounds.

MARSDENIA CONDURANGO
(Asclepiadaceae)

CONDURANGO

Parts Used: Bark, latex.

Constituents: Condurango bark contains glycosides (based on condurangogenins), a volatile oil and phytosterols.

Medicinal Actions & Uses: The bark's main effect is to stimulate stomach secretions. It is often used in South American folk medicine as a bitter and digestive tonic. Condurango is a specific treatment for nervous indigestion and anorexia nervosa, since it bitterness slowly increases the appetite as well as the stomach's ability to process more food. The herb is also thought to stimulate the liver and pancreas, and may be taken for liver disorders. Condurango also encourages menstruation. The caustic white latex is applied to remove warts.

MARTYNIA ANNUA
(Pedaliaceae)

Leaves: given in epilepsy, applied to tuberculosis glands of the need.

Juice: used as gargle for sore throat.
Fruit: alextric, useful in inflamm.

MATRICARIA AUREA
(Asteraceae)

Plant, chiefly **flowers**: used in Spain as tonic, diaphor. anthelm., antipyr., antilystenic, and for pain in bowels.

MATRICARIA CHAMOLILLA
(Asteraceae)

Flowers: stim., attenuant, discutient, carmin, uscd in constitutional debil ity, in hysteria, dyspep, and intermittant fevers.

Oil: used externally in rheum. effective in flatulence and colic.

Essential oil, chamazulene, apignine, α-heteroside, salicylic acid, and non-erystalline, β-heteroside, pure azulene for essential – anti inflammatory.

MATTHIOLA INCANA
(Brassicaceae)

Seeds: bitter, tonic, stim., expect. diur., stomach, aphrodis; mixed with wine given as antid. to poisonous bites, used in infusion in cancer.

MECANOPSIS ACULEATA
(Papaveraceae)

Plant: especially the root: considered narcotic and poisonous.

MEDICAGO SATIVA
(Leguminosae)

ALFALFA, LUCERNE

Parts Used: Aerial parts, sprouting seeds.

Constituents: Alfalfa contains isoflavones, coumarins, alkaloids, vitamins and porphyrins. The isoflavones and coumarins are oestrogenic.

Medicinal Actions & Uses: Alfalfa is perhaps more therapeutically useful as a food than a medicine. It is given to convalescents who require easily assimilated nutrients. In view of alfalfa's oestrogenic activity, it could prove useful in treating problems relating to menstruation and the menopause.

MELALEUCA ALTERNIFOLIA
(Myrtaceae)

Tea Tree an evergreen reaching 7 m, with layers of papery bark, pointed leaves and white flower spikes.
Key Constituents:
Volatile oil (percentages are variable), terpinen 4-01 40% gammaterpinene 24%, alpha-terpinene 10%, cineol 5%.
Key Actions:
Antiseptic
Antibacterial
Antifungal
Antiviral
Immune stimulant

MELALEUCA LEUCADENDRON
(Myrtaceae)

Oil distilled from fresh leaves and twigs: used internally and also as external application in rheum; stimulant and antisp. in choleraic diarr, rubft., and in psoriasis, chr. pityriasis, acne and eczema; mosquito repellant. Essential oil, cajupatol, identified with encalystol; bark yields resinol melaleucin.

MELALEUCA LEUCADENDRON
(Myrtaceae)

CAJUPUT

Part Used: Essential oil.

Constituents: The volatile oil contains terpenoids, mainly cineole (50-60%), beta-pinene, alpha terpineol and others. Cineole is strongly antiseptic.
Medicinal Actions & Uses: Cajuput is normally combined with other essential oils such as eucalyptus (*Eucalyptus globulus*). Its antiseptic properties treat colds, sore throats, coughs and, especially, chest infections. The diluted oil may either be steam inhaled or applied to the chest or throat to treat laryngitis, tracheitis and bronchitis. As cajuput stimulates the circulation and is antispasmodic, it is used as a friction rub for rheumatic joints and neuralgia.

MELASTOMA DECEMFIDUM
(Melastomaceae)

Root, leaf and fruit: astrin., used in diarh, and diseases of uterus.
Root: used in Cambodia in treatment of liver complaints, with jaundice, considered stim. tonic, in form of infusion prescribed in malaise and vertigo.

MELASTOMAA MALABATHRICUM
(Melastomaceae)

Leaves: used in diarh. and dysent.
Leaves and flower-tops: given as astrin. in leveor. and chr. diarh. in Indo-China.

MELIA AZEDARACH
(Meliaceae)

Root, Bark, Fruit, flowers and leaves: deobstruent, resolv., alexipharmic.
Flowers and leaves: applied as poultice to relieve nervous headaches.

Juice of leaves: used internally as anthelm., antilithic, diur., emmen.

Seeds: prescribed in rheum.

Oil: properties similar to **neem oil**.

Leaves and bark: used internally and externally in leprosy and scrophula. Alkaloid azaridine, resin, tannin, meliotannic acid, benzoic acid. Bakayamin sterol, bitter principle mayosine, fixed oil contains sulphur.

MELIA COMPOSITAE
(Meliaceae)

Fruit pulp: useful for colic.

Juice of green fruit: with a third of its weight of sulphur and equal quantity of curds, heated and applied to skin diseases and to sores infested with maggots.
Glued.

MELIANTHUS MAJOR
(Sapindaceae)

Root: poisonous, emetic, remedy against snake-bite.

MELILOTUS OFFICINALIS
Fabaceae
(Ribbed Melilot)

The flowering stems or just the flowers alone are used medicinally. When dry they have a bitter taste and a hay-like smell due to coumarin. The constituents include melilotin and other coumarin (Hydroxy-coumarin & Hydrocoumarin), di-coumarol, a powerful anticoagulant, glycosides, tannins, an essential oil and flavonoid pigments. These substances give Ribbed Melilot aromatic, expectorant, antispasmodic, antithrombic, astringent and anti-inflammatory properties. It is contained in some pro-

Melilotus officinalis

prietary preparations for use in the treatment of thrombosis and varicose veins and is a component of medicinal cigarettes for bronchial asthma. More often, however, Ribbed Melilot is used in plasters and salves, compresses or bath preparations. **Ribbed Melilot should always be taken internally under professional supervision.** Large doses may cause bleeding, headache, vertigo and vomiting.

MELILOTUS INDICA
(Fabaceae)

Seeds: useful in bowel complaints and infantile diarh. given as grnel.

Plant: used as a discutient and emol. externally as a formentation, poultice or plaster for swellings.

MELISSA OFFICINALIS
Lamiaceae
(Balm)

The leaves are used medicinally. Their constituents include 0.1 to 0.25 per

Melissa officinalis

cent of an essential oil with citral, linalool, geraniol and citronellal as its main components, plus tannins, a bitter compound and hydroxyterpenic acid. It is the oil that imparts the lemon scent. Balm has carminative, antispasmodic, stomachic, diaphoretic and sedative properties. It is used in infusions for digestive disorders, nausea, flatulence, nervous anxiety, headache and insomnia. It can be used in potpourris, herb pillows and in herb mixtures for aromatic baths and cosmetic waters.

With their delicate lemon flavour the leaves have a variety of uses in cooking and they also make a refreshing addition to salads, cold drinks and wine cups.

MELISSA PARVIFLORA
(Lamiaceae)
Leaves and Stems: antipyr. used in brain, liver and heart diseases and in bites of venomous insect.
Fruit: brain tonic., useful in hypochondriac conditions.
Plant: strengthens the gums and removes bad taste from the mouth.

MELOTHRIA MADERASPATAMA
(Cucurbitaceae)
Root: in decoct. useful in flatulence and masticated for relief of toothache.
Tender shoots and leaves: used as a gentle aper. recommended in vertigo and biliousness.
Seeds: in decoct. sudorific, crushed and applied on aching bodies; specially on strained backs.

MEMECYLON AMPLEXICAULE
(Melastomaceae)
Decoct. of flowers and shoots: used in skin diseases.
Roots: ecbolic.

MEMECYLON UMBELLATUM
(Cucurbitaceae)
Leaves: used as a cooling astrin; in conjunctivitis as a lotion, internally given in leucor. and gonor.
Decoct. of root: useful in excessive menstrual discharges.

MENTHA AQUATICA VAR. CRISPA
Lamiaceae
(Water Mint)
The flowering stems or the leaves alone are used medicinally. Their constituents include as essential oil with carvone as the main component,

Mentha aquatica var. Crispa

tannins and bitter compounds. These substances give Water Mint stomachic, carminative, cholagogic, antispasmodic and slight astringent properties. It is used fresh or dried in an infusion to stimulate the appetite, to treat digestive and gall bladder disorders, flatulence, diarrhoea and abdominal spasms. The essential oil obtained by distillation from the fresh leaves and stems is also used medicinally. **Large doses of Water Mint may cause vomiting.**

MENTHA HAPLOCALYX
(Labiatae)

BO HE (CHINESE), CORN MINT

Parts Used: Aerial parts.
Constituents: *Bo he* contains a volatile oil comprising mainly menthol (up to 95%) with menthone, menthyl acetate, camphene, limonene and other terpenoids.
Medicinal Actions & Uses: In Chinese herbal medicine *bo he* is a popular treatment for colds, sore throats, sore mouth and tongue, and a host of other conditions ranging from toothache to measles. Like peppermint (*M. X piperita*), it helps to lower the temperature, has anticatarrhal properties, and may be taken for dysentery and diarrhoea. The juice has also been used to treat earache. *Bo he* is often combined with *ju hua* (*Chrysanthemum X morifolium*), to treat headaches and bloodshoot or sore eyes.

MENTHA X PIPERATA
Lamiaceae
(Peppermint)

The leaves are used medicinally but they must be free of mint rust (*Puccinia menthae*). The constituents include an essential oil (peppermint oil) with, as its main components, menthol (up to 50 per cent), 5-10 per cent esters (mainly menthyl acetate) and menthone (15-30 per cent), plus tannins and bitter compounds. These substances give Peppermint stomachic, carminative, cholagogic, milk antispasmodic, expectorant, antiseptic and local anaesthetic properties. It is used in proprietary medicines and in infusions to stimulate the appetite, to treat respiratory infections, digestive and gall bladder disorders, diarrhoea, flatulence and abdominal spasms. Because of the methol it contains Peppermint is added to aromatic waters, drops and spirits for embrocations and as analgesics to relieve sprains and headaches. The

Mentha X Piperata

essential oil obtained by steam distillation from the fresh or partially dried herb is used for flavouring toothpastes and mouthwashes, pharmaceutical preparations, liqueurs and confectionery. The fresh leaves make a refreshing tea substitute.

MENTHA PULEGIUM
(Labiatae)

PENNYROYAL

Parts Used: Aerial parts.
Constituents: Pennyroyal's volatile oil contains pulegone (27-92%), isopulegone menthol and other terpenoids. Pennyroyal also contains bitters and tannins.
Medicinal Actions & Uses: Similar to many respects to peppermint (*M. X piperita*), pennyroyal is a good digestive tonic. It increases the secretion of digestive juices, relieves flatulence and colic, and occasionally is used as a treatment for intestinal worms. It makes a good remedy for headaches and for minor respiratory infections, helping to check fever and reduce catarrh. Pennyroyal powerfully stimulates the uterine muscles and encourages menstruation. An infusion of pennyroyal is used externally to treat itchiness and fornication (a sensation of ants crawling over the body), inflamed skin disorders such as eczema, and rheumatic conditions including gout.

MENTHA SPICATA
(Labiatae)

Seeds: mucilaginous
Leaves: given in fever and broncht. and in decoct. used as a lotion in apthae.
Herb: considered stim. carmin., and antisp.
Essential oil.

MENYANTHES TRIFOLIATA
(Menyanthaceae)

BOGBEAN

Parts Used: Leaves.
Constituents: Bogbean contains iridoid glycosides, flavonol glycosides, coumarines, phenolic acids, sterols, triterpenoids, tannin and very small amounts of pyrrolizidine alkaloids. The iridoids are strongly bitter and stimulate digestive secretions.
Medicinal Actions & Uses: Bogbean is a strongly bitter herb that encourages the appetite and stimulates digestive secretions. It is taken to improve an underactive or weak digestion, particularly if there is abdominal discomfort. This herb is also used as an aid to weight gain. Bogbean is

Menyanthes trifoliata

thought to be an effective remedy for rheumatoid arthritis, especially when this condition is associated with weakness, weight loss and lack of vitality. In the main, bogbean is prescribed in combination with other herbs such as celery seed (*Apium graveolens*) and white willow (*Salix alba*).

MERIANDRA BENGALENSIS
(Lamiaceae)

Infusion of leaves: useful application to aphthae and sore throats, diminishes or arrests the secretion of milk.
M. Strobilifera
Decoct. When made strong forms a useful lotion for ulcers and heals raw abrasions of the skin, dries up breast milk.

MERREMIA TRIDENTATA
(Convolvulaceae)

Plant: used in rheum. piles and urinary disorders; tonic, and laxat.

MERREMIA VITIFOLIA
(Convolvulaceae)

Plant: given in stranguary and urethral discharges.
Juice: cooling, diur, and a preparation from it is applied to inflamed eyes.
Root: eaten raw as stomch.
Mesua Ferrea
(Guttiferae)
Flowers: astrin., stomch, used in cough attended with expectoration; made into paste with butter and used in bleeding piles and burning of the feet.
Flower buds: used in dysen.
Unripe fruits: arom., sudorific.
Bark: astrin., arom., combined with gingere used as sudorific.
Leaves and flowers: in snake-bite and scorpion stings.
Essential oil, bitter substances, mesuol.

MICHELIA CHAMPACA
(Magnoliaceae)

Bark: febge, stim., expect. astrin.
Dried root and root bark: purg., in form of infusions useful emmen. mixed with curdled milk useful application to abscesses.
Flowers and fruits: considered stim., antisp. tonic, stomach. Carmin, bitter and cooling, used in dyspep., nausea and fever, useful as diur., in renal diseases and in gonor., mixed with sesamum oil forms an external application in vertigo.
Juice of leaves: given with honey in colic.
Seeds and fruit: used for healing cracks in feet.
Flowers contain essential oil.

MILLETTA RETICULATA
(Leguminosae)

Jɪ Xᴜᴇ Tᴇɴɢ

Parts Used: Root, vine.

Constituents: Little is known about the active constituents.

Medicinal Actions & Uses: In Chinese herbal medicine, pain is often thought to be due to poor is often thought to be due to poor or obstructed blood flow. In this tradition, *ji xua teng* is classified as a herb that invigorates the blod, and is mainly used to treat menstrual problems. *Ji xue teng* is used to relieve period pain or normalise an irregular or absent cycle, especially where this may be due to anaemia. This herb is also prescribed for certain types of arthritic pain, as well as for numbness of the hands and feet.

MICHELIA NILAGIRICA
(Magnoliaceae)

Bark: febge.

Essential oil, bitter substance.

MICROGLOSSA VOLUBILIS
(Asteraceae)

Plant: given in Gold Coast as an enema to cure fever in babies; in Liberia used as a vermifuge; in Cameroon Mountain are used as a remedy for severe cough.

MILLETIA AURICULATA
(Fabaceae)

Roots: applied to sores on cattle to kill vermin; used as fish poison.

MILLETIA PACHYCARPA
(Fabaceae)

Root: used as fish poison.

Saponin, resin, rotenone,

MIMOSA PUDICA
(Mimosaceae)

Decoct. of root: useful in gravellish complaints.

Leaves and roots: used in piles and fistula

Leaves: rubbed into a paste applied to hydrocele.

Leaf and stem: in scorpion sting.

Alk. mimosine.

MIMOSA RUBICAULIS
(Mimosaceae)

Leaves: in form of infusion prescribed in piles; bruised are applied to burns.

Root: to powder form given when from weakness the patient vomits his food.

MIMUSOPS ELENGI
(Sapotaceae)

Bark: astrin., tonic, useful in fevers.

Leaves: in snake-bite

Pulp of ripe fruit: astrin., used in curing chr. dysen.

Seeds: bruised and locally applied within the anus of children in cases of constipation.

Seeds contain saponin, kernel oil.

MIRABILIS JALAPA
(Nyctaginaceae)

Root: aphrodis, purg.

Leaves: naturant, lesser inflam. applied to boils, phlegmous and whillow.

Alkaloid trigonelline.

MITCHELLA REPENS
(Rubiaceae)

Sǫᴜᴀᴡ Vɪɴᴇ

Parts Used: Aerial parts, berries.

Constituents: Squaw vine is believed to contain tannins, glycosides and saponins.

Medicinal Actions & Uses: Squaw vine is still extensively used to aid labour and childbirth, and is also considered to have a tonic action on the uterus and the ovaries. It is taken to normalise menstruation and to relieve heavy periods and period pain. This herb has also been recommended for stimulating breast-milk production, but other herbs with a similar action, such as fennel (*Foeniculum vulgare*), are preferred. The berries, crushed and mixed with tincture of myrrh (*Commiphora molmol*), are helpful for sore nipples. An astringent herb, squaw vine has also been prescribed for diarrhoea and colitis.

Caution: Do not take during the first six months of pregnancy.

MOLLUGO CERVIAMA
(Ficoidaceae)

Plant: febge., used for promoting the flow of lochial discharges, and as a cure for gonor.

MOLLUGO LOTOIDES
(Ficoidaceae)

Dried plant: in diarh., purg., cure for boils, bilious attacks and for wounds and pains in the limbs.
Juice: given internally to weak children.

MOLLUGO NUDICAULIS
(Ficoidaceae)

Leaves: applied to boils to draw out pus.
Plant: bitter, considered pectoral, used in atheram, and whooping cough.

MOLLUGO OPPOSITIFOLIA
(Ficoidaceae)

Plant: stomach, aper., antisep., administered for suppression of the lochia, applied warm moistened with little castor oil as cure for earache.
Juice: applied in skin diseases and itch.

MOLLUGO PENTAPHYLLA
(Ficoidaceae)

Plant: stomach, aper., antisep.
Infusion of plant: emmen.
Leaves: bitter, antiper.

MOMORDICA CHARANTIA
(Cucurbitaceae)

Cerasee

Parts Used: Leaves, fruit, seeds, seed oil.

Constituents: Cerasee contains a fixed oil, an insulin-like peptide, glycosides (mormordin and charantin), and an alkaloid (mormordicine). The peptide is known to lower sugar levels in the blood and urine.

Medicinal Actions & Uses: The unripe fruit is mainly used to treat late-onset diabetes. The ripe fruit is a stomach tonic, and induces menstruation. In Turkey, it is used to treat ulcers. The fruit is much used in the West Indies for worms, urinary stones and fever. The fruit juice is taken as a purgative, and is prescribed for colic. A decoction of the leaves is taken for liver problems and colitis, and it may be applied to skin conditions. The seed oil is used to help heal wounds.

Fruit and leaves: anthelm, useful in piles, leprosy, jaundice, and as vermifuge.

MOMORDICA DIOICA
(Cucurbitaceae)

Root: toasted and used to stop bleeding from piles; used in urinary complaints; ground to paste smeared over the body as a sedative in high fever with delirium; used in snake-bite and scorpion sting; juice used as antisep. **Powder or infusion of dried fruits**: if introduced into nostrils, produces a powerful errhine effect, and prorokesa copious discharge from the schneiderian mucous membrane. Alkaloid.

MONARDA PUNCTATA
(Labiatae)

HORSEMINT

Parts Used: Aerial parts.
Constituents: Horsemint's volatile oil has thymol as the main constituent.
Medicinal Actions & Uses: Having a strong volatile oil, horsemint is primarily used for digestive and upper respiratory problems. It is taken as an infusion to relieve nausea, indigestion, flatulence and colic. It is also employed to reduce fevers and upper respiratory catarrh. The herb has an antiseptic action within the chest. Taken internally or applied externally, horsemint reduces fever by encouraging sweating. It also strongly stimulates menstruation. Applied externally, the plant is a counter-irritant. It helps to lessen the pain in arthritic and rheumatic joints by increasing the flow of blood in the affected area, and thereby hastening the flushing out of toxins.

MONSONIA OVATA
(Geraniaceae)

MONSONIA

Parts Used: Aerial parts.

Medicinal Actions & Uses: Monsonia is used throughout southwestern Africa as a treatment for diarrhoea, acute and chronic dysentery, and ulcerative colitis. The plant's astringent properties act to tighten and protect the inner linings of the intestinal tract. Given monsonia's long traditional use for intestinal disorders and infections, it is possible – but as yet unsubstantiated by research – that the plant has a direct antimicrobial effect.

MONTIA PERFOLIATA
(Portulacaceae)

MINER'S LETTUCE

Parts Used: Aerial parts.
Constituents: Miner's lettuce is rich in vitamin C.
Medicinal Actions & Uses: Apart from its value as a nourishing vegetable, miner's lettuce, like its relative purslane (*Portulaca oleraceae*), may also be taken as an invigorating spring tonic and an effective diuretic.

MORINDA CITRIFOLIA
(Rubiaceae)

Root: cath.
Leaves: administered internally as tonic and febge; used as a healing application to wounds and ulcers.
Baked fruit: in Indo-China used as emmen., and given in asthma and dysent.
Unripe berries: charred and mixed with salt applied successfully to spongy gums.
Juice of leaves: applied to gout externally.
Root contains glued. morindin anthraquinone derivation.

MORINDA TINCTORIA
(Rubiaceae)
Root: used internally as an astrin. Calycoside morindin from wood.

MORINDA OFFICINALIS
(Rubiaceae)
BA JI TIAN

Part Used: Root.

Constituents: *Ba ji tian* contains morindin and vitamin C.

Medicinal Actions & Uses: The pungent, sweet-tasting *ba ji tian* is an important Chinese herb. It is a kidney tonic, and therefore strengthens the *yang*. It is also used as a sexual tonic, treating impotence and premature ejaculation in men, infertility in both men and women, and a range of other, often-hormonally linked conditions, such as an irregular menstrual cycle. *Ba ji tian* is also prescribed for conditions affecting the lower back or pelvic region, including pain, cold and urinary weaknes – especially frequent urination or incontinence.

MORINGA CONCANENSIS
(Moringaceae)
Roots: used as subst. for *M. oleifera* properties same as *M. oleifera*.

MORINGA OLEIFERA
(Moringaceae)
Root: used as stim. in paralytic affections and intermittent fever, used in epilepsy; rubft. in palsy and chr. rheum; carmin, stomach; abortif; as cardiac and circulatory tonic; in form of a compound spirit useful in fainting, giddiness, nervous debilitym spasmodic affections of the bonds, hysteria and flatulence.

Root bark: used as formentation to relieve spasm.

Bark: abortif.

Fruit: used as diseases of liver and spleen, articular pains, tetanus and paralysis.

Flowers: stim., aphrodis.

Oil from seeds: used as external application in rheum.

Gum: used for dental caries, mixed with sesame oil poured into the ears for relief of otalgia.

Seeds: used in venereal affections. Alkaloid, gum, moringine and moringinine, antibiotic pterygospernum.

MORUS ALBA
(Moraceae)
WHITE MULBERRY, SANG YE (CHINESE)

Parts Used: Leaves, twigs, fruit, root, bark.

Constituents: The leaves contain flavonoids, anthocyanins and artocapin. The fruit contains the vitamins A, B^1, B^2 and C.

Medicinal Actions & Uses: White mulberry leaves are expectorant, encouraging the loosening and coughing up of catarrh, and are prescribe in China as a treatment for coughs. The leaves are also taken to treat fever, sore and inflamed eyes, sore throats, headaches, dizziness and vertigo. The fruit juice is cleansing and tonic, and has often been used as a gargle with mouthwash. The root bark may be used for toothache, and is considered laxative. An extract of the leaves has been given by injection for elephantiasis. The twigs are used to combat excess fluid retention and joint pain.

The fruit is taken to prevent premature greying of the hair, and to treat dizziness, ringing in the ears, blurred vision and insomnia.
Related Species: The black mulberry (*M. nigra*) native to Iran, is cultivated for its sweet, deep red fruit.

MOSCHOSMA POLYSTACHYUM
(Lamiaceae)

Juice of plant: squeezed into the nostrils of children to cure headache in Gold Coast.

MUCUNA GIGANTEA
(Fabaceae)

Bark: used in rheum. complaints it is pulverised, mixed with dry ginger and rubbed over the affected parts.
Bristles of pods: used as poison.

MUCUNA MONOSPERMA
(Fabaceae)

Seeds: used as expect. in cough and asthma and applied externally as a sedative.

MUCUNA PRURITA
(Fabaceae)

Seeds: aphrodis., nerve tonic, in scorpion-sting.
Pods: anthelm.
Root: purg. prescribed as remedy for delirium in fever; powdered and made into paste applied to the body in dropsy; strong infusion mixed with honey given in cholera.
Reddish viscous oil, and alkls. mucunine and mucunadine.

MUNDULEA SERICA
(Fabaceae)

Seeds: fish poison.
Bark and root contain toxic glycoside.

MURRAYA KOENIGII
(Rutaceae)
CURRY PATTA

Parts Used: Leaves, berries.
Constituents: Curry patta contains a glycoside (koenigin), volatile oil and tannin.
Medicinal Actions & Uses: Curry patta leaves increase digestive secretions and relieve nausea, indigestion and vomiting. They also treat diarrhoea and dysentery. The leaves are considered a hair tonic in India and are thought to prevent greying. They may also be used as a poultice for burns and wounds. The berry juice may be mixed with lime juice (*Citrus aurantiifolia*) and applied to soothe insect bites and stings.

MUSA PARADISIACA
(Musaceae)
VAR. SAPIENTURY

Root and Stem: tonic, antiscor. useful in blood and venereal diseases.
Root: anthelm.
Unripe fruit: in combination with other drugs used in diabetes.
Ripe fruit: antiscor. used as a mild demulc. astrin. diet in cases of dysent.
Juice of flowers: mixed with curds used in dysent. and menor.
Sap of the stem: used in nervous affections like hysteria and epilepsy; drunk in dysen. and diarh., forms a valuable drink and mouthwash to allay thirst in cholera.
Young leaves: used as a cool dressing for blisters and burns.

MUSSAENDA GLABRATA
(Rubiaceae)

Root: half a tole given with cow's urine in white leprosy.

White leaves: two tolas given in milk in jaundice.

Flowers: pectoral, diur., given in asthma, intermittent fevers, and dropsy; externally applied as detergent to ulcers.

MYRICA CERIFERA
(Myricaceae)
BAYBERRY

Part Used: Root bark.

Constituents: Bayberry contains triterpenes (including traxerol, taraxerone and myricadiol), flavonoids, tannins, phenols, resins and gums. Myricadiol has a mild effect on potassium and sodium levels. Myricitrin is antibacterial.

Medicinal Actions & Uses: Bayberry is used to increase circulation, stimulate perspiration and keep bacterial infections in check. Colds, flu, coughs and sore throats benefit from treatment with this herb. It helps to strengthen resistance to infection and to tighten and dry mucous membranes. An infusion is helpful for spongy gums, and a gargle is used for sore throats. Bayberry's astringency is beneficial for irritable bowel syndrome and mucous colitis. An infusion can help treat excess vaginal discharge. A paste of the powdered root bark may be used externally on ulcers and sores.

MYRISTICA FRAGRANS
(Myristicaceae)
NUTMEG & MACE ROU DOU KOU (CHINESE)

Nutmeg tree growing to 12 m, with aromatic leaves and clusters fo small yellow flowers.

Key Constituents: Nutmeg;
Volatile oil (up to 15%), including alpha-pinene, beta-pinene, alpha-terpine, beta-terpinene, mysristicin, elincin, safrole. Fixed oil (nutmeg better:), myristine, butyrin
Mace;
Volatile oil (similar to nutmeg but with a higher concentration of myristicin).
Key Actions:
Nutmeg Mace
Carminative
Stimulant
Relieves muscle spasms
Carminative
Prevents vomiting
Stimulant

MYRSTICA MALABARICA
(Myrsticaceae)

Seeds: in form of **lep** used as an external application.
Fat: mixed with little oil applied to indolent ulcers, allays pain, cleanses the surface and establishes healthy action.
Essential oil.

MYRSINE AFRICANA
(Myrsinaceae)

Fruit: used as anthelm. specially for tape worm, laxt. in dropsy and colic.
Gum: warm remedy for dysent.
Decoct. of leaf: used as a blood purifier.
Berries yield embelic acid and quercitol.

MYROXYLON PEREIRAE
(Leguminosae)
PERUVIAN BALSAM

Part Used: Oleo-resin.
Constituents: The oleo-resin contains 50-65% volatile oil (mainly benzyl

benzoate and benzyl cinnamate) and resins.

Medicinal Actions & Uses: Peruvian balsam is strongly antiseptic and stimulates repair of damaged tissue. It is most commonly taken internally as an expectorant and anticatarrhal remedy to treat bronchitis, emphysema and bronchial asthma. It may also be taken to treat sore throats and diarrhoea. Externally, the balsam is applied to skin afflictions.

MYRTUS COMMUNIS
(Myrtaceae)

MYRTLE

Parts Used: Leaves, essential oil.
Constituents: Myrtle contains tannins, flavonoids and a volatile oil (mainly alpha-pinene, cineole and myrtenol). ˙
Medicinal Actions & Uses: Myrtle leaves are astringent, tonic and antiseptic. They may be used externally to heal wounds or internally to remedy disorders of the digestive and urinary systems. The essential oil is antiseptic and anticatarrhal and is used to treat chest ailments.

NARDOSTACHYS JATMANSI
(Valerianaceae)

Root: arom., bitter, tonic, stim., antisp., employed for treatment of epilepsy, hysteria and convulsive affections; used in palpitation of heart; subst. for valerian; useful in intestinal colic.
Essential oil, crystalline acid jatmansic acid. Oil antibacterial.

NAREGAMIA ALATA
(Meliaceae)

Root: emetic, cholag., expect., useful in acute dysen.
Leaves and stems: in decoct. given with bitters and aromatics for biliousness.
Plant: used in rheumatism and itch.
Root bark contains alk. naregamin.

NASTURTIUM FONTANUM
(Brassicaceae)

Plant: appetizer, antiscor., stim. used in troubles of the chest.
Glucd. essential oil.

NASTURTIUM OFFICINALE
(Water Cress)

BRASSICACEAE

The flowering stems, collected before flowering, are used medicinally. Their constituents include a glucosinolate (gluconasturtiin), which decomposes into a pungent essential oil, also provitamin A, vitamins B complex,

Nasturtium officinale

C, D and E, iodine, and various minerals with iron, manganese and calcium. These substances give Watercress tonic, stomachic, diuretic and irritant properties. It has numerous uses in herbal medicine, for example, as an appetizer, for digestive and gall bladder disorders, and for coughs and asthma. The fresh diluted juice or an infusion of the dried herb can be taken internally but, fresh or dried, **Watercress should always be used with care**; large doses may cause inflammation of the mucosa of the bladder and gasrtointestinal tract. Externally the fresh juice is used to treat some skin disorders.

Being rich in vitamin C, fresh Watercress is a useful addition to the diet in winter and early spring; it can be eaten raw in salads or used to flavour soups.

NAUCLEA MISSIONIS
(Rubiaceae)

Powdered Bark or decoct.: used in leprosy, ulcers, rheum. and constipation.

NELUMBO NUCIFERA
(Nymphaeaceae)

Flowers: cooling, used as astrin. in diar., also cholera, in fever and diseases of the liver; recommended as a cardiac tonic.

Seeds: used to check vomiting, given to children as diur., and refrig.; form a cooling medicine for skin diseases and leprosy; considered as antid. to poisons.

Filaments: considered astrin. and cooling, useful in burning sensation of the body, bleeding piles and menor.

Root: in powder form prescribed for piles as demulc., also for dysen. and dyspep.; used as a paste in skin affections and ringworm.

Leaf, pedicles and embryo contain alk. nelumbine .

NEPETA CATARIA
Lamiaceae
(Catmint)

The flowering stems, without the woody parts, are used medicinally. When dry they have a sharp, balsam-like taste and a strong, pungent aroma reminiscent of Balm (*Melissa officinalis*). The constituents include 0.5-0.7 per cent of an essential oil with carvacrol and thymol, plus tannins and bitter compounds. These substances give Catmint mild sedative, stomachic, carminative, antidiarrhoeal, diuretic, antipyretic and emmenagogic properties. It is used in tea mixtures

Nepeta cataria

in the treatment of nervous disorders, neuroses and migraine and, on its own in infusions, for gastrointestinal disorders, chills, colds, ammenorrhoea and other menstrual complaints. Externally it is used in ointments for haemorrhoids.

NEPETA HINDOSTANA
(Lamiaceae)
Plant: largely used in fevers and as cardiac tonic, internally taken in gonor.
Decoct.: used as gargle in sore throat.

NEPHELIUM LAPPACEUM
(Sapindaceae)
Fruit: considered stomach. and anthelm. in china; used as astrin., and febge. in Cambodia.

NERIUM INDICUM
(Apocynaceae)
Plant: poisonous.
Root: powerful resolv. and attenuant, used externally; beaten in to a paste with water applied to chancres and ulcers on the penis.
Decoct. of leaves: used to reduce swellings.
Oil prepared from root-bark: used in skin diseases of a scaly nature, and in leprosy.
Glucd. root, bark and seeds contain the toxic principles neriodorin, nerioderin and karabin.
All are powerful cardiac poisons.

NERIUM OLEANDER
(Apocynaceae)
Plant: poisonous.
Leaves contain the glucds. neriin and oleandrin; leaves yield another glucd. folinerin; used like digitalis; 1 mg. cor.

NEURACANTHUS
(Acanthaceae)
Sᴘʜᴀᴇʀᴏsᴛᴀᴄʜʏᴜs
Root: powdered and made into a paste used as cure for ringworm; administered in that form of indign. in which fatty or saponaceous grape-like masses are observed in the stools.

NICANDRA PHYSALOIDES
(Solanaeae)
Plant: diur.
Alcoholic extract of the fresh plant yielded 0.65% of a bitter principle named nicandrin.

NICOTIANA TABACUM
(Solanaceae)
Tᴏʙᴀᴄᴄᴏ
Parts Used: Leaves.
Constituents: Tobacco contains alkaloid (notably nicotine) and a volatile oil.
Nicotine is stimulant and addictive.
Medicinal Actions & Uses: Tobacco is no longer used medicinally. The dried leaves make a good insecticide, but external application should be avoided as nicotine, is readily absorbed through the skin.
Caution Tobacco should not be taken in any form.
Principal alk. is nicotine; others in lesser amount are nicoteine, nicrotimine, anabasine, nor-nicotine, etc. nicotine 2-4%.

NIGELLA SATIVA
Ranunculaceae
(Black Cumin)
The ripe seeds have a camphor-like scent and a bitter, later aromatic taste.

Nigella sativa

Their constituents include saponin, an essential oil, a bitter compound (nigelline) and tannins. These substances give Black Cumin diuretic, cholagogic, carminative, anthelmintic, antispasmodic, galactagogic & emmenagogic properties. The seeds are mainly used for digestive and menstrual disorders, and for bronchitis. They are, however, slightly poisonous; **large doses should not be taken**.

In cooking the seeds can be used as a substitute for pepper and can be sprinkled on bread and cakes. The distilled essential oil is used as a flavouring in confectionery. In France the seeds are sometimes called *quatre-epices* (four spices). Black Cumin seeds should not be confused with those of true Cumin (*Cuminum cyminum*), which taste quite differently. The two plants are unrelated.

NIPA FRUTICANS
(Arecaceae)

Pounded leaves: used as remedy for bites of centipedes and as a cure for ulcers in the Phillippine Islands. Saccharose.

NOTONIA GRANDIFLORA
(Asteraceae)

Fresh stems: used in the form of an extract as a preventive of hydrophobia.

NOTOPTERYGIUM INCISIUM
(Umbelliferae)

QIANG HUO

Part Used: Root.
Constituents: *Qiang huo* contains a volatile oil, including angelical.
Medicinal Actions & Uses: *Qiang huo* is taken mainly for colds and chills, fevers, headache, general aches and pains, and malaise. The herb is warming and pungent and promotes sweating. It is also prescribed for neck and back pain.

NYCTANTHES ARBORTRISTIS
(Oleaceae)

Leaves: useful in fever and rheumatism; fresh juice given with honey in chr. fever.
Decoct. of leaves: prepared over a gentle fire, recommended as a specific for obstinate sciatica.
Expressed juice of leaves: cholag., laxt., mild bitter tonic, given with a little sugar to children as remedy for intestinal worms.
Flowers yield crystalline nyctanthin, leaves alk., resins, peppermint-like oil, amorphous glucd.

NYMPHAEA ALBA
(Nymphaceae)
WHITE WATER LILY

Medicinal Actions & Uses: The rhizome of the white water lily is astringent and antiseptic. A decoction treats dysentery, or diarrhoea due to irritably bowel syndrome. White water lily has also been employed to treat bronchial catarrh and kidney pain, and taken as a gargle for sore throats. The rhizome may be used to make a douche for vaginal soreness and discharge, or to make a poultice – often in combination with slippery elm (*Ulmus rubra*), or linseed (*Linum usitatissimum*), for boils and abscesses. White water lily flowers have long been reputed to reduce sexual drive. Their generally calming and sedative effect on the nervous system makes them useful in the treatment of insomnia, anxiety and similar disorders.
Root and Stock: astrin., slightly narcotic, administered in dysent.
Flowers: anti-aphrodis.
Infusion of flower and fruit: given in diarh. and as diaphor.
Alk. nupharine, nymphacine, cardiac glued. nymphalin.

OCHNA PUMILA
(Ochnaceae)
Root: used as antid. to snake-bite; its decoct. given in menstrual complaints also for consumption and asthma.

OCHRADENUS BACCATUS
(Resedaceae)
Baluchistan – *Kalirram, Kirmkush.*

Twigs, leaves and flowers – fried, ground to a powder, mixed with little *neshar* and applied to dry wounds and sore to kill maggots, etc., in Baluchistan.

OCHROCARPUS
(Guttiferae)
LONGIFOLIUS
Flower buds: astrin., arom.
Flowers: stim., carmin., useful in some forms of dyspep. and in haemorrhoids.
Essential oil.

OCIMUM AMERICANUM
(Lamiaceae)
Leaves: made into paste used in parasitical skin diseases and applied to finger and toe-nails during fever when the extremeties are cold.
Essential oil: Camphor, cintronethol, methyl-cinnamate, citronellic acid, eugenol, borneol, citral, methyl heptone, canaphene, limonene, dipentene, Sabinene, geraniol, d-α-pinene.

OCIMUM BASILICUM
Lamiaceae
(Basil)
The flowering stems are the medicinal parts. Their constituents include an essential oil with linalool as the main component, also tannins, glycosides and saponin. These substances give Basil stomachic, carminative, expectorant, antispasmodic, mild sedative and galactogogic properties. An infusion is sometimes used for chronic gastritis, stomach pains, flatulence, constipation, respiratory disorders such as cough and whooping cough, and for urinary infections. It is an excellent preventive for travel sickness. Externally Basil can be used

Ocimum basilicum

for invigorating baths, in compresses for slow-healing wounds, and in gargles.

The essential oil, obtained by steam distillation from the fresh herb, is used in perfumery, in making incense, and in the food industry. In cooking Basil leaves can be used, preferably fresh, to flavour soups, salads and meat and fish dishes.

The herb has been traditionally taken to increase breast-milk production. Used as insect repellent, antibacterial in action.

Essential oil yields same compounds as *O. americanum*.

OCIMUM GRATISSIMUM
(Lamiaceae)

Plant: arom. baths of fumigations prepared with it recommended in the treatment of rheumatism and paraly-sis; a strong decoct. effectual in the aphthae of children.

Decoct. of leaves: useful in seminal weakness, remedy in gonor.

Seeds: given in headache and neutralgia.

Essen. oil, thymol, eugenol, methyl chavicol, myrene, citral, geraniol.

OCIMUM SANCTUM
(Labiatae)

HOLY BASIL, TULSI (HINDI)

An aromatic annual growing to about 70 cm, with small, purple-red or white flowers.

Key Constituents:
Volatile oil (1%) including eugenol (70-80%), methyl-chavicol, methyl eugenol, caryophyllene.
Flavonoids (apigenin, uteolin)
Triterpene (ursolic acid)

Key Actions:
Lowers blood sugar levels
Antispasmodic
Analgesic
Lowers blood pressure
Reduces fever
Adaptogenic
Anti-inflammatory

OENOTHERA BIENNIS
(Onagraceae)

EVENING PRIMROSE

Parts Used: Leaves, stem, bark, flowers, seed oil.

Constituents: Evening primrose oil is rich in essential fatty acids – cis-linoleic (about 70%) and cis-gammalinolenic acid (about 9%) in particular. Its action mostly depends on the gammalinolenic acid (GLA), a precursor of prostaglandin E^1. The oil is often combined with vitamin E to prevent oxidation.

Medicinal Actions & Uses: The flowers, leaves and stem bark of evening primrose have astringent and sedative properties. All three parts have been employed in the treatment of whooping cough. Evening primrose has also been taken for digestive problems and asthma, and used as a poultice to ease the discomfort of rheumatic disorders. The oil, applied, externally, is beneficial in the treatment of eczema, certain other itchy skin conditions and breast tenderness. Taken internally, the oil has an effect in lowering blood pressure and in preventing the clumping of platelets. The oil is now commonly taken for premenstrual problems, including tension and abdominal bloating. Multiple sclerosis may benefit from internal treatment with the oil, as may rheumatoid arthritis, intermittent claudication (a cramp-like pain in the leg) and other problems relating to the circulation.

OLAX SCANDENS
(Olacaceae)

Bark: used in a preparation given in anaemia.

OLDENLANDIA AURICULATA
(Rubiaceae)

Plant: emol., used in dysen. and cholera.
Alk. hedyotine, auricularine.

OLDENLANDIA BIFLORA
(Rubiaceae)

Plant: used in remittent fever, gastric irritation and nervous depression.
Alk.

OLDENLANDIA CORYMBOSA
(Rubiaceae)

Decoct. of plant: given in remittent fever with gastric irritation and nervous depression.
Plant: given in jaundice and diseases of liver and used as anthelm.
Juice: applied in burning of the palms of hands and soles of feet from fever.

OLDENLANDIA SPS.
(Rubiaceae)

O. Diffusa

Decoct. of plant: used in biliousness, impure blood, fever and gonor.

O. herbacea:

Decoct. of plant: used in mild cases of malarial fever of a low remittent type.

OLDENLANDIA UMBELLATA
(Rubiaceae)

Leaves: expect., given in consumptive and asthmatic affections.
Leaves and root: expect., prescribed in bronchial catarrh, broncht. and asthma.
Root: in snake-bite.
Alizarine.

OLEA CUSPIDATA
(Oleaceae)

Oil from fruit: rubft.
Leaves and bark: bitter astrin., used as antiper. in fever and debility.
Leaves: considered as a cure for gonor.
Gum: used mixed with antimony in applications to the eye.

OLIGOMERIS SUBULATA
(Resedaceae)

Plant: pounded, and the juice thus

extracted is used by women to apply to their breasts to keep them soft in Kalat.

OLEA EUROPAEA
(Oleaceae)

OLIVE

Parts Used: Leaves, oil.

Constituents: Olive leaves contain oleoropine, oleasterol and leine. Olive oil contains about 75% oleic acid, a mono-unsaturated fatty acid.

Medicinal Actions & Uses: Olive leaves lower blood pressure and help to improve the function of the circulatory system. They are also mildly diuretic and may be used to treat conditions such as cystitis. Possessing some ability to lower blood sugar levels, the leaves have been taken for diabetes. The oil is nourishing and improves the balance of fats within the blood. It is traditionally taken with lemon juice in teaspoonful doses to treat gallstones. The oil has a generally protective action on the digestive tract and is useful for dry skin. Externally, it is a good, although sticky, carrier oil for essential oils.

ONONIS SPINOSA
Fabaceae
(Spiny Restharrow)

The root is used medicinally. Its constituent include an essential oil, tannins, fatty oil and a flavonoid glycoside (ononin). These substances give Spiny Restharrow diuretic, antiseptic, cholagogic, tonic and hypotensive properties. It is used to treat dropsy and inflammations of the bladder and kidneys, to lower high blood

Ononis spinosa

pressure and to alleviate rheumatic and arthritic pain. **Restharrow should be used only under supervision of a qualified medical or herbal practitioner;** it should not be taken often, in large doses and preferably not on its own. For these reasons it is generally prescribed in herbal tea mixtures.

ONOSMA BRACTEATUM
(Boraginaceae)

Plant: tonic, alter., in decoct. much used in rheumatism, syphilis and leprosy; good refrig. and demulc., useful for relieving excessive thirst and restlessness in febrile excitement and also useful in relieving functional palpitation of the heart, irritation of the bladder and stomach and strangury.

ONOSMA ECHIOIDES
(Boraginaceae)

Leaves: alter., in powder given to children as purg.

Flowers: used as cordial and stim. in rheumatism and palpitation of heart.

Root: bruised and used as application to eruptions.

OPERCULINA TURPETHUM
(Convolvulaceae)

TURPETH

Part Used: Root.

Constituents: Turpeth root contains turpethin resin (approximately 4%) and a volatile oil.

Medicinal Actions & Uses: Turpeth root is chiefly used in small doses to clear the bowels. It is sometimes known as "Indian jalap", and is used in much the same way as this plant (*Ipomoea purga*), though its action is slower and less drastic. Turpeth should be taken with care with and combined with herbs that ease griping pain and flatulence, such as ginger (*Zingiber officinale*). In Ayurvedic medicine, turpeth is prescribed with picrorrhiza (*Picrorrhiza kurroa*) to treat jaundice.

OPHIOGLOSSUM VULGATUM
(Ophioglossaceae)

Plant: used as vulnerary and as remedy for wounds in England and Spain.

OPHIORRHIZA MUNGOS
(Rubiaceae)

Root: bitter, tonic, considered to be remedy against the bites of venomous snakes, mad dogs, etc.

OPUNTIA DILLENI
(Cactaceae)

Fruit: refrig., useful in gonor.; baked and given in whooping cough; in form of a syrup given to control spasmodic cough and expectoration.

Milky juice: purg.

Leaves: mashed up and applied as poultice to allay inflam. and heat; made into a pulp applied to the eyes in ophthalmia; heated and applied to boils to hasten suppuration.

Plant: in snake-bite.

OPUNTIA INDICA
(Cactaceae)

PRICKLY PEAR

Parts Used: Flowers, fruit, stems.

Constituents: The fruit of prickly pear contains mucilage, sugars, vitamin C and other fruit acids. The flowers contain a flavonoid.

Medicinal Actions & Uses: Prickly pear flowers are astringent and reduce bleeding, and are used for problems of the gastro-intestinal tract – particularly diarrhoea, colitis and irritable bowel syndrome. The flowers are also taken to treat an enlarged prostrate gland. The fruit is nutritious.

ORCHIS MASCULA
(Orchidaceae)

PURPLE ORCHID, SALEP

Part Used: Tuber.

Constituents: Purple orchid contains about 48% mucilage.

Medicinal Actions & Uses: Once believed to have aphrodisiac powers, salep is now seen as a nourishing vegetable somewhat similar to the potato (*Solanum tuberosum*). Its current medicinal use is generally confined to the treatment of diarrhoea and irritated gastro-intestinal tracts in children.

ORIGANUM MAJORANA
(Sweet Marjoram)

LAMIACEAE

The flowering stems are the medicinal parts. Their constituents include 1-2 per cent of an essential oil with a spicy fragrance containing terpinines and terpineol, plus tannins, bitter compounds, carotenes and vitamin C. These substances give Sweet Marjoram stomachic, carminative, choleretic, antispasmodic and weak sedative properties. In herbalism it is used mainly for various gastrointestinal disorders and to aid digestion. It is also an ingredient of ointments and bath preparations used to alleviate rheumatism.

Mostly, however, Sweet Marjoram is used as a culinary herb. Of all the marjorams it has the best flavouring for cooking and is an excellent addition to soups, sauces and meat dishes.

ORIGANUM VULGARE
Lamiaceae
(Marjoram)

The flowering stems are used medicinally. The constituents include 0.4 per cent of an essential oil with thymol as its main component, also bitter compounds and tannins (8 per cent). These substances give Marjoram as-

Origanum majorana

Origanum vulgare

tringent, expectorant, antispasmodic, antiseptic, mild tonic, stomachic and carminative properties. It is used in herbal tea mixtures to treat stomach and gall bladder disorders, diarrhoea, coughs, asthma, nervous headache, general exhaustion and menstrual pain. Externally Marjoram is used in gargles, bath preparations, liniments and inhalents.

Marjoram is a favourite kitchen herb, especially in Italy where it is used to flavour pizzas and spaghetti dishes. It has a stronger taste than Sweet Marjoram (*Origanum majorana*).

ORMOCARPUM SENNOIDES
(Fabaceae)
Root: tonic, stim., used in paralysis and lumbago.

OROXYLUM INDICUM
(Bignoniaceae)
Root bark: astrin., tonic, useful in diar. and dysen.

Bark: made into powder alongwith *haldi* useful cure for sorebacks of horses; in powder or infusion diaphor., useful in acute rheumatism; bitter tonic.

Tender fruits: grateful carmin., stomch.

Seeds: purg.

Stem: in scorpion-sting.

Crystalline bitter oroxylin; alk. (J.)

ORTHOSIPHON ARISTATA
(Labiatae)
JAVA TEA

Parts Used: Leaves.

Constituents: Java tea contains flavones (including sinenstein), a glycoside (orthosiphonin), a volatile oil and large amounts of potassium.

Medicinal Actions & Uses: Java tea is listed in the French, Indonesian, Dutch and Swiss pharmacopoeias. It is thought to increase the kidney's ability to eliminate nitrogen-containing compounds. The herb is much used as a diuretic and as a treatment for kidney infections, stones and poor renal function resulting from chronic nephritis. It is also used to treat cystitis and urethritis.

ORYZA SATIVA
(Poaceae)
RICE

Rice gruel: in disorganised digestion, in bowel complaints, in diar. & dysen.

Rice-water: demulc., refrig., soothing, nourishing drink in febrile diseases & inflammatory states of intestines.

OSBECKIA CRINITA
(Melastomaceae)
Decoct. of Dried leaves: used for tooth ache in Tongking.

OSMUNDA REGALIS
(Osmundaceae)
Plant: tonic and styptic used for rickets in England.

OSYRIS ARBOREA
(Santalaceae)
Infusion of leaves: powerful emetic.

OTHONNOPSIS INTERMEDIA
(Asteraceae)
Plant: used as cure for headaches, boils and pimples.

OUGENIA DALBERGIOIDES
(Fabaceae)
Bark: febge.; when incised furnishes a kino-like exudation which is used in dysen. and diar.; in decoct. given when the urine is highly coloured; used as fish poison.

OXALIS CORNICULATA
(OXALIDACEAE)

Leaves: cooling, refrig., stomach, anti scorb.
Plant: used as cure for scurvy.
Acid potassium oxalate.

OXYSTELMA ESCULENTUM
(Asclepiadaceae)

Decoct. of plant: used as gargle in aphthous ulcerations of mouth and in sore throat.
Roots: considered specific for jaundice.
Milky sap: used as a wash for ulcers.

P

PACHYGONE OVATA
(Menispermaceae)

Dried fruit: used for destroying vermin and stupefying fish.

PAEDERIA FOETIDA
(Rubiaceae)

Plant: considered specific for rheum. affections, administered both internally and externally.
Roots: emetic.
Juice of leaves: astrin., given to children in diar.
Herb contains essen. oil, alk.

PAEONIA EMODI
(Paeoniaceae)

Tubers: useful medicine for uterine diseases, colic, bilious obstructions, dropsy, epilepsy, convulsions and hysteria; given to children as blood-purifier.

Seeds: emetic, cath.
Infusion of dried flowers: useful in diar.

PAEONIA LACTIFLORA
(Paeoniaceae)

WHITE PEONY, BAI YAO (CHINESE)

An epright perennial, growing to 2 m, with large white flowers and divided, dark green leaves.
Key Constituents:
Monoterpenoid glycosides (paeoniflorin, albiflorin)
Benzolic acid
Pentagalloyl glucose
Key Actions:
Antispasmodic
Tonic
Astringent
Antgesic

PAEONIA OFFICINALIS
Paeoniaceae
(Peony)

The flowers are the main medicinal parts but the roots and seeds are also effective. Only the petals of double, red varieties are used. The constituents include glycosides, an alkaloid (peregrinine), tannins, sugars and mucilage and, in the flowers, the anthocyanidin pigment paeonidin. These substances given Peony antispasmodic, diuretic, vasoconstrictive, sedative and emmenagogic properties. It was once used internally to relieve spasm of the smooth muscles, asthmatic attacks, epileptic seizures, to treat gout, kidney stones and haemorrhoids, and as an abortifacient. Because of its toxicity **Peony should be used only under strict medical su-**

Paeonia officinalis

arasaponin A and arasaponin B) and a flavonoid (dencichine).

San qi also helps to improve blood flow through the coronary arteries, thus finding use as a treatment for arteriosclerosis, high blood pressure and angina. *San qi* treats internal bleeding of almost any kind. The herb may also be applied externally as a poultice as a means to help speed the healing of wounds and bruises.

PANAX QUINQUEFOLIUM *(Araliaceae)*

AMERICAN GINSENG

Part Used: Root.

Constituents: American ginseng contains steroidal saponins, including panaquilon.

Medicinal Actions & Uses: The action of American ginseng is presumed to be similar to, but milder than, that of its Chinese cousin, ginseng (*P. ginseng*). American ginseng increases the ability to tolerate stress of all kinds. In traditional Chinese medicine, American ginseng is employed as a *yin* tonic, treating weakness, fever, wheezing and coughs.

pervision; it should never be used for self-medication. Herbalists, if they use it at all, would prescribe it only in herbal mixtures, never on its own. In homeopathy a tincture of the fresh plant is used for the relief of haemorrhoids.

PANAX NOTOGINSENG *(Araliaceae)*

SAN QI

Medicinal Actions & Uses: Like ginseng (*P. ginseng*), *san qi* is a tonic that supports the function of the adrenal glands, in particular the production of corticosteroids and male sex hormones.

Part Used: Root.

Constituents: *San qi* contains steroidal saponins (including

PANDANUS TECTORIUS *(Pandanaceae)*

Leaves: bitter, pungent, arom., used in leprosy, small-pox, syphilis, scabies and leucoderma.

Oil from bracts: stim., antisp., administered for headache and rheumatism.

Essen. oil outer part of the flower yields an essen. oil containing 70% methyl ether of β-phenylethyl alcohol; blossoms yield 0.1-0.3% essen.

oil containing benzyl benzoate, benzyl salicylate, benzyl acetate, benzyl alcohol, geraniol, linalool, linalyl acetate, bromostyrene, guaiacol, phenylethyl alcohol and aldehydes.

PANICUM ANTIDOTALE
(Poaceae)

Smoke of the burning plant: used to fumigate wounds, and as disinfectant in small-pox.

PANICUM MILIACEUM
(Poaceae)

Plant: used as cure for gonor.

PANICUM MILIARE
(Poaceae)

Plant: nervine stim., tonic sometimes used as subst. for *P. miliaceum*.

PAPAVER ARGEMONE
(Papaveraceae)

Infusion or syrup of petals: much esteemed as sudorific in Spain.

PAPAVER BRACTEATUM
Papeveraceae
(Iranian Poppy)

The roots are used medicinally. Their constituents include the toxic alkaloids thebaine, alpinigenine and oripavine. It is possibly to derive codeine and other pain-killing substances from thebaine. Unlike opium alkaloids, thebaine does not have additive narcotic properties, it cannot be used directly and it thus poses no danger of drug addiction: morphine, the precursor of the addictive drug heroin, can be obtained only with great difficulty from it. For pharmaceutical purposes, therefore, there

Papaver bracteatum

may be considerable social and economic benefits in introducing this poppy into cultivation in place of Opium Poppy. Crop scientists have discovered that Iranian Poppy can provide up to 37 kg of codeine per hectare compared with Opium Poppy's much lower yield of 3 kg per hectare. **Iranian Poppy should never be used for self-medication.**

PAPAVER DUBIUM
(Papaveraceae)

Petals: sudorific.
Two alkls. aporeine and aporeidine isolated from the plant, resembling that of the baine in action.

PAPAVER NUDICAULE
(Papaveraceae)

Petals: sudorific
Contains alks. narcotine, thebaine, isothebaine, thebaine, isothebaine, glaucidine, two phenolic bases and one non-phenolic base isolated from

the plant does not contain morphine; contains 0.16% total alks. yielding thebaine, protopine and oripavine.

PAPAVER RHOEAS
Papeveraceae
(Common Poppy)

The red petals are used medicinally. Their constituents include traces of rhoeadane alkaloids (for example, rhoeadine and rhoeagenine), a red anthocyanin pigment, and mucilage. These substances give Common Poppy sedative, hypnotic, demulcent and mild expectorant properties and the dried petals are used in herbal medicine to treat irritable coughs and hoarseness, bronchitis, and to induce sleep. The flowers are also used in the pharmaceutical industry to colour medicines. The dried petals are not poisonous but the plant can poison like-stock if they eat it in large quantities.

The seeds, which have a pleasant nutty flavour, are used sprinkled on bread and cakes, especially on the Continent, and they are processed for their oil.

PAPAVER SOMNIFERUM
Papaveraceae (Opium Poppy)

The pharmaceutical industry processes the dried latex obtained from the unripe seed capsules to make

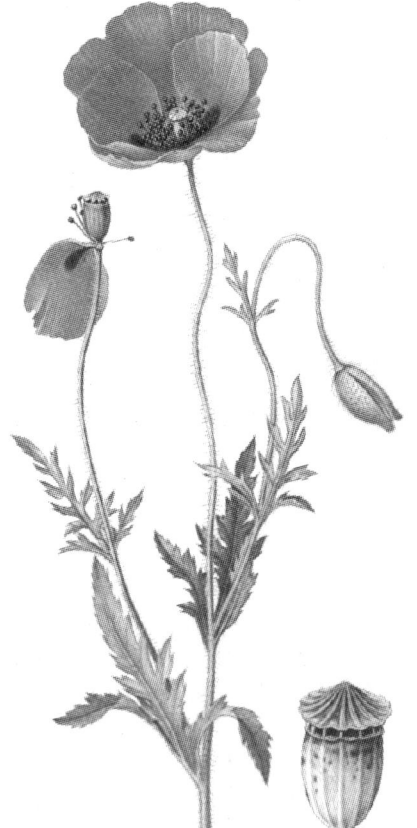

Papaver rhoeas   Papaver somniferum

opium; it also uses the dried ripe capsules and the empty dry ripe capsules. About 25 alkaloids have been isolated from opium but 6 (morphine, thebaine, codeine, narcotine, papaverine and narceine) account for about 98 per cent of the total. Morphine, the most potent constituent, is present in the greatest amount. Although opium is still occasionally used in its crude state, the alkaloids are now mostly used in purified form by the medical profession. These drugs have not yet been superceded by any synthetic product. Morphine is a narcotine, analgesic and antispasmodic. Codeine (about 3 per cent) has similar effects to morphine but is much milder. All opium alkaloids are narcotics and long-term use causes physical deterioration and eventually death. Needless to say **Opium Poppy should never be collected and used for self-medication**.

PARMELIA PERIATA
(Parmeliaceae)
Astrin., laxt., tonic, carmin., aphrodis., used in dyspep., amenor., calculi, in scorpion-sting and snake-bite.

PARIETARIA OFFICINALIS
(Urticaceae)
PELLITORY OF THE WALL
Parts Used: Aerial parts.
Constituents: Pellitory of the wall contains flavonoids and tannins.
Medicinal Actions & Uses: Pellitory of the wall is chiefly employed as a diuretic, demulcent and stone-preventing herb. In European herbal medicine, it is regarded a having a restorative action on the kidneys supporting and strengthening their function.

It has been prescribed for conditions such as nephritis, pyelitis (inflammation of the kidney), kidney stones, renal colic (pain caused by kidney stones), cystitis and oedema (fluid retention). It is also occasionally taken as a laxative.

PARSONSIA HELICANDRA
(Apocynaceae)
Juice of plant: given internally in insanity.

PASPALUM SCROBICULATUM
(Poaceae)
Plant: used in scorpion-sting, occasionally develops narcotic properties.

PASSIFLORA FOETIDA
(Passifloraceae)
Decoct. of leaves: used in biliousness and asthma.
Fruit: emetic.
Leaves: applied on the head in giddiness and headache
HCN.

PASSIFLORA INCARNATA
(Passifloraceae)
Passion Flower, Passiflora Maypop
A climbing vine growing to 9 m, with 3-lobed leaves, ornate flowers and egg-shaped fruit.
Key Constituents:
Flavonoids (apigenin)
Meltrol
Cyanogenic glycosides (gynocardin)
Indole alkaloids (harman)
Key Actions:
Sedative
Antispasmodic
Tranquilling

PAULLINIA CUPANA
(Sapindaceae)

GUARANA

Parts Used: Seeds.

Constituents: Guarana contains xanthine derivatives (including up to 70% caffeine, together with theobromine and theophylline), tannins and saponins. The xanthines are stimulant, diuretic and reduce fatigue over the short term.

Medicinal Actions & Uses: Guarana's medicinal uses are similar to those of coffee (*Coffea arabica*) it is taken for headache and migraine, for mild depression and to boost energy levels. The problems that apply to long-term or excessive coffee consumption also apply to guarana – both stimulate over the short term but tend to inhibit the body's restorative processes over the longer term. In view of guarana's high tannin content, long-term use is even less advisable, because tannins impair the intestine's ability to absorb nutrients. Nevertheless, guarana is a useful short-term remedy for boosting energy, or for treating a tension headache that cannot be treated with rest. Guarana's astringency also treats chronic diarrhoea.

PAUSINYSTALIA YOHIMBE
(Rubiaceae)

YOHIMBE

Part Used: Bark.

Constituents: Yohimbe contains about 6% indole alkaloids (including yohimbine), pigments and tannins. The alkaloids have a cerebral stimulant action at moderate doses, but are highly toxic in large doses.

Medicinal Actions & Uses: Yohimbe is little used in herbal medicine owing to its potential toxicity. In Western Africa, it is often employed as a stimulant and as a means to counter impotence. Yohimbe has been used in conventional medicine in the treatment of impotence.

PAVETTA INDICA
(Rubiaceae)

Root: bitter, aper., prescribed in visceral obstructions; pulverised and mixed with ginger and rice-water given in dropsy.

Leaves: boiled in water and a formentation made which is used in haemorrhoidal pains.

Root contains glucd.

PAVONIA ODORATA
(Malvaceae)

Root: astrin., tonic, dysent., cooling, demuk, carmin, febge.

Plant: used in rheumatism.

PEDALIUM MUREX
(Pedaliaceae)

Fruit: demulc., diur., antisp., aphrodis.; in decoct. given for incontinence of urine, spermatorrhoae, nocturnal emission and impotency.

Infusion of leaves and stems: used in gonor. and dysuria.

Juice of fruit: emmen., used in puerperal diseases and to promote lochial discharges.

Decoct. of root: antibil.

Young branches contain mucil. and root alk.

PEGANUM HARMALA
(Rutaceae)

Plant: aphrodis., emmen., galact., abortif.

Seeds: narcotic, given in fever and colic; used as a remedy for tapeworm in man.
Decoct. of leaves: given for rheumatism.
Root: applied to kill lice.
Seeds and roots contain 4 alks.: harmine, harmaline, harmalol and peganine.

PELLAEA CALOMELANOS
(Polypodiaceae)
The Sutos use the rhizome as an anthclm. and smoke the leaf for colds in the head and chest.

PENNISETUM GLAUCUM
(Poaceae)
Plant: tonic, useful in diseases of heart, appetizer.
Amylase isolated from malted seed.

PENTAPETES PHOENICEA
(Sterculiaceae)
Plant: emol., demuk. used in snake-bite.
Fruit: mucilaginous.Pentatropis Cynanchoides
(Asclepiadaceae)
Dry roots: given in decoct. as astrin. and cooling alter., used in gonor.

PERGULARIA EXTENSA
(Asclepiadaceae)
Plant: expect., emetic.
Juice of leaves: used as expect. in catarrhal affections, in infantile diar., given in asthma; applied to rheum. swellings in combination with lime or ginger; in snake-bite.
Fresh leaves: made into pulp used as a poultice in carbuncle with good effect.
Root bark: mixed with cow's milk used as a purg. in rheum. cases.

Plant contains a bitter resin, two bitter principles and a glucd. possessing physiological action similar to pituitrin and several sterols.

PERICAMPYLUS GLAUCUS
(Menispermaceae)
Root: antid. to snake poisons narcotic alk.

PERILLLA OCIMOIDES
(Lamiaceae)
Leaves, stems and seeds: considered resolv., diaphor, and cephalic in China and Indo-China.
Plant: used as sedative, antisp. and antid. in Annan.

PERIPLOCA APHYLLA
(Asclepiadaceae)
Milky juice: used as application for swellings and tumours.
Plant: stated to be used in cerebral fever and as stomch.
Decoct. of bark: purg.
Resin alcohol, bitter substance, etc.

PERISTROPHE BICALYCULATA
(Acanthaceae)
Plant: macetated in an infusion of rice said to be antid. to snake poison.

PEIOVSKIA ABROTANOIDES
(Lamiaceae)
Flowers: soaked in water applied to the body of a patient suffering from fever as a cooling medium.

PERSEA AMERICANA
(Lauraceae)
Avocado

An evergreen tree, growing too 20 m, with dark green, leathery leaves and white flowers.

Key Constituents:
Leaves & bark
Volatile oil (methylchvicol, alpha-pinene)
Flavonoids
Tannins
Fruit pulp
Unsaturated fats
Protein (about 25%)
Sesquiterpenes
Vitamins A, B_1 and B_2
Key Actions:
Leaves & bark
Astringent
Carminative
Relieve coughs
Promote menstrual flow

PETASITES HYBRIDUS
Asteraceae (Butterbur)
The rhizomes and leaves have medicinal action. When dry the rhizomes have an unpleasant smell and a bitter taste. The constituents include an essential oil, a bitter compound, an alkaloid, mucilage, tannins and inulin. These substances give Butterbur antispasmodic, diuretic, diaphoretic and anthelmintic properties. An infusion is used to treat coughs, hoarseness, urinary disorders and to expel intestinal parasites. A poultice of fresh leaves can be applied externally to swellings, rashes, swollen veins and glands and rheumatic joints. In homeopathy a tincture of the fresh plant is used for neuralgia.

PETROSELINUM CRISPUM
(Umbelliferae)
PARSLEY
Parts Used: Leaves, root, seeds.
Constituents: Parsley contains a volatile oil (including about 20% myristicin, about 18% apiole, and many other terpenes), flavonoids, phthalides, coumarins (including bergapten), vitamins A, C and E and highly levels of iron. The flavonoids are anti-inflammatory and antioxidant. Myristicin and apiole have diuretic properties. The volatile oil relieves griping pain and flatulence and is a strong uterine stimulant.
Medicinal Actions & Uses: The fresh leaves are highly nutritious and can be considered a natural vitamin and mineral supplement in their own right. The seeds have a much stronger diuretic action than the leaves, and may be substituted for celery seeds (*Apium graveolens*) in the treatment of gout, rheumatism and arthritis. Both plants act by encouraging the flushing out of waste products from the inflamed joints, and the waste's subsequent

Petasites hybridus

Petroselinum crispum

Peucedanum ostruthium

elimination via the kidneys. Parsley root is more commonly prescribed than the seeds or leaves in herbal medicine. It is taken as a treatment for flatulence, cystitis and rheumatic conditions. Parsley is also valued as a promoter of menstruation, being helpful both in stimulating a delayed period and in relieving menstrual pain.

PEUCEDANUM OSTRUTHIUM
Apiaceae (Materwort)

The rhizomes are the medical parts. When dry they have a penetrating aroma and burning taste and cause salivation. The constituents include a large amount of essential oil with limonene, phellandrene, and pinene as its main components, plus a coumarin glycoside (imperatorin), bitter compounds and tannins. These substances give Masterwort diuretic, diaphoretic, stomachic and carminative properties. It is used in an infusion or in powder form for anorexia, digestive disorders, flatulence and enteritis. **In large doses it is toxic**. A tincture of the fresh rhizome is used in homeopathy for similar complaints.

PEUMUS BOLDUS
(Apiaceae)

BOLDO

Parts Used: Leaves.
Constituents: Boldo contains 0.7% isoquinoline alkaloids (including boldine), well as a volatile oil and flavonoids.
Medicinal Actions & Uses: Boldo stimulates liver activity and bile flow

and is chiefly valued as a remedy for gallstones at liver or gallbladder pain. It is normally taken for a few weeks at a time, either as a tincture or infusion. Boldo is also a mild urinary antiseptic and demulcent and may be taken for infections such as cystitis. In the Anglo-American tradition, boldo is combined with barberry (*Berberis vulgaris*) and fringe tree (*Chionanthus virginicus*) in the treatment of gallstones.

PHASEOLUS ACONITIFOLIUS
(Fabaceae)
Root: narcotic
Seeds: used as diet in fever.

PHASEOLUS ADEMANTHUS
(Fabaceae)
Decoct: used in bowel complaints and stricture.

PHASEOLUS LUNATUS
(Fabaceae)
Seeds: astrin., used as diet in fever. Seeds contain HCN glued. Plant sometimes exhibits poisonous properties.

PHASEOLUS MUNGO
(Fabaceae)
Seeds: cooling, astrin., used as diet in fever and to strengthen the eye.

PHASEOLUS RADIATUS
(Fabaceae)
Seeds: used both internally and externally, in paralysis, rheumatism and affections of the nervous system, considered hot and tonic, useful in piles, affections of the liver and cough, and in fever.
Root: narcotic.

PHASEOLUS TRILOBUS
(Fabaceae)
Leaves: tonic, sedative, used in cataplasms for weak eyes, administered in decoct. in irregular fever.

PHASEOLUS VULGARIS
(Leguminosae)
French Bean, Haricot Bean
Parts Used: Beanpods, beans.
Constituents: French beans contain allantoin, sugars, leucine, tyrosine, arginine and inositol. Fruit contain Rubidium.
Medicinal Actions & Uses: In addition to being an important food in many parts of the world, French beans have two principal medicinal uses. The pods are a medium-strength diuretic, stimulating urine flow and the flushing of toxins from the body. Powdered or infused, they are also

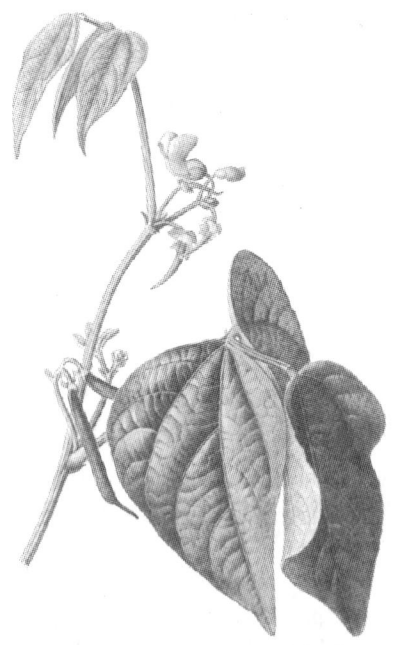

Phaseolus vulgaris

hypoglycaemic, reducing blood glucose levels in the treatment of diabetes. Powdered beans may be dusted on to eczema to ease itching and dry the skin.

PHELLODENDRON AMURENSE
(Rutaceae)

HUANG BAI

Part Used: Bark.
Constituents: *Huang bai* contains isoquinoline alkaloids (including berberine), sesquiterpene lactones and plant sterols. Due to its alkaloid content, *huang bai* is antimicrobial and antibiotic.
Medicinal Actions & Uses: A strongly bitter remedy, *huang bai* is used within Chinese herbal medicine to "drain damp heat". It is prescribed for conditions such as acute diarrhoea and dysentery, jaundice, vaginal infection (including trichomonas), and certain skin conditions. It is also given for urinary system disorders such as frequent urination, pain and infection.

PHLOGACANTHUS THYRSIFOLIUS
(Acanthaceae)

Plant: used like *Adhatoda vasica*.

PHOENIX DACTYLIFERA
(Arecaceae)

Fresh juice: cooling, laxt.
Gum: useful in diar. and diseases of the genito-urinary system.
Fruit: demulc., expect., nutrient, laxt., aphrodis., prescribed in asthma, chest complaints and cough; also in fever, gonor., etc.

Vitamin B and antiscor. vitamin; fruit contains vitamin A, B and D.

PHOENIX PUSILLA
(Arecaceae)

Fruit juice: cooling, laxt.
Gum: used in diarh. and genito-urinary diseases.

PHOENIX SYLVESTRIS
(Arecaceae)

Fruit: tonic and restor.
Juice of tree: used as a cooling beverage.
Root: used in toothache.
Kernels: made into a paste with the root of *Achyranthes aspera*, eaten with betel leaves as remedy for ague.

PHRAGMITES MAXIMA
(Poaceae)

Root: regarded as cooling, diur; and diaphor.

PHYLLANTHUS NIRURI
(Euphorbiaceae)

Plant: used as a diur. in dropsical affections, gonor. and other troubles of the genito-urinary tract.
Infusion of young shoots: given in dysen.
Fresh root: remedy for jaundice.
Leaves: stomch.
Milky juice: used as applicattion to offensive sores.
Powdered leaves and roots: pulverised and made into poultice with rice-water used to lessen oedematous swellings and ulcers.
Leaves contain bitter substance phyllanthin. – hyphophyllanthin and phyllanthin.

PHYLLANTHUS RETICULATUS
(Euphorbiaceae)

Leaves: diur., cooling.
Bark: alter. and attenuant.
Juice of leaves: made into pills with camphor and cubebs and allowed to dissolve in the mouth act as remedy for bleeding gum; used for diar. in infants.

PHYLLANTHUS SIMPLEX
(Euphorbiaceae)

Fresh leaves: bruised and mixed with butter milk used as a wash for itch in children.
Root: used as external application for mammary abscess.
Fresh leaves, flowers and fruits: with cumin seeds and sugar made into an electuary used for the cure of gonor.

PHYLLANTHUS URINARIA
(Euphorbiaceae)

Plant: used as diur. in dropsical affections, also in gonor. and other genito-urinary troubles; fish poison.
Root: given to sleepless children. Alkaloid principle.

PHYLLANTHUS ALKEKENGI
Solanaceae (Cape Gooseberry)

The ripe fruits, without the calyx, contain a bitter compound (physalin), alkaloids, organic pigments and vitamin C. They are diuretic and can be used to treat kidney and urinary disorders, gravel, gout, arthritis and rheumatism. Because of their high concentration of vitamin C the fresh berries are also beneficial during conva-

Physalis alkekengi

lescence. They can be eaten on their own (in small quantities, otherwise they may cause diarrhoea), in jam or made into a tea.

The dried Chinese lanterns make an attractive winter decoration as they retain their shape and colour for a long time.

PHYSALIS MINIMA
(Solanaceae)

Fruit: considered tonic, diur., aper., used for horses and gonor.
Juice of leaves: mixed with water and mustard oil used as a remedy against earache.

PHYSALIS PERUVIANA
(Solanaceae)

Juice of leaves: given in worms and bowel complaints.
Plant: diur.

PHYSOCHLAINA PRAEALTA
(Solanaceae)

Leaves: applied to boils; poisonous. Leaves contain 1.02% alk. hyoscyamine, hyoscine, and potassium nitrate, potassium chloride and potassium sulphate; roots contain 0.64% alk. hyoscyamine and sucrose.

PHYTOLACCA DECANDRA
(Phytolaccaceae)

POKE ROOT

Part Used: Root.

Constituents: Poke root contains triterpenoid saponins, lectins, proteins, resin and mucilage. The triterpenoid saponins are strongly anti-inflammatory, the proteins are antiviral and the lectins are mitogenic (break up chromosomes).

Medicinal Actions & Uses: Poke root is taken internally as a tincture in small amounts to treat rheumatic and arthritic conditions. The root has also been used to treat respiratory tract infections, such as sore throats and tonsillitis, as well as swollen glands and chronic infections. The herb is sometimes prescribed for pain and infection of the ovaries or testes, and as a lymphatic "decongestant", stimulating the clearance of waste products. As a poultice or ointment, it is applied to sore and infected nipples and breasts, acne, folliculitis, fungal infections and scabies.

PICRASMA EXCELSA
(Simaroubaceae)

QUASSIA

Part Used: Bark.

Constituents: Quassia contains quassinoid bitter principles (including quassin), alkaloids, a coumarin (scopoletin) and vitamin B^1. Some of the quassinoids have been shown to have cytotoxic (cell-killing) and anti-leukaemic actions.

Medicinal Actions & Uses: The strongly bitter quassia supports and strengthens weak digestive systems. It increases bile flow, the secretion of salivary juices and stomach acid, and improves the digestive process as a whole. Quassia is commonly used to stimulate the appetite, especially in the treatment of anorexia. Its bitterness has led to its being used for malaria and other fevers, and it is given in the Caribbean for dysentery. The bark has been used in the form of an enema to expel threadworms and other parasites. A decoction of the bark may be used as an insect repellant.

PICRORRHIZA KURROA
(Scrophulariaceae)

PICRORRHIZA

Part Used: Rhizome.

Constituents: Picrorrhiza contains the bitter glycoside kutkin (composed of picrosides I to III and kutkoside), cucurbitacins and apocynin. Apocynin is powerfully anti-inflammatory and reduces platelet aggregation.

Medicinal Actions & Uses: In India, picrorrhiza is used as a bitter tonic, equivalent in many respects to gentian equivalent in many respects to gentian (*Gentiana lutea*), and given for a wide range of digestive and liver troubles, including insufficient stomach acid secretion, indigestion, jaundice, hepatitis, liver cirrhosis and constipation. In China, picrorrhiza is chiefly employed to treat chronic diarrhoea and dysentery. Picrorrhiza also helps treat asthma, acute and chronic infections, conditions where the immune system is compromised, and auto-immune diseases, including psoriasis and vitiligo. The herb's traditional use for liver disorders is well founded and picorrhiza may play an important role in treating liver disease. Glucd. picrorrhizin; roots contain glucosidic principle kutkin, vanillic acid, kutkiol and kutkisterol.

PIMENTA OFFICINALIS
(Myrtaceae)

ALLSPICE

Parts Used: Berries, essential oil.
Constituents: Allspice contains about 4% volatile oil (comprising up to 80% eugenol), proteins, lipids, the vitamins A, C, B^1 and B^2 and minerals.
Medicinal Actions & Uses: A digestive stimulant, allspice is taken to relieve flatulence and indigestion. It may also be used to treat diarrhoea. Allspice is often combined with herbs that have a tonic or laxative effect. The herb has an action that is similar to that of cloves (*Eugenia caryophyllata*), both are stimulant, stomach-settling and antiseptic. Allspice essential oil is also stomach-settling.

PIMPINELLA ANISUM
(Umbelliferae)

ANISE

Parts Used: Seeds, essential oil.
Constituents: Anise contains a volatile oil (comprising 70-90% anethole,

Pimpinella anisum

together with methyl chavicol and other terpenes), furanocoumarins, flavonoids, fatty acids, phenylpropanoids, sterols and proteins. Anethole has an observed oestrogenic effect and the seeds as a whole are mildly oestrogenic. This effect may substantiate the herb's use as a stimulant of sexual drive and of breast-milk production
Medicinal Actions & Uses: Anise seeds are known for their ability to reduce wind and bloating and to settle the digestion. They are commonly given to infants and children to relieve colic, and to people of all ages to ease nausea and indigestion. Anise seeds' antispasmodic properties make them helpful in countering period pain, asthma, whooping cough and

bronchitis. The seeds' expectorant action justifies them use for these respiratory ailments. Anise seeds are thought to increase breast-milk production and may be beneficial in treating impotence and frigidity. Anise essential oil is used for similar complaints, and is also used externally to treat lice and scabies.

PIMPINELLA SAXIFRAGA
Apiaceae (Burnet Saxifrage)

The roots are the medicinally active parts. When dry they have a strong aroma – of billy-goat – and absorb moisture easily. The constituents include an essential oil, furanocoumarins and their derivative pimpinellin, tannins, saponins and

Pimpinella saxifraga

resin. These substances give Burnet Saxifrage expectorant, antispasmodic, stomachic, diuretic and antiseptic properties. It is used in an infusion or powdered form to treat asthma, infections of the upper respiratory tract, digestive disorders, flatulence and diarrhoea. **Large doses can harm the kidneys.** Externally Burnet Saxifrage is used in compresses and bath preparations to treat slow-healing wounds and in gargles.

PINGUICULA VULGARIS
(Lentibulariaceae)

Butterwort

Parts Used: Leaves.
Constituents: Butterwort contains mucilage, tannins, benzoic acid, cinnamic acid and valeric acid. Cinnamic acid has antispasmodic properties.
Medicinal Actions & Uses: Butterwort is little employed in European herbal medicine today. Its main use is as a cough remedy, with properties similar to those of sundew (*Drosera rotundifolia*), another insect eating plant. Butterwort may be used to treat chronic and convulsive coughs.

PICRASMA QUASSIOIDES
(Simarubaceae)

Wood: bitter, subst. for Quassia.
Bark, wood and root – febge.
Leaves: applied to itch.
Stemwood contains bitter principle quassin.

PINUS GERARDIANA
(Pinaceae)

Seeds: anodyne, stim.
Oil from seeds: applied as a dressing to wounds and ulcers; external appli-

cation in head diseases.
Essential oil.

PINUS ROXBURGII
(Pinaceae)

Resin: stim., used internally as stomch., and as a remedy for gonor., externally as a plaster applied to buboes and abscesses for suppuration.
Wood: stim., diaphor., useful in burning of the body, cough, fainting and ulcerations.
Wood and oleo-resin: used in snakebite and scorpion-sting.
Essential oil.
α-and-β pinene.

PINUS SYLVESTRIS
(Pinaceae)

SCOTS PINE

Parts Used: Leaves, branches, stems, seeds, essential oil.
Constituents: The leaves of Scots pine contain a volatile oil (consisting mainly of alpha-pinene, but also including beta-pinene, d-limonene and other constituents).
Medicinal Actions & Uses: Scots pine leaves, taken internally, have a mildly antiseptic effect within the chest, and may also be used for arthritic and rheumatic problems. Essential oil from the leaves may be taken for asthma, bronchitis and other respiratory infections, and for digestive disorders such as wind. Scots pine branches and stems yield a thick resin, which is also antiseptic within the respiratory tract. The seeds yield an essential oil with diuretic and respiratory-stimulant properties. The seeds are used for bronchitis, tuberculosis and bladder infections. A decoction of the seeds may be applied to help suppress excessive vaginal discharge.

PIPER ANGUSTIFOLIA
(Piperaceae)

MATICO

Parts Used: Leaves.
Constituents: Matico contains a volatile oil (including camphor, borneol and azulene), tannins, mucilage and resins.
Medicinal Actions & Uses: Matico is an aromatic stimulant, diuretic and astringent used extensively for gastric and intestinal problems, including peptic ulcers, diarrhoea and dysentery. It is commonly used in South American herbal medicine for internal bleeding, particularly within the digestive tract – for example, rectal bleeding and haemorrhoids. It is also taken for bleeding in the urinary tract. Applied externally, a decoction of matico makes a valuable remedy for minor wounds, insect stings and inflamed skin, and it may also be used either as a mouthwash or a douche.

PIPER BETLE
(Piperaceae)

BETEL

Parts Used: Leaves, root, fruit.
Constituents: Betel leaves contain up to 1% volatile oil (including cadinene, chavicol, chavibetol and cineole). As with many volatile oils, the percentages are variable. Malaysian samples have been shown to contain up to 69% chavibetol.
Medicinal Actions & Uses: Betel leaves are chiefly used as a gentle stimulant, since they apparently induce a mild sensation of well-being. They also affect the digestive system, stimulating salivary secretions, relieving flatulence, and preventing worm

infestation. In many Asian traditions, including Ayurvedic medicine, betel leaves are thought to have aphrodisiac and nerve tonic properties. In Chinese herbal medicine, betel root, leaves and fruit are sometimes used as a mild tonic and stomach-settling herb. The root has been used with black pepper (*P. nigrum*) or jequirity (*Abrus precatorius*) to produce sterility in women.

PIPER CUBEBA
(Piperaceae)
CUBEB

Part Used: Fruit.

Constituents: Cubeb contains a volatile oil (up to 20%), a bitte principle (cubebin), an alkaloid (piperidine), resin and fixed oil.

Medicinal Actions & Uses: Like other members of the pepper family, cubeb has a significant antiflatulent and antiseptic action. The fruit is used medicinally as a means to counter infections of the urinary tract, and has been taken in the past as a treatment for gonorrhoea. In addition, the fruit is helpful in relieving digestive problems such as flatulence and bloating. Cubeb is occasionally employed as an expectorant in the treatment of chronic bronchitis.

PIPER METHYSTICUM
(Piperaceae)
KAVA KAVA

An evergreen shrub climbing to 3 m, with fleshy stems and heart-shaped leaves.

Key Constituents:
Resin containing kava lactones, including kawain
Piperidine alkaloid (piper-methylsticine)

Key Actions:
Stimulant
Tonic
Reduces anxiety
Urinary antiseptic
Analgesic Induces sleep

PIPER NIGRUM
(Piperaceae)
PEPPER

Parts Used: Fruit, essential oil.

Constituents: Pepper contains a volatile oil (including beta-bisabolene, camphene, beta caryophyllene and many other terpenes and sesquiterpenes), up to 9% alkaloids (especially piperene, which is largely responsible for the herb's acrid taste), about 11% proteins, and small quantities of minerals. White pepper contains very small amounts of volatile oil.

Medicinal Actions & Uses: The familiar sharp taste of pepper reflects the stimulant and antiseptic effect it has on the digestive tract and the circulatory system. Pepper is commonly taken, either alone or in combination with other herbs and spies, to warm the body, or to improve digestive function in cases of nausea, stomach ache, flatulence, bloating, constipation or lack of appetite. The essential oil eases rheumatic pain and toothache. It is antiseptic and antibacterial and reduces fever.

PISCIDIA ERYTHRINA
(Leguminosae)
JAMAICA DOGWOOD

Part Used: Root bark.

Constituents: Jamaica dogwood contains isoflavones (including lisetin, jamaicin and icthyone), organic acids (such as piscidic acid), beta-sitosterol and tannins.

Medicinal Actions & Uses: Jamaica dogwood is a useful and undervalued remedy that acts both as a sedative and as a painkiller. It is chiefly employed in the treatment of insomnia and over excitability, as it calms mental activity. It is also prescribed for nerve pain, toothache and period pain. As an antispasmodic, it is useful for treating muscle spasms, especially in the back, and spasmodic respiratory ailments such as asthma and whooping cough.

Cautions: Do not take Jamaica dogwood during pregnancy or if suffering from heart problems.

PISONIA ACULEATA
(Nyctaginaceae)

Bark and leaves: used as counter-irrit. for swellings and rheum. pains.
Juice: mixed with pepper and other ingredients given in pulmonary complaints of children.

PISTACIA INTEGERRIMA
(Anacardiaceae)

Galls: tonic, expect., used in cough, phthisis, asthma, etc.; powdered and fried in ghee given internally in dysen.; antid. to snake venom and scorpion-sting.

Essen. oil galls gave 1.3% essen. oil and two crystalline acids essen. oil contains α-pinene 25, camphene 27, *dl*-limonene 4-5, 1:8-cineole 10, α-terpeneol 20, aromadendren 4-5% and caprylic acid.

PISTACIA LENTISCUS
(Anacardiaceae)
MASTIC TREE
Part Used: Resin.

Constituents: The resin contains alpha -and beta-masticoresins, a volatile oil (comprising mainly alpha-pinene), tannins, masticin and mastic acid. Pinenes are strongly antiseptic.
Medicinal Actions & Uses: Mastic resin is little used today, though it could be employed as an expectorant for bronchial troubles and coughs, and as a treatment for diarrhoea. It has also been used to treat boils, ulcers and similar skin conditions. The resin has been mixed with other compounds as a temporary filling for decayed teeth.

PISTIA STRATIOTES
(Araceae)

Plant: demulc., refrig., given in dysuria; used to destroy bugs.
Root: emol., laxt., diur.
Leaves: made into poultice applied to haemorrhoids; mixed with rose water and sugar given in asthma and cough and with rice and coconut milk in dysen.
Ashes: applied to ringworm of the scalp.
Juice of leaves: boiled in coconut oil and the preparation used externally in chr. skin diseases.

PISUM SATIVUM
(Fabaceae)
GARDEN PEA
Seeds: believed to cause dysen. when eaten raw; in Spain their flour considered emol. and resolv., and applied in form of a cataplasm.

Alk. trigonelline; As 0.026 mg. in 100 g. ash of the seeds oil from ripened seeds has anti-sexharmonic effect; produces sterility and antagonises effect of male sexhormone.

PITHECELLOBIUM BIGEMINUM
(Minosaceae)

Plant: fish poison.
Seeds: prescribed in diabetes mellitus.
Decoct. of leaves: applied externally; used as a nostrum for leprosy and for promoting growth of hair.

PITTOSPORUM FLORIBUNDUM
(Pittosporaceae)

Bark: bitter, arom., given in broncht., expect., febge., narcotic, antid. to snake poison.
Oil: alter., tonic, stim., specific for certain skin diseases, local application in rheumatism, chest affections and phthisis, ophthalmia, sprains and bruises, sciatica and leprosy; internally prescribed in cutaneous diseases, secondary syphilis, chr. rheumatism and leprosy.
Bitter glucd., essen. oil

PLANTAGO OVATA
(Plantaginaceae)

PSYLLIUM, ISPAGHULA, (HINDI), FLEA SEED

An annual, growing to 40 cm high, with narrow leaves and clusters of minure white-brown flowers.
Key Constituents: Mucilage
Fixed oil (2.5%)-mainly linoleic, oleic and palmitic fatty acids, Starch.
Key Action:
Seeds: demulc., cooling, diur., used in inflammatory conditions of the mucous membrane of gastro-intestinal and genito-urinary tracts, in chr. dysen., diar. and constip.

PLANTAGO LANCEOLATA
Plantaginaceae
(Ribwort Plantain)

The leaves and seeds are used medicinally. The most important constituent of the leaves is an iridoid glycoside (aucubin). It is unstable and causes the dried leaves to darken. Other constituents include mucilage, carotenes, tannins, enzymes and silicic acid. These substances give the leaves expectorant, emollient, demulcent, vulnerary and astringent properties. An infusion of them is used for cough, whooping cough, hoarseness, bronchitis and other respiratory disorders. Coughing in children is alleviated by a thickened syrup made from the leaves sweetened with honey.

Plantago lanceolata

The seeds contain abundant mucilage, fatty oil, aucubin and enzymes. Swallowed whole they are an effective and harmless laxative. Crushed fresh leaves can be applied externally to swellings, bruises and inflamed wounds.

PLANTAGO MAJOR
(Plantaginaceae)

COMMON PLANTAIN

Parts Used: Leaves.

Constituents: Common plantain contains iridoids (such as aucubin, also found in *Euphrasia species*), flavonoids (including apigenin), tannins, plant acids and mucilage. Aucubin increases uric acid excretion by the kidneys, apigenin is anti-inflammatory.

Medicinal Actions & Uses: Common plantain quickly staunches blood flow and encourages the repair of damaged tissue. It may be used instead of comfrey (*symphytum-officinale*) in treating bruises and broken bones. An ointment or lotion may be used to treat haemorrhoids, fistulae (abnormal passages in the skin) and ulcers. Taken internally, common plantain is diuretic, expectorant and anticatarrhal. It is commonly prescribed for gastritis, peptic ulcers, diarrhoea, dysentery, irritable bowel syndrome, respiratory catarrh, loss of voice and urinary tract bleeding.

PLATANUS ORIENTALIS
(Platanaceae)

Fresh leaves: bruised and applied to the eyes in ophthalmia.
Bark: boiled in vinegar given in diar., dysen., hernia and toothache. Allantoin, asparagin.

PLATYSTOMA AFRICANUM
(Lamiaceae)

Used for fever, feverish chills or rheum. symptoms in Northern Nigeria.
Leaves and seeds: used for children's cough in Gold Coast.

PLESMONIUM MARGARITIFERUM
(Araceae)

Crushed seeds: used as local anaesthetic to cure toothache, a small quantity is placed in the hollow tooth and covered with cotton.

PLUMBAGO INDICA
(Plumbaginaceae)

Root: acrid., vesic., stim., when tempered with little bland oil used as external application in rheum. and paralytic affections; also prescribed internally for these complaints; powerful sialogogue, remedy for secondary syphilis and leprosy.
Milky juice: useful in ophthalmia and application to scabies.
Plumbagin.

PLUMBAGO ZEYLANICA
(Plumbaginaceae)

CEYLON LEADWORT

Parts Used: Leaves, root.

Constituents: Ceylon leadwort contains plumbagin, which stimulates sweating,

Medicinal Actions & Uses: Ceylon leadwort root is acrid and stimulates sweating. In West Africa, the root is traditionally mixed with okra (*Hibiscus esculentus*) to treat leprosy. In Nepal, a decoction of the root is used to treat baldness. In Indian herbal medicine, the leaves and root are used to treat infections and digestive problems such as dysentery. Externally, a paste of the leaves and root is applied to painful rheumatic areas or to chronic and itchy skin problems. The paste acts as a counter-irritant. By raising blisters and increasing circulation, it speeds the clearing of toxins from the affected area.

Root: appetizer, used in skin diseases, diar., dyspep., piles, anasarca; made into a paste with vinegar, milk or salt and water applied externally in leprosy and other skin diseases.

Tincture of root bark: powerful sudorific, antiper.

Milky juice: used as application in scabies and unhealthy ulcers.

Contains plumbagin which externally is a strong irrit. and a powerful germicide, stimulates muscular tissue in smaller doses and paralyses in larger ones, stimulates the contraction of the muscular tissue of the heart, intestines and worms, stimulates the secretion of sweat, urine and bile and has stimulant action on the nervous system.

PLUMERIA ALBA
(Apocynaceae)

Latex: applied to ulcers, herpes and scabies in Guiana.

Seeds: considered haemostatic.

Root bark: purg., alter., detergent, given for blennorrhagia and herpes; in form of an extract used internally and externally for syphilitic ulcers.

PLUMERIA RUBRA
(Apocynaceae)

Root bark: drastic purg., used mostly in blennorhagia in Guiana.

Latex: given in toothache and for carious teeth.

Flowers: arom., bechic and used as a pectoral syrup.

PLUMERIA RUBRA. *VAR.* ACUTIFOLIA
(Apocynaceae)

Root bark: purg., antiherpetic, useful in gonor. and venereal sores.

Bark: given with coconut, ghee and rice in diar.

Milky juice: employed as a rubft. in rheumatism, purg.

Bitter glucd., essen. oil; plumeric acid.

PODOPHYLLUM PELTATUM
(Berberidaceae)

AMERICAN MANDRAKE

Part Used: Rhizome.

Constituents: The rhizome of American mandrake contains lignans (especially podophyllotoxin), flavonoids, resin and gums. The lignans are responsible for the rhizome's purgative action.

Medicinal Actions & Uses: Despite 19th century beliefs in its safety. American mandrake is no longer taken internally on account of its cytotoxic (cell-killing) action. However, applied externally as a poultice, lotion or ointment, the root can be an effective treatment for all kinds of warts.

POGOSTEMON CABLIN
(Labiatae)
PATC CHOULI

Parts Used: Young leaves and shoots, essential oil.

Constituents: Patchouli contains a volatile oil comprising mainly the sesquiterpenes patchoulol (35%) and bulnesene.

Medicinal Actions & Uses: Patchouli is used in herbal medicine in Asia as an aphrodisiac, antidepressant and antiseptic. It is also employed for headaches and fever Patchouli essential oil is used in aromatherapy to treat skin complaints. It is thought to have a generative effect on skin tone and to help clear conditions such as eczema and acne. The oil may also be used for varicose veins and haemorrhoids.

POGOSTEMON PARVIFLORUS
(Labiatae)

Fresh leaves: styptic, bruised and applied as a cataplasm to clean wounds and promote healthy granulation.

Root: remedy for haemor., useful in uterine haemor.; antid. to scorpion-sting and snake-bite.

Alk., essen. oil.

POLIANTHES TUBEROSA
(Amaryllidaceae)

Flowers: diur., emetic.

Bulbs: dried and powdered used as a remedy for gonor.

Essential oil.

POLYCARPEA CORYMBOSA
(Caryophyllaceae)

Herb: administered both internally and externally as remedy for venomous bites from reptiles.

Pounded leaves: used cold or warm as poultice over boils and inflammatory swellings; used for bites from animals and given with molasses in form of a pill in jaundice.

POLYCARPON INDICUM
(Caryophyllaceae)

Infusion of roasted leaves: given for cough following upon a fever, more particularly measles in Indo-China. Hotter parts of India in fields and waste places.

POLYGALA AMARA
Polygalaceae (Dwarf Milkwort)

The flowering stems, sometimes with the roots, are used medicinally. When dry they have a distinctive bitter taste (the specific epithet *amara* means

Polygala amara

bitter). The constituents include important triterpenoid saponins (senegins), a bitter compound (polygamarin), the glycoside gaultherin, tannins and an essential oil. These substances give Dwarf Milkwort expectorant, stomachic and diuretic properties. It is used in the form of a decoctioon or powder to treat coughs, bronchitis and other infections of the upper respiratory tract, and digestive disorders. It is also include in proprietary expectorant medicines. In folk medicine it is still recommended for nursing mothers but it has not yet been established whether the plant really is a galactagogue.

POLYGALA CHINENSIS
(Polygalaceae)

Root: given in cases of fever and dizziness.

POLYGALA CROTALARIOIDES
(Polygalaceae)

Used medicinally in catar. affections. **Root**: chewed or ground and drunk with water to expel phlegm from the throat; provokes coughing; used as cure for snake-bite.

POLYGALA ELONGATA
(Polygalaceae)

Plant: used in biliousness and coustip. Specific for snake-poison.

POLYGALA SENEGA
(Polygalaceae)

SENECA SNAKEROOT
Part Used: Root.
Constituents: Seneca snakeroot contains triterpenoid saponins (including sengins), phenolic acids, methyl salicylate, polygalitol and plant sterols. The triterpenoid saponins promote the clearing of phlegm from the bronchial tubes.

Medicinal Actions & Uses: In North American and European herbal medicine. Seneca snakeroot is used as an expectorant to treat bronchial asthma, chronic bronchitis and whooping cough. The root has a stimulant action on the bronchial mucous membranes, promoting the coughing up of phlegm from the chest and thereby easing wheeziness. In large doses, the root is emetic. Seneca snakeroot is also thought to encourage sweating and to stimulate the secretion of saliva.

POLYGALA SIBERICA
(Polygalaceae)

Roots: given as a subst. for senega in colds and coughs in Japan, China and Malaya; in Indo-China used as diur. and also given in broncht., amnesia, sexual impotency and seminal losses.

POLYGALA VULGARIS
(Polygalaceae)

MILKWORT

Parts Used: Aerial parts, root.
Constituents: Milkwort contains triterpenoid saponins, a volatile oil, gaulthern and mucilage.
Medicinal Actions & Uses: The bitter tasting milkwort still has a reputation for increasing milk production in nursing mothers, but this attribute is in fact unfounded. While milkwort is infrequently used in European herbal medicine today, it – like Seneca snakeroot (P. Senega) – is a valuable herb for the treatment of respiratory

troubles such as chronic bronchitis, bronchial asthma and convulsive coughs, including whooping cough. Milkwort is also thought to have sweat-inducing & diuretic properties.

POLYGONUM AVICULARE
Polygonaceae (Knotgrass)

The flowering stems are used medicinally. They are usually collected in autumn when they have the greatest concentration of silicic acid. They also contain tannins, a flavonoid glycoside (avicularin) and perhaps also saponins. Knotgrass, like other members of the dock (Polygonaceae) family, has strong astringent and haemostatic properties and it is used in an infusion to check external and internal bleeding, gastritis and enteritis and severe diarrhoea. The infusion also acts as a tonic. Knotgrass is also a component of diuretic tea mixtures taken for kidney disorders, uroliths and kidney stones.

POLYGONUM BISTORTA
Polygonaceae
(Common Bistort)

The rhizomes are used medicinally their constituents include abundant tannins (15-20 per cent), starch, a bitter compound (catechin) and silicic acid. These substances (especially the tannins) have Common Bistort strong astringent, antidiarrhoeal, haemostatic and anti-inflammatory properties. The dried rhizome is used in infusions, decoctions or powder form to check internal and external bleeding, to treat gastritis and enteritis, severe diarrhoea, dysentery and incontinence of urine. The high concentration of starch in the rhizome, which produces mucilage, is also beneficial. Exter-

Polygonum aviculare

Polygonum bistorta

nally Common Biswort is used in gargles, mouthwashes, compresses and bath separations.

The young leaves can be eaten in salads cooked like spinach and the root is edible after it has been soaked and roasted.

POLYGONUM HYDROPIPER
Polygonaceae (Water Pepper)

The flowering stems are used medicinally. The most important constituents are tannins. The plant also contains bitter compounds, an essential oil, a glycoside, formic acid, acetic acid, polygonic acid and vitamin C (in fresh material). These substances give Waterpepper strong astringent, haemostatic, diuretic and anti-inflammatory properties. It is used in an infusion or powder form for uterine haemorrhage and menstrual disorders for dropsy and other urinary complaints, rheumatism, diarrhoea and haemorrhoids. Fresh bruised leaves can be applied to bleeding or slow-healing wounds. Waterpepper is also used in homeopathy and in veterinary medicine. **Large doses of Waterpepper should not be taken** because they may cause gastrointestinal irritation. The fresh juice may also irritate the skin.

POLYGONUM LAPTHIFOLIUM
Polygonaceae
(Pale Persiacaria)

The flowering stems are used medicinally. Their constituents include tannins, essential oil, organic acids and a large amount of vitamin C (in fresh material). These substances give

Polygonum hydropiper

Polygonum lapthifolium

Pale Persicaria astringent, haemostatic and diuretic properties and it is used to check internal and external bleeding, diarrhoea and various urinary disorders.

POLYGONUM MULTIFLORIUM
(Polygonaceae)

He Shou Wu (Chinese)? Flowery Knotweed

A perennial climber, growing to 10 m, with red stems, light green leaves and white or pink flowers.

Key Constituents:
Crysophanic acid
Anthraquinones (emodin, rhein)
Lecithin
Key Actions:
Mildly sedative
Nourishes the blood
Tonic

POLYGONATUM MULTIFLORUM
(Liliaceae)

Solomon's Seal

Part Used: Rhizome
Constituents: Solomon's seal contains saponins (similar to diosgenin), flavonoids and vitamin A.
Medicinal Actions & Uses: Like arnica (*Arnica montana*), Solomon's seal is believed to prevent excessive bruising and to stimulate tissue repair. Used mainly in the form of a poultice, the rhizome has astringent and demulcent actions that undoubtedly contribute to its ability to accelerate healing. Solomon's seal has also been recommended for tuberculosis, as a remedy for menstrual problems, and as a tonic. In Chinese herbal medicine, it is considered a *yin* tonic, and

is thought to be particularly applicable to respiratory system problems – sore throats, dry and irritable coughs, bronchial catarrh and chest pain.

POLYGONATUM ODORATUM
Liliaceae
(Angular Solomon's Seal)

The roots and rhizomes are the medicinal parts. Their constituents include starch, saponins, mucilage, tannins, sugar and organic pigments. These substances give Angular Solomon's Seal diuretic, hypoglycaemic, astringent and vulnerary properties. It has been used internally as a diuretic but mostly it is used only externally in compresses or bath preparations for treating rheumatism, bruises, eczema and other skin disorders. **All species of Solomon's seal should be taken internally only under the supervi-**

Polygonatum odoratum

sion of a qualified medical or herbal practitioner; large doses can be harmful.

POLYMNIA UVEDALIA
(Compositae)

BEARSFOOT

Part Used: Root.

Medicinal Actions & Uses: Bearsfoot is perhaps best known for its use as a hair tonic, having traditionally been a popular ingredient in hair lotions. It is still used in this way, but today the root is more often taken internally as a treatment for non-malignant swollen glands and especially for mastitis. The root is thought to have a beneficial effect on the stomach, liver and spleen, and may be taken to relieve indigestion and counteract liver malfunction. The herb has laxative properties, and it may also act to relieve pain.

POLYPODIUM VULGARE
(Polypodiaceae)

POLYPODY

Part Used: Rhizome.

Constituents: Polypody rhizome contains saponins (based on polypodosapogenin), ecdysteriods, pholoroglucins, volatile oil, fixed oil and tannins.

Medicinal Actions & Uses: Polypody stimulates bile secretion and is a gentle laxative. Traditionally, it has been used in European herbal medicine as a treatment for hepatitis and jaundice, and as a remedy for indigestion and loss of appetite. Polypody makes a safe treatment for constipation in children. The rhizome

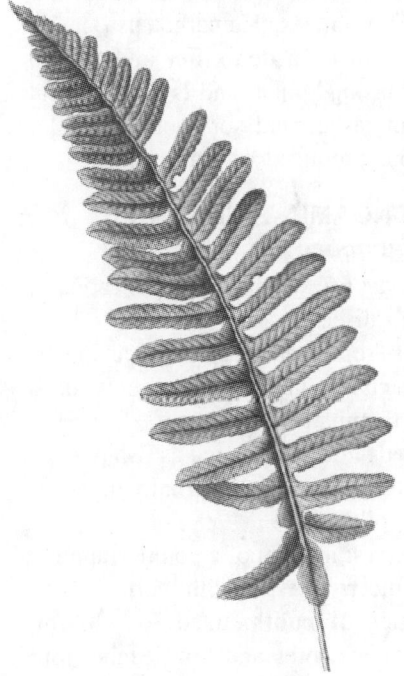

Polypodium vulgare

is also expectorant, having a supportive and mildly stimulating effect on the respiratory system. It may be taken for relief of catarrh, bronchitis, pleurisy and dry irritable coughs. The rhizome combines well with marshmallow (*Althaea officinalis*).

POMADERRIS ELLIPTICA
(Rhamnaceae)

KUMARHOU

Parts Used: Aerial parts.

Medicinal Actions & Uses: Kumarhou is a traditional Maori remedy that has been used to treat a wide range of illnesses. Its most common use is as a remedy for problems of the respiratory tract, such as asthma and bronchitis. However, it has also been used in the treatment of indigestion and heartburn, diabetes and kidney problems. Kumarhou is considered to be a detoxifier and "blood cleansing" plant, and is used to treat skin rashes and sores, including lesions produced by skin cancer.

PONGAMIA PINNATA
(Fabaceae)

Seeds: used as external application in skin diseases.

Oil from seeds: useful in cutaneous affections, herpes and scabies; used in rheumatism.

Seeds and roots: used as fish poison.

Fresh bark: used internally in bleeding piles.

Leaves: in form of a poultice applied to ulcers infested with worms.

Juice of roots: used for closing fistulous sores and for cleaning foul ulcers; given internally with equal quantities for coconut milk and lime water for gonor.

Seeds contain 27 to 36.7% of a bitter fatty oil and traces of an essen. oil, seeds yield fixed oil, three crystalline substances, karanjin, pongamol and glabrin.

POPULUS × CANDICANS
(Salicaceae)

BALM OF GILEAD

Parts Used: Buds, stem bark.

Constituents: Balm of Gilead buds contain flavonoids, phenolic glycosides (including salicin) and fatty acids. Salicin's analgesic, anti-inflammatory and fever-reducing actions resemble those of aspirin.

Medicinal Actions & Uses: Balm of Gilead is a common ingredient of cough mixtures. Its expectorant, antiseptic and analgesic properties make it an excellent remedy for sore throats, dry irritable coughs, bronchitis and other respiratory ailments. In France and Germany, balm of Gilead is applied as a salve to grazes, small wounds, chapped and itchy skin, sunburn, chillblains and haemorrhoids. A preparation of balm of Gilead, applied externally, may also help relieve the pain of rheumatic joints and strained muscles. As Culpeper noted, the plant is also thought to reduce breast-milk production.

POPULUS NIGRA
Saliaceae (Black Poplar)

The leaf buds are used medicinally. Their constituents include as essential oil, tannins, the phenolic glycosides salicin and populin, and resin. These substances give Black

Populus nigra

Poplar pronounced diuretic, anti-inflammatory and antiseptic properties. It is used in an infusion for infections of the upper respiratory tract, for gout (it reduces the amount of uric acid in the blood), and to relieve rheumatic pain. An ointment from the dried or fresh buds – or from the dried or fresh bark of young twigs – is applied externally to treat skin rashes, cut, haemorrhoids and arthritic and rheumatic joints. White Poplar and Aspen have similar medicinal uses.

POPULUS TREMULA
Salicaceae (Aspen)

The leaf buds, occasionally the young bark and the leaves of Aspen, are used medicinally. Like the buds of Black Poplar (*P. nigra*), Aspen buds contain the phenolic glycosides salicin and

Populus Tremula

populin, also an essential oil and bitter compounds. These substances give Aspen strong diuretic, anti-inflammatory and antiseptic properties and the buds are used in an infusion for infections of the urinary tract, enlarged prostrate gland, for gout and rheumatism. Externally compresses, bath preparations and ointments are used for haemorrhoids and for the treatment of burns. Preparations from fresh leaves are used in homeopathy.

POPULUS TREMULOIDES
(Salicaceae)

QUAKING ASPEN

Parts Used: Bark.

Constituents: The bark contains phenolic glycosides (including salicin and populin) and tannins. Salicin and populin are salicylates, substances that have fever-reducing, pain-relieving and anti-inflammatory properties that are similar to those of aspirin.

Medicinal Actions & Uses: Like willow bark (*Salix alba*), quaking aspen bark has widely recognised anti-inflammatory and pain-relieving properties. It is often taken to treat arthritic and rheumatic aches and pains. It is also used to lower fever, especially when this condition is associated with rheumatoid arthritis. Being stimulant, quaking aspen bark acts as a tonic remedy in the treatment of anorexia and another debilitated states. The bark's significant astringent and antiseptic qualities make it useful for treating diarrhoea and the symptoms of irritable bowel syndrome. It is also used to treat urinary tract infections.

PORIA COCOS
(Polyporaceae)

Fu Ling, Indian Bread

Parts Used: Inner mass of the fungus.
Constituents: *Fu ling* contains beta-pachyman, beta-pachymanase and pachymic acid.
Medicinal Actions & Uses: *Fu ling* is much used as a diuretic and tonic in Chinese herbal medicine, being classified as a herb that "drains dampness". It is prescribed for a variety of conditions affecting the urinary system, including fluid retention and difficulty in passing urine. *Fu line* has a soothing and tranquillizing effect upon the nervous system, and can be most helpful in treating stress-related problems, for example anxiety, tension headaches, palpitations and difficulty in sleeping. In common with many other tonic herbs, *fu ling* plays a useful role in supporting convalescence after long-term illness.

PORTULACA OLERACEA
(Portulacaceae)

Purslane

Parts Used: Aerial parts.
Constituents: Purslane contains mucila, plant acids, sugars, vitamins A, B_1 and C, a calcium. Chinese research (unconfirmed in the West) also lists noradrenaline and dopamine as constituents.
Medicinal Actions & Uses: Purslane has long been considered valuable in the treatment of urinary and digestive problems. The diuretic effect of the juice makes it useful in the alleviation of bladder ailments, for example difficulty in passing urine. The plant mucilaginoous properties also makes it a soothes remedy for gastro-intestinal problems such as dysentery and diarrhoea. In Chinese herbal medicine, purslane is employed for similar problems and, additionally, for appendicitis. The Chinese also use the plant as an antidote for wasp stings and snake bite. Used as an external wash, the juice or a decoction relieves skin ailments such as boils and carbuncles, & also helps to reduce fever.

POTENTILLA ANSERINA
Rosaceae
(Silverweed)

The flowering stems and leaves, and sometimes the rhizomes are used medicinally. The constituents of the aerial parts include tannins, bitter compounds, mucilage, organic pigments and mineral compounds. These

Potentilla anserina

substances give Silverweed strong astringent, anti-inflammatory, haemostatic, antiseptic and stomachic properties. An infusion is used for gastritis and enteritis, severe diarrhoea, indigestion, colic, internal haemorrhage, bladder and kidney complaints and excessive menstrual bleeding. Externally Silverweed is used in gargles to relieve painful gums and toothache and in bath preparations and compresses to treat wounds, swellings and skin disorders. In homeopathy a tincture prepared from the fresh plant is used for stomach disorders and for painful menstruation.

POTENTILLA ERECTA
(Rosaceae)
TORMENTIL

Parts Used: Aerial parts, root.
Constituents: Tormentil contains 15-20% tannins, catechins, ellagitannins and a phlobaphene.
Medicinal Actions & Uses: Containing even more tannins that oak bark (*Quercus robus*), all parts of tormentil are strongly astringent, finding use wherever this action is required. The plant makes a beneficial gargle for throat infections, and an effective mouthwash for treating mouth ulcers and infected gums. Tormentil may be taken for conditions that give rise to diarrhoea, such as irritable bowel syndrome, colitis, ulcerative colitis and dysentery, and for rectal bleeding. Applied externally as a lotion or ointment, tormentil helps relieve haemorrhoids (especially those that are bleeding). In the form of a lotion, tormentil is used to help staunch wounds and protect areas of damaged or burned skin.

POTHOS SCANDENS
(Araceae)
Stem and leaves: used in snake-bite.
Powdered leaves: applied to the body as a cure for small-pox.
Stem: cut up with camphor smoked like tobacco for asthma.

POUZOLZIA ZEYLANICA
(Urticaceae)
Plant: used in syphilis, gonor., and snake-poison.

PRANGOS PABULARIA
(Apiaceae)
Fruit: stim., carmin., stomch., diur., emmen., promotes the expulsion of the foetus; in decoct. used to cure the rot in sheep.
Seeds: stomch., aphrodis.
Roots: used to cure itch., as diur. and emmen.
Plant: considered heating.
Fruit contains essen. oil, alk., valeric acid ester; fresh plant contains 2% essen. oil consisting of myrcene 48, α-pinene 4, camphene traces, borneol, dihydro-cuminol (free and as acetate) 17.5, aldehyde traces and resinous residue.

PREMNA INTEGRIFOLIA
(Verbenaceae)
Decoct. of root: cordial, stomach., good for liver complaint.
Decoct. of plant: used in rheumatism and neuralgia.
Leaves: rubbed along with pepper administered in colds and fever; in decoct. given for flatulence; in form of soup used as stomch. and carmin.
Stem bark contains alk. premnine; decreases force of contraction of heart

and produces dilation of the pupils; another alk. ganiarine.

PREMNA LATIFOLIA
(Verbenaceae)

Leaves: diur., given internally and applied externally in dropsy.
Milk of the bark: applied to boils.
Juice of bark: given to cattle in colic.

PRIMULA VERIS
(Primulaceae)

COWSLIP

Parts Used: Flowers, leaves, root.
Constituents: Cowslip contains triterpenoid saponins, flavonoids, phenols, tannins and a trace of vola-

Primula veris

tile oil. The flavonoids, mainly in the flowers, are antioxidant, anti-inflammatory and antispasmodic. The triterpenoid saponins, which are concentrated in the root (5-10%) are strongly expectorant.

Medicinal Actions & Uses: Cowslip is an under-used but valuable plant. The root is strongly expectorant, stimulating a more liquid mucus and so easing the clearance of phlegm. It is given for chronic coughs, especially those associated with chronic bronchitis and catarrhal congestion. The root is also thought to be mildly diuretic and antirheumatic and to slow blood clotting. The leaves have similar properties to the root, but are weaker in action. The flowres are believed to be sedative and are recommended for overactivity and sleeplessness, particularly in children. Cowslip flowers' antispasmodic and anti-inflammatory properties make them potentially useful in the treatment of asthma and other allergic conditions.

PRUNELLA VULGARIS
(Labiatae)

SELF-HEAL, XU KU CAO (CHINESE)

A creeping perennial, growing to 50 cm, with pointed oval leaves and violet-blue or pink flowers.

Key Constituents:
Pentacyclic triterpenes (based on ursolic, betulinic and oleanolic acids)
Tannins
Caffeic acid
Vitaminss B_1, C, K

Key Actions:
Heals wounds
Astringent
Stops internal bleeding
Gently lowers blood pressure

Prunella vulgaris

PRUNUS ARMENIACA
(Rosaceae)
APRICOT

Parts Used: Fruit, seeds, bark.
Constituents: Apricot fruit contains fruit sugars, vitamins and iron. The seeds contain up to 8% amygdalin, the cyanogenic glycoside that yields laetril and hydrocyanic (prussic) acid. The bark contains tannins.
Medicinal Actions & Uses: Apricot fruit is nutritious, cleansing and mildly laxative. A decoction of the astringent bark soothes inflamed and irritated skin conditions. Though the seeds contain highly toxic prussic acid, they are prescribed in small amounts in the Chinese tradition as a treatment for coughs, asthma and wheezing, and for excessive mucus

production and constipation. An extract from the seeds, laetril, has been used in Western medicine as a highly controversial treatment for cancer. The seeds also yield a fixed oil, similar to almond oil (from *P. amygdalus*), that is often used in the formulation of cosmetics.

PRUNUS AVIUM
(Rosaceae)
SWEET CHERRY

Parts Used: Stems, fruit.
Constituents: The stems of sweet cherry contain phenols, including salicylic acid, and tannins. Cherry fruit contains small quantities of salicylates and cyanogenic glycosides, and vitamins A, B_1 and C. The seeds also contain amygdalin, a cyanogenic glycoside.
Medicinal Actions & Uses: In Euporean herbal medicine, cherry stems have long been used for their diuretic and astringent properties. They have been prescribed for cystitis, nephritis, urinary retention and for arthritic problems, notably gout. Cherries can be helpful part of an overall regime treating arthritic problems. Their high sugar content makes them mildly laxative.
Caution: The seeds are toxic and should not be consumed.

PRUNUS DULCIS
Rosaceae (Almond)

From the medical viewpoint bitter almonds are more useful than sweet almonds. They contain up to 50 per cent fatty oil, proteins, enzymes, vitamins, sugar and the cyanogenic glycoside amygdalin, which on hy-

Prunus dulcis

drolysis produces the poisonous prussic acid. A medicinal water made from bitter almonds is used to treat cough, nausea, vomiting and retching. It was once used to flavour other natural substances or by synthesis extracts. More important is the extracted oil, from both bitter and sweet almonds. Sweet almond oil is used to prepare emulsions in which other remedies may be suspended, especially cough medicines, and it is used in massages. The oil from both almonds is widely used in cosmetic and toilet preparations and in the confectionary, perfumery and liqueur industries.

Sweet almonds are the familiar nuts used in sweet and savoury dishes.

PRUNUS MUME
(Rosaceae)

Part Used: Fruit.

Constituents: *Wu mei* contains fruit acids and sugars, vitamin C and plant sterols.

Medicinal Actions & Uses: The sour-tasting astringent *wu mei* is used in Chinese medicine to counter diarrhoea and dysentery, to stop bleeding and to ease coughs. It may also be effective in expelling hookworms. Externally, a plaster of the fruit is applied to the sites of removed corns and warts to hasten healing.

PRUNUS PADUS
Rosaceae (Bird Cherry)

The bark from young twigs is the medicinally active part. Its constituents include the cyanogenic glycosides laurocerasinn and isoamygdalin, an essential oil and tannins. These substances give Bird Cherry diuretic, sedative and anti-pyretic properties but it is used only rarely in folk medicine because of the danger of prussic acid poisoning. It has been used to

Prunus padus

treat rheumatic and arthritic pain and fever. **Bird Cherry should be used internally only under strict medical supervision.** It is also used in homeopathy.

PRUNUS SEROTINA
(Rosaceae)

WILD CHERRY

Part Used: Inner bark.

Constituents: Wild cherry contains prunasin (a cyanogenic glycoside that yields hydrocyanic acid), benzaldehyde, eudesmic acid, coumarins and tannins. Prunasin reduces the cough reflex.

Medicinal Actions & Uses: Figuring in official pharmacopoeias and much used in the Anglo-American tradition, wild cherry bark is an effective remedy for chronic dry and irritable coughs. It combines well with coltsfoot (*Tussilago farfara*) as a treatment for asthma and whooping cough. The astringent bark also helps treat indigestion and the symptoms of irritable bowel syndrome, especially when these conditions are of nervous origin.

PRUNUS SPINOSA
Rosaceae (Blackthorn)

The flowers and ripe fruits are used medicinally. The constituents of the flowers include traces of flavonoid and cyanogenic glycosides (including amygdalin, but this is lost on drying), sugars, tannins and vitamin C (in fresh material). These substances give the flowers diuretic, tonic and mild laxative properties. They are used in an infusion and, in homeopathy, in tincture form. The fruits contain tannins, organic acids, pectin, sugars and vi-

Prunus spinosa

tamin C. When dried they are astringent and are used to treat bladder, kidney and stomach disorders, including diarrhoea.

The fresh fruits can be eaten raw or in a compote, or made into juice, syrup, jam and wine.

PSALLIOTA CAMPESTRIS
(Agaricaceae)

TONIC, LAXT., APHRODIS.

Juice contains a thermostable substance which increases the strength of contraction of the heart of the toad and acts on the rate of the heart of the dog.

PSEUDARTHRIA VISCIDA
(Fabaceae)

Plant: used in biliousness, rheumatism, excessive heat and fever, diar., asthma, heart diseases, worms and piles.

PSIDIUM GUAJAVA
(Myrtaceae)

Fruit: laxt.

Leaves: used as astrin. for bowels and for wounds and ulcers; their decoct. used in cholera for arresting vomiting and diar.
Leaves contain essen. oil, eugenol.

PROSOPIS SPICIGERA
(Mimosaceae)

Pod: astrin.

Bark: used as remedy in rheumatism and scorpion-sting.

Flowers: pounded and mixed with sugar eaten by women during pregnancy as a safeguard against miscarriage.

Ashes: rubbed over the skin to remove hair.

PSORALEA CORYLIFOLIA
(Fabaceae)

Bu Gu Zhi

Parts Used: Seeds.

Constituents: *Bu gu zhi* psoraline, isopsorlin and bavachin.

Medicinal Actions & Uses: Valued as a *yang* tonic, *bu gu zhi* is taken in China to treat impotence and premature ejaculation and to improve vitality. The seeds are also used to counter debility and other problems reflecting "kidney *yang* deficiency" such as lower back pain, frequent urination, incontinence and bed-wetting. *Bu gu zhi* is used externally to treat skin conditions such as psoriasis, alopecia (loss of hair) and vitiligo (loss of skin pigmentation). In Vietnam, a tincture of the seeds is used in the treatment of rheumatism.

PTEROCARPUS MARSUPIUM
(Fabaceae)

Kino

Part Used: Sap.

Constituents: Kin contains tannins, flavonoids and marsupin.

Medicinal Actions & Uses: Kino is a strongly astringent herb that tightens the mucous membranes of the gastro-intestinal tract. Kino relieves chronic diarrhoea and the irritation caused by intestinal infection and colitis. Though its taste is unpleasant, this herb makes a good mouthwash and gargle. It is widely used in Asia as a douche for excessive vaginal discharge.

PTEROCARPUS SANTALINUS
(Fabaceae)

Wood: astrin., tonic, used as cooling external application for inflam. and headache, in bilious affections and skin diseases, in fever, boils, and to strengthen the sight, diaphor., in scorpion-sting.
Fresh shoots yield glucd., colouring matter (*J. chem.* Soc., 1912, 1061;)

PTEROSPERMUM ACERIFOLIUM
(Sterculiaceae)

Flowers: used as a general tonic.

Flowers and bark: charred and mixed with kamala applied in suppurating small-pox.

Down on the leaves: used as haemostatic.

PUERARIA LOBATA
(Fabaceae)

Ge Gen (Chinese), Kudzu

Parts Used: Root.
Constituents: *Ge gen* contains isoflavonoids, purerarin, daidzein and plant sterols. Daidzein is oestrogenic.
Medicinal Actions & Uses: In China, *ge gen* is frequently used as a remedy for measles, often in combination with *sheng ma (Cimicifuga foetida)*. *Ge gen* is also given for muscle aches and pains, especially when they are linked with fever or are affecting the neck are upper back. The root may be taken to treat symptoms of headache, dizziness or numbness caused by high blood pressure. *gen* also treats diarrhoea and dysentery. The root is prescribed with *ju hua (Chrysanthen × morifolium)* to treat alcohol intoxication, hangovers and alcoholism.

PUERARIA TUBEROSA
(Fabaceae)

Root: given as demulc. and refrig. in fevers; peeled and bruised into a cataplasm used to reduce swellings; crushed and rubbed on the body in fever and rheumatism; used as emetic, tonic and lactag.

PULMONARIA OFFICINALIS
(Boraginaceae)

LUNGWORT

Parts Used: Leaves.
Constituents: Lungwort contains allantoin, flavonoids, tannins, mucilage, saponin and vitamin C. Unlike many of its relatives in borage family, it does not contain pyrrolizidine alkaloids.
Medicinal Actions & Uses: Given its high mucilage content, lungwort is indeed useful remedy for chest conditions, and it is particularly beneficial in cases of chronic bronchitis. It combines well with herbs such as

Pulmonaria officinalis

coltsfoot (*Tussilago farfara*) as a treatment for chronic coughs (including whooping cough), and it can be taken for asthma. Lungwort can also be used as a treatment for sore throats and catarrh. In the past, lungwort was given for the coughing up of blood arising from tubercular infection. Lungwort leaves are astringent, and have been applied externally to stop bleeding.

PULSATILLA CHINENSIS
(Ranunculaceae)

CHINESE ANEMONE
BAI TOU WENG (CHINESE)

Part Used: Root.
Constituents: Chinese anemone contains lactones (including protoanemonin and anemonin), pulsatoside and anemonol. Protoanemonin is antibacterial and irritant. It is absent from the dried root.

Medicinal Actions & Uses: Chinese anemone is thought to clear toxicity and to lower fever. It is most commonly taken as a decoction to counter infection within the gastro-intestinal tract. The root is also used to treat malarial fever and vaginal infections.

PUNICA GRANATUM
(Lythraceae)

POMEGRANATE

Parts Used: Rind, bark, fruit pulp.

Constituents: Pomegranate fruit rind and bark contain pelletierene alkaloids, elligatannins (up to 25%) and triterpenoids. The alkaloids are highly toxic.

Medicinal Actions & Uses: Both the rind and bark of the pomegranate are considered to be specific remedies for tapeworm infestation. The alkaloids present in the rind and bark cause the worm to release its grip on the intestinal wall. If a decoction of pomegranate rind or bark is immediately followed by a dose of a strong laxative or purgative, the worm will be voided. The rind and bark are also strongly astringent and occasionally have been used to treat diarrhoea. In Spain, the juice of pomegranate fruit pulp is taken to comfort an upset stomach and as a remedy to relieve wind.

PYGEUM AFRICANUM
(Boraginaceae)

PYGEUM

Parts Used: Bark.

Constituents: Pygeum contains phytosterols (beta-sitosterol), triterpenes (ursolic and oleanolic acids), long chain alcohols (n-tetracosanol) and tannins.

Medicinal Actions & Uses: In conventional medicine in France, the fat-soluble extract of pygeum bark has become the primary treatment for an enlarged prostate gland. A decoction of the bark may reduce the severity of chronic prostate inflammation, and it may also help reverse male sterility when this is due to insufficient prostate secretions. In combination with other plants, pygeum may be valuable in the treatment of prostatic cancer.

QUERCUS PETRAEA
Fagaceae (Sessile oak)

Quercus petraea

The bark and leaves, sometimes the leaves and galls ('oak apples') are used medicinally. The acorns are separated from the cupule, dried, divested of the seed coat and roasted. Roasting converts the starch in them into dextrins and eliminates the bitter tannins. Acorns are then a suitable food in cases of severe diarrhoea and inflammation of the lymph nodes. Roasted acorns can be ground to make a coffee substitute and mixed with cocoa and sugar to make acorn cocoa. These beverages will check diarrhoea and act as a general tonic. With their astringent, antiseptic and anti-inflammatory properties fresh oak leaves will promote the healing of wounds. They contain a flavonoid glycoside (quercitrin) besides tannins. Galls (tumours caused by a gall wasp) are a raw material for the production of pure tannins, which have many uses in medical practice in powders, gargles, ointments and other preparations for checking bleeding.

Quercus robur

QUERCUS ROBUR
(Fagaceae)

COMMON OAK

Parts Used: Bark, galls (growths produced by insects or fungi).

Constituents: Common oak bark contains 15-20% tannins (including phlobatannin, ellagitannins and gallic acid). Oak galls contain about 50% tannins.

Medicinal Actions & Uses: Common oak bark, prepared as a decoction, is used as a gargle to treat sore throats and tonsillitis. It may also be applied as a wash, lotion or ointment to treat haemorrhoids, anal fissures, small burns and other skin problems. Less commonly, a decoction of the bark is taken in small doses to treat diarrhoea, dysentery and rectal bleeding. Powdered oak bark may be sniffed to treat nasal polyps or sprinkled on eczema to dry the affected area. Oak galls are very astringent. They are used, in small quantities, in place of bark.

QUILLAJA SAPONARIA
(Rosaceae)

SOAP BARK

Part Used: Inner bark.

Constituents Soap bark contains up to 10% triterpenoid saponins, calcium oxalate and tannins. The saponins are strongly expectorant and can cause inflammation of the digestive tract.

Medicinal Actions & Uses: Soap bark has a long tradition of used as a treatment for chest problems. Its strong expectorant effect is beneficial in the treatment of bronchitis, especially in the early stages of the illness. Like other medicinal plants that contain saponins, soap bark stimulates the production of a more fluid mucus in the airways, facilitating the clearing of phlegm through coughing. Soap bark is useful for treating any condition featuring congested catarrh within the chest, but it should not be used for dry irritable coughs. Soap bark is also used externally, appearing in the formulations of dandruff shampoos.

RADERMACHERA XYLOCARPA
(Bignoniaceae)
Oil from wood – used in cutaneous affections

RANDIA DENSIFLORA
(Rubiaceae)
Bark – very bitter and given in Indochina in the so-called forest fever.
Wood – presented in Cambodia in the treatment of paludism.

RANDIA DUMETORUM
(Rubiaceae)
Fruit – irritating emetic; used as fish-poison.
Pulp of fruit – used in dysen., anthelm., abortif.; ground to coarse powder applied to the tongue and palate for fevers and incidental ailments of children during teething.

Bark – astrin., given internally and also applied externally when bones ache during fever; externally applied as anodyne in rheumatism.
Aqueous extract of root bark actively insecticidal.
Fruits contain neutral and acid saponin, essen. oil, and acid resin; neutral saponin is the active constituent and lead in seeds.

RANDIA ULIGINOSA
(Rubiaceae)
Unripe fruit – roasted in wood ashes used as a remedy in dysen. and diar., the central portion consisting of the stone and seeds being rejected; astrin.
Root – boiled in ghee given in dysen. and diar.

RANUNCULUS FICARIA
(Lesser Celandine)
RANUNCULACEAE

The flowering stems are used medicinally. Their constituents include the toxic protoanemonin and anemonin, a saponin (ficarin), tannins and abundant vitamin C (in fresh material). These substances give Lesser Celandine astringent and anti-scorbutic properties. In the past it was used internally to treat scurvy, a disease caused by lack of vitamin C in the diet. Nowadays Lesser Celandine is mostly used only externally in ointments and bath preparations for treating haemorrhoids, warts and scab. It has an acrid burning taste and is a severe irritant, especially the fresh flowering plant. Grazing cattle avoid it. **The fresh plant should never be taken internally.** Even handling the

Ranunculus ficaria

Raphanus sativus

the plant may cause skin blistering. It is rendered non-toxic by drying and heat-processing. The first leaves and very young top parts are also not toxic. They have been eaten like spinach, but there are safer ways of adding vitamin C to the diet.

RANUNCULUS SCLERATUS (Ranunculaceae)

Plant – emmen, galact, poisonous
Leaves – Vesic, applied to the skin to raise blisters.
Plant contains anemonin and protoanemonin, deadly poisonous.

RAPHANUS SATIVUS
Brassicaceae (Garden Radish)

The black roots contain antiseptic thioglucosinolates, vitamins C and B complex and mineral salts. They are generally eaten raw and unpeeled and have antiseptic, tonic, carminative, choleretic and stomachic actions. They are used sliced or grated on bread and butter, or the pressed juice alone is taken for hepatitis and gall bladder disorders (but not for inflammation of the gall bladder), gallstones and digestive disorders. Radish is also used in homeopathy. The popular red, red and white or white radishes (*R. sativus var. radicula*) are less potent but are excellent and wholesome salad vegetables. The leaves are also edible.

RAUVOLFIA SERPENTINA (Apocynaceae)

INDIAN SNAKEROOT, SARPAGANDHA (HINDI)
Parts Used: Root.

Constituents: Indian snakeroot contains a complex mixture of indole alkaloids including reserpine, rescinnamine, ajmaline and yohimbine. Ajmaline has been used to regulate heart beat.

Medicinal Actions & Uses: Indian snakeroot is useful in the treatment of high blood pressure and anxiety. The root has a pronounced sedative and depressant effect on the sympathetic nervous system. By reducing the system's activity, the herb brings about the lowering of blood pressure. It may also be used to treat anxiety and insomnia, as well as more serious mental health problems such as psychosis. Indian snakeroot is a slow-acting remedy and it takes some time for its effect to become fully established.

REMIREA MARITIMA
(Cyperaceae)
Infusion of leaves: given as sudorific and diur. in Brazil and Guiana.

REMUSATIA VIVIPARA
(Araceae)
Root: made into an ointment with turmeric used as remedy for itch.
Juice: with cow's urine considered alexipharmic.

RESEDA ODORATA
(Resedaceae)
Root: acrid, used as laxat. diaphor, and diur. in Spain.
Seeds: applied externally as resolv.

RHAMNUS CATHARICUS
Rhamnacaee (Buckthorn)
The ripe fruits are used medicinally. Their constituents include anthraquinone glycosides, flavonols,

Rhamnus catharicus

pectin and vitamin C. Dried or fresh, the fruits are strongly laxative, also diuretic. An infusion of the macerated fruits or a syrup is used (for adults only) for cases of chronic constipation. On the Continent the fruits are still used on medical practice, but in Britain now only in herbal medicine. **Buckthorn fruits must always be taken with great care** because strong doses irritate the gastrointestinal mucosa to the point of bleeding and cause vomiting and severe diarrhoea. The dried bark (at least one-year old) has also been used as a laxative. **Fresh bark should never be used.**
The fruit and bark yield a yellow dye.

ROSA ALBA
(Rosaceae)
Flowers: used as a cooling medicine in fever and in palpitation of heart.

RHAMNUS FRAGULA
Rhamnaceae
(Alder Buckthorn)

The bark from young twigs is used medicinally. After drying it is either heat-treated or stored for at least a year before using to destroy the anthraquinone glycosides and derivatives in the fresh material. These substances cause severe vomiting, diarrhoea and abdominal cramps. Other constituents in the bark include tannins and bitter compounds. The bark or an extract from it is used as a strong laxative for chronic constipation usually in cases where other, milder agents have failed to produce the desired result. Alder Buckthorn bark is also a choleretic and very small doses are used to treat liver, gall bladder and spleen disorders. On the Continent it is still widely used in conventional medicine; in Britain it is now mostly used only in herbal medicine. **Alder Buckthorn should be used only under the supervision of a qualified medicinal or herbal practitioner;** strong doses are toxic. **Fresh bark and the fruits should never be used.**

RHAMNUS VIRGATA
(Rhamnacaee)
Fruit: bitter, emetic, purg., given in affections of the spleen.
Oxymethyl-anthraquinones, rhamnose.

RHAPHIDOPHORA LACINIATA
(Araceae)
Juice of the plant: used in snake-bite and scorpion sting.

RHEUM EMODI
(Polygonaceae)
Rhizome and roots: purg. astrin. tonic.
Rhizomes yield glucd. rhaponticin, chrysophanic acid and certain anthraquinone derivatives.

RHEUM PALMATUM
Polygonaceae
(MEDICINAL RHUBARB)

The rhizomes of 5 - to 7-year old plants are used medicinally. When dried they have a bitter taste. The constituents include two types of glycosides: tannin glycosides with free gallic acid, cinnamic acid and glucose; and anthraquinone glycosides based on the aglycones chrysophanol, emodin, aloe-emodin and rhein (up to 10 per cent). The rhizomes also contain starch and calcium oxalate. In small doses Rhubarb is astringent and is used to treat diarrhoea, and to stimulate the appetite. Stronger doses are laxative after 8 to

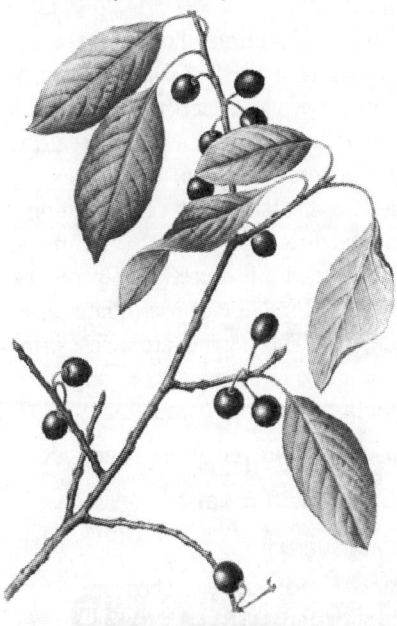

Rhamnus fragula

Rheum palmatum

diabetes.

RHODODENDRON CAMPANULATUM
(Ericaceae)

Leaves: mixed with tobacco and made into snuff used in colds and homicrenia; used in chronic rheumatism, sciatica and syphilis.
Dried twigs and woods: used in phthisis and chronic fevers in Nepal.

RHUS CORIARIA
(Anacardiaceae)

Astrin., styptic, tonic, diur. used in dysent., haemoptysis, conjunctivitis.

RHUS GLABRA
(Anacardiaceae)

Smooth Sumach

Parts Used: Root bark, berries.
Constituents: Smooth sumach contains tannins. Its other constituents are unknown.
Medicinal Actions & Uses: The astringent root bark of smooth sumach is often used as a decoction. It is taken to alleviate diarrhoea and dysentery, applied externally to treat excessive vaginal discharge and skin eruptions, and used as a gargle for sore throats. The berries are diuretic, help reduce fever, and may be of use in late-onset diabetes. The berries are also astringent and can be used as a gargle for mouth and throat complaints.

RIBES NIGRUM
(Grossulariaceae)

Blackcurrant

Parts Used: Leaves, berries.
Constituents: Blackcurrant leaves

10 hours and are used to treat chronic constipation. Rhubarb is included in some proprietary preparations and is also a component of herbal tea mixtures and digestive powders. **Medicinal Rhubarb must not be used by individuals with urinary disorders, uroliths, kidney stones, arthritis and rheumatism, or by children or nursing mothers.**

RHINACANTHUS COMMUNIS
(Acanthaceae)

Root, leaves and seeds: useful remedy for ringworm and other skin diseases.
Roots: boiled in milk used as aphrodis. antid. to snakebite.

RHIZOPHORA MUCRONATA
(Rhizophoraceae)

Bark: astrin. used as cure for

Ribes nigrum

contain a volatile oil, tannins and vitamin C; the berries contain anthocyanosides (about 0.3%) flavonoids, pectin, tannins, vitamin C and potassium.

Medicinal Actions & Uses: In Europe, blackcurrant leaves are used for their diuretic effect. By encouraging the elimination of fluid, the leaves help to reduce blood volume and thereby to lower blood pressure. The leaves are also used as a gargle for sore throats and mouth ulcers. According to French investigators, blackcurrant leaves increase the secretion of cortisol by the adrenal glands, and thus stimulate the activity of the sympathetic nervous system. This action may prove useful in the treatment of stress-related conditions. Blackcurrant berries and

their juice are high in vitamin C. They help improve resistance to infection and make a valuable remedy for treating colds and flu. According to the herbal authority R.F. Weiss, the juice is "as good as, if not better than, lemon juice (*Citrus limon*) for patients with pneumonia, influenza, etc." The juice also helps to stem diarrhoea, and calms indigestion. Juice that is fresh or vacuum-sealed is more effective than concentrate.

RICINUS COMMUNIS
(Euphorbiaceae)

CASTOR OIL, PLANT

Parts Used: Seed oil, seeds.
Constituents: The seeds contain 45-55% fixed oil, which consists mainly of glyceride of ricinoleic acid, ricin (a highly toxic protein), ricinine (an alkaloid) and lectins. The seeds are

Ricinus communis

highly poisonous — 2 are sufficient to kill an adult — but the toxins do not pass into the expressed oil.

Medicinal Actions & Uses: Castor oil is well known for its strongly laxative (and in higher doses, purgative) action, prompting a bowel movement about 3-5 hours after indigestion. The oil is so effective that it is regularly used to clear the digestive tract in cases of poisoning. Castor oil is well tolerated by the skin, and it is sometimes used as a vehicle for medicinal and cosmetic preparations. In India, the oil is massaged into the breasts after childbirth to stimulate milk flow. Indian herbalism uses a poultice of castor oil seeds to relieve swollen and tender joints. In China, the crushed seeds are used to treat facial palsy.

Robinia pseudoacacia

ROBINIA PSEUDOACACIA
Fabaceae
(FALSE ACACIA)

The flowers and bark have medicinal action but they are rarely used therapeutically. The flowers contain flavonoid glycosides and an essential oil. They relax muscular and nervous spasms. The bark contains the toxic albumins robin and fazin, which are emetic and purgative. There have been cases of poisoning from the seeds and bark. Symptoms of poisoning include drying of the throat, vomiting, diarrhoea, dilated pupils, weak and irregular pulse and breathing difficulties. **False Acacia should never be collected and used for self-medication.**

ROSA CANINA
Rosaceae
(Dogrose)

The hips are the medicinally active parts. Their constituents include vitamin C (upto 1 per cent), carotenes, vitamin B complex, sugars, pectin, tannins and malic and citric acids. The fruits also contain fatty oil. The best-known and widely used medicine is rosehip tea, which has tonic, astringent, mild diuretic and mild laxative effects. When fresh the hips are an excellent source of vitamin C. Fresh or dried, they are beneficial for convalescents, against exhaustion and colds. The tea is best made by macerating the crushed hips (without the hard acnes), not by lengthy boiling. A decoction from the hips can be used

Rosa canina

Rosa centifolia

as a gargle for bleeding gums and it will relieve toothache.

Fresh hips can also be made into jam, syrup and medicinal wine.

ROSA CENTIFOLIA
Rosaceae

(CABBAGE ROSE)

Rose oil is obtained by steam distillation or solvent extraction from the petals. Its main constituent is geraniol. Besides essential oil, rose petals also contain tannins, glycosides and pigments. All these substances give the petals astringent and anti-inflammatory properties and they are used in infusions to treat diarrhoea and externally to bath wounds. Rose oil is the basis of many perfumes and it is also used to give a pleasant taste and smell

to medicines, chiefly ointments, lotions and creams. The familiar rose water, which can be used as a face tonic, is prepared by water distillation from the petals or by impregnation of distilled water with rose oil.

ROSA DAMASCENA
(Rosaceae)

Petals: applied externally as astrin. made into a conserve with equal parts of white sugar, known as gulland, used as tonic and fattening.

Buds: astrin. considered aper., cardiacal, tonic cephalic, removing bile and cold humours.

Essential oil.

ROSA GALLICA
(Rosaceae)

Parts Used: Flowers, essential oil.
Constituents: Rose contains a vola-

tile oil consisting of geraniol, nerol, citronellol, geranic acid and other terpenes, and many other substances. **Medicinal Actions & Uses:** The rose is currently little used in herbal medicine, but it is probably time for a re-evaluation of its medicinal benefits. The essential oil, called "attar of rose", is used in aromatherapy as a mildly sedative, antidepressant and anti-inflammatory remedy. Rose petals and their preparations have a similar action. They also reduce high cholesterol levels. Rosewater is mildly astringent and makes a valuable lotion for inflamed and sore eyes.

ROSMARINUS OFFICINALIS
Lamiaceae

(ROSEMARY)

The leaves and young shoots are used medicinally. The constituents include an essential oil (up to 2 per cent) with cineole, camphor and borneol as the main components, plus tannins, saponin and organic acids. These substances give Rosemary a pronounced rubefacient action and the dried herb and rosemary oil, obtained by steam distillation from the fresh parts, are components of anti-rheumatic liniments and ointments. Rosemary also has sedative, diuretic, stomachic, cholagogic, tonic, aromatic, antispasmodic and antiseptic properties. It is especially beneficial for fatigue and neuralgia. **Very strong doses are poisonous. Rosemary oil should never be taken internally.**

Fresh or dried Rosemary leaves have many culinary uses and they can also be used in soothing herbal bath preparations. Infusions and oil extracts are excellent hair tonics. The distilled

Rosmarinus officinalis

oil is used in toilet preparations, disinfectants and extensively in perfumery.

ROYLEA ELEGANS
(Labiatae)

Infusion of leaves: bitter, tonic, febge, drunk for contusions produced by blows.

RUBIA TINCTORUM
Rubiaceae

(MADDER)

The rootstock of two-to three-year old plants is used medicinally. It remains red when dried. The constituents include anthraquinone glycosides and derivatives such as alizarin (madder red) and purpurin (madder purple), organic acids, tannins, pectins and sugars. The medicinal action is mainly

Rubia tinctorum

Rubus fruticosus

due to the glycosides, which are isolated and included in proprietary preparations used on the Continent. Madder has diuretic, antispasmodic, antiseptic, choleretic, emmenagogic and laxative properties. It is used in powder form or in an infusion for kidney and bladder disorders, and it has a marked ability to disintegrate kidney stones and uroliths. In homeopathy a tincture of the fresh roots is used for amenorrhoea, spleen disorders and various other complaints. When taken internally the colouring matter in the rootstock gives urine, mucus, sweat and milk a pink tinge.

RUBUS FRUTICOSUS
Rosaceae
(BLACK BERRY)

The leaves contain tannins, organic acids, sugars and vitamin C and have mild astringent, antiseptic, antifungal, diuretic and tonic properties. An infusion is used in herbalism to treat stomach disorders, enteritis and diarrhoea. The fragment tea is also prescribed for flue, colds and coughs and it is a pleasant substitute for ordinary tea. Externally a decoction is used as a gargle for sore throats and pharyngitis, as a component of mouthwashes and is bath preparations for treating skin rashes, wounds, ulcers and fungal infections. The fully ripe fruits contain vitamin C, organic acids, pectins and sugars and can be eaten raw or cooked and made into jam, syrup and medicinal cordial and wine.

RUBUS IDAEUS
Rosaceae
(RASPBERRY)

The leaves contain tannins, pectin,

Rubus idaeus

vitamin C, organic acids and fragarin and other substances that have oxytocic properties (they act on the uterine muscles). Raspberry leaves also have astringent, diuretic, expectorant, stomachic cholagogic and tonic properties. They are used in infusions on their own or in herbal mixtures to treat diarrhoea and stomach disorders, and for chest complaints. Externally an infusion can be used as a mouthwash and gargle. Fresh or fermented, the leaves make a pleasant substitute for ordinary tea. The fresh fruits contain organic acids, sugars, pectin and vitamin C and with the addition of sugar and wine vinegar and diluted with water they make a gargle for fevers and sore throats and a base for summer drinks.

RUMEX ACETOSELLA
(Polygonaceae)

SHEEP'S SORREL

Parts Used: Aerial parts.

Constituents: Sheep's sorrel contains oxalates and anthraquinones (including chrysophanol, emodin and physcion). In isolation, the anthraquinones are irritant and has a laxative effect.

Medicinal Actions & Uses: Sheep's sorrel is a detoxifying herb, the fresh juice having a pronounced diuretic effect. Like other members of the dock family, it is mildly laxative, and holds out potential as a long-term treatment for chronic disease, in particular that of the gastro-intestinal tract.

RUMEX CRISPUS
(Polygonaceae)

Root: mildly laxat. and astrin.
Emodin, chrysophanic acid, essential oil.

RUMEX NEPALENSIS
(Polygonaceae)

Roots: purg. subst. for rhubarb.
Chrysophanic acid.

RUMEX VESICARIUS
(Polygonaceae)

Leaves: cooling. aper. diur. astrin. used in snake-bite.
Seeds: cooling, prescribed roasted in dysent. and scorpion sting.
Juice: cooling, useful in heat of the stomach and to allay the pain of toothache, and by its astrin. properties to check nausea.

RUNGIA REPENS
(Acanthaceae)

Plant: dried and pulverised given in fevers and coughs and considered vermifuge.
Fresh leaves: bruished and mixed

with castor oil applied to scalp in cases of tinea capitis.

RUSCUS ACULEATUS
(Liliaceae)
BUTCHER'S BROOM

Parts Used: Aerial parts, rhizome.
Constituents: Butcher's broom contains saponin glycosides, including ruscogenin and neoruscogenin. These constituents have a structure similar to that of diosgenin, found in wild yan (*Dioscorea villosa*). They are anti-inflammatory and cause the contraction of blood vessels, especially veins.
Medicinal Actions & Uses: Butcher's brooms is not widely used today, but, in view of its positive effect on varicose veins and haemorrhoids, it could be due for a revival. In the European tradition, both the aerial parts and the rhizome are considered to be diuretic and mildly laxative.

RUTA GRAVEOLENS
(Rutaceae)
RUE

Parts Used: Aerial parts.
Constituents: Rue contains about 0.5% volatile oil (including 50-90% 2-undecanone), flavonoids (including rutin), furanocoumarins (including bergapten), about 1.4% furoquinoline alkaloids (including fagarine, arborinine, skimmianine, and others). Rutin has the effect of supporting and strengthening the inner lining of blood vessels and reducing blood pressure.
Medicinal Actions & Uses: Rue is chiefly used to encourage the onset of menstruation. It stimulates the muscles of the uterus and promotes menstrual blood flow. In European

Ruta graveolens

herbal medicine, rue has also been taken to treat conditions as varied as hysteria, epilepsy, vertigo, colic, intestinal worms, poisoning and eye problems. The latter use is well founded, as an infusion used as an eyewash brings quick relief to strained and tired eyes, and reputedly improves the eyesight. Rue has been used to treat many other conditions, including multiple sclerosis and Bell's palsy.

S

SACCHARUM ARUNDINACEUM
(Poaceae)

Root: demulc., diur.

SACCHARUM MUNJA
(Poaceae)

Stem: refrig. aphrodis. useful in burning sensations, blood troubles, erysipelas, thirst and urinary complaints.
Root: burnt near women after delivery, and burns and scalas, its smoke being considered beneficial.

SACCHARUM OFFICINARUM
(Poaceae)

Stems: sweet, laxat. diur., cooling, aphrodis.
Calcium oxalate

SACCHARUM SPONTANEUM

Plant: laxat., aphrodis, useful in burning sensations, strangury, phthisis, vesical calculi, diseases of blood, biliousness, haemorrhagic diathisis.

SAGERARIA LAURIFOLIA
(Annonaceae)

Leaves: bitter, astrin., and pungent, used for fermentation.

SALACIA OBLONGA
(Celastraceae)

Root bark: used in gonor. rheumatism, and skin diseases.

SALICORNIA BRACHIATA
(Chenopodiaceae)

Ashes: used for mange and itch and considered emmen. and abortif.

SALIX ALBA
Salicaceae (White Willow)

The bark from two to three year old twigs is used medicinally. After drying it has a bitter taste. The constituents include the phenolic glycoside salicin and tannins (up to 14 per cent).

Salix alba

Because of the salicylic compounds it contain willow bark of White Willow and other willow species was once used in medicinal practice but nowadays it has been replaced by synthetic preparations, such as aspirin (acetosalicylic acid); however, it still has a place in herbal medicine for fevers and neuralgia. Externally a decoction is used in bath preparations for rheumatic and arthritic pains and in ointments and compresses for cuts, skin ulcers and burns.

SALIX BABYLONICA
(Salicaceae)

Leaves and bark: tonic, astring. used in intermittent ad remittent fevers.
Bark: anthelm.
Leaves contain enzyme salicinase.

SALIX CAPREA
(Salicaceae)

Decoct. of leaves: given in fevers.
Distilled water from the flowers: cordiac, stim., aphrodis., externally applied in headache and ophthalmia.
Ashes of the wood: useful in haemoptysis.
Stem and leaves: astrin.
Gum and juice: used to increase visual powers.
Fresh bark contains glucd. salicine, enzyme salicinase and two more glucosidases.

SALMALIA MALABARICA
(Bombacaceae)

Root: stim., tonic, form the chief ingredient in the *musla-semul*, a medicine which is aphrodis; given in impotence.
Root and bark: emetic.
Gum: aphrodis., demulc., haemostatic, astrin., tonic, alter., used in diar., dysen. and menor.
Flowers and fruits: used in snakebite.
Gum contains catechutannic acid, seeds contain 22.3% crude fat with 0.5% stearin.

SALVADORA OLEOIDES
(Salvadoraceae)

Root bark: vesic.
Leaves: used as purg. and as a cure for cough.
Fruit: aphrodis.
Oil from seeds: used a stimulating application in painful rheum. affections and after child-birth.
Leaves and root contain alk., trime-thylamine; seeds contain fatty oil and ethereal oil.

SALVADORA PERSICA
(Salvadoraceae)

Leaves: used as external application in rheumatism; their juice given in scurvy.
Shoots and leaves: pungent, used as antid. to poisons of all sorts.
Fruit: carmin., diur., deobstruent.
Stem bark: used in decoct. in low fever and as a stim. and tonic in amenor.
Root bark: acrid.
Alk. trimethylamine.

SALVIA OFFICINALIS
Lamiaceae

(SAGE)

The leaves are used medicinally. Their constituents include an essential oil (upto 2.5 per cent) with thujone (15-35 per cent), borneol, cineole and camphor, also bitter compounds (salvin and picrosalvin), oestrogenic substances, resin and tannins. These substances give Sage antiseptic, antifungal, astringent, diuretic, carminative, antidiarrhoeal, antispasmodic and antidiaphoretic properties. It has a wide variety of medicinal uses. In herbal medicine, for example, it is used in an infusion to reduce sweating and lactation, and to treat colds and coughs, nervous conditions and gastrointestinal disorders. A tincture prepared from the fresh leaves is also used in homeopathy. **Sage should not be taken in large doses for a long period because of the thujone it contains.** The essential oil, obtained by steam distillation of the partially

Salvia officinalis

dried leaves, is used by the pharmaceutical, perfumery, liqueur and food
industries.

In cooking Sage is usually used with
pork, but it is also good with other
meat dishes and in salads and spreads.

SALVIA SCLAREA
Lamiaceae

(CLARY)

The flowers, flowering stems and
leaves are used medicinally. When
dried all parts are strong-smelling and
have a bitter taste. Their constituents
include linalool and linalyl acetate as
the main components. and bitter compounds. These substances give Clary
tonic, astringent, antidiaphoretic, carminative, antispasmodic and
emmenagogic properties. In herbal
medicine an infusion is used to treat
digestive upsets, cramps flatulence,
diarrhoea, as a general tonic, to inhibit perspiration and in the treatment
of menstrual disorders. The decoction
and Clary vinegar can be applied ex-

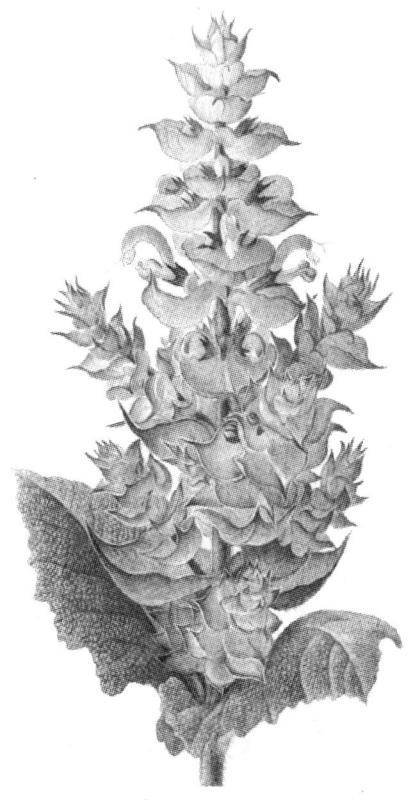

Salvia sclarea

ternally in compresses or bath preparations to wounds and ulcerous conditions. The essential oil, which is
known as sage clary or Muscatel oil,
is obtained by steam distillation from
the fresh or partially dried flowering
stems and leaves. It is used in herbal
medicine, but more widely in toilet
waters, perfumes and soap and to flavour wine vermouths and liqueurs.

SAMADERA INDICA
(Simarubaceae)

Bark: used in fever.
Leaves: bruished and externally applied in erysipelas; their infusion used
as an insecticide.

Infusion of wood: taken as a tonic.
Oil from kernels: used as an application in rheumatism.
Glucd. samaderin, bitter substance in all parts.

SAMBUCUS EDULIS
Caprifoliaceae
(DWARF ELDER)

The rhizomes, sometimes the flowers and fruits, are used medicinally. Their constituents include a glycoside, a bitter compound, anthocyanins, an essential oil, tannins and a saponin. These substances give Dwarf Elder diuretic, marked diaphoretic and strong laxative properties. It is occasionally prescribed for dropsy, kidney disorders and rheumatism in herbal medicine and in homeopathy. **It should be taken internally only under strict medical supervision**; large doses cause vertigo, vomiting

and diarrhoea. **The berries should never be eaten whole or used for self-medication.** The alcohol extract from macerated rhizomes is said to promote hair growth and it is sometimes used to treat dandruff.

SAMBUCUS NIGRA
Caprifoliaceae
(ELDER)

Nowadays primarily the flowers and fruits are used medicinally. The constituents of the flowers include the flavonoid glycosides rutin and quercitrin, the cyanogenic glycoside sambunigrin, essential oil, mucilage, tannins and organic acids. These substances are principally diaphoretic and an infusion is used to treat colds and other respiratory infections and mild nervous disorders. Cosmetically the

Sambucus edulis

Sambucus nigra

flowers are very beneficial and the wine made from them is a safe hypnotic. The fruits contain organic pigments (anthocyanins), amino acids, sugar, rutin and a large amount of vitamin C (when fresh). They are mildly laxative, also diuretic, diaphoretic and mildly sedative. They are used in tea mixtures for colds and weight loss and are also beneficial for nervous disorders such as insomnia and migraine. **The leaves are purgative and should not be used.**

SAMBUCUS RACEMOSA
Caprifoliaceae

(RED-BERRIED ELDER)

The fully ripe fruits are used medicinally. They are bitter tasting and are used fresh or dried. Their constitu-

Sambucus racemosa

ents include vitamin C, essential oil, sugar, pectins, organic pigments (anthocyanins), organic acids and trace amounts of glycosides. These substances give Red-berried Elder diaphoretic, nutritive, and antiseptic properties, which are more pronounced in the fresh fruits. The fruits have been used for constipation and gastrointestinal infections, but **they cause vomiting and diarrhoea in large amounts** and it is best not to use them for medical purposes or in cooking.

SANSEVIERIA ROXBURGHIANA
(Haemodoraceae)

Root: prescribed in the form of an electuary in consumptive complaints and coughs of long standing.

Juice of tender shoots: administered to children to clear their throats of viscid phlegm.

Plant: alk. sanservierine, Aconitic acid, alk and resin.

SANGUINARIA CANADENSIS
(Papaveraceae)

BLOOD ROOT

Part Used: Rhizome.

Constituents: Blood root contains isoquinoline alkaloids, notably sanguinarine (1%), and many others, including berberine. Sanguinarine is a strongly expectorant substance that also has antiseptic and local anesthetic properties.

Medicinal Actions & Uses: In contemporary herbal medicine, blood root in chiefly employed as an expectorant, promoting coughing and the clearing of mucus from the respiratory tract. The plant is prescribed for chronic bronchitis and — as it also has an antispasmodic effect — for asthma and whooping cough. Blood root may also be used as a gargle for sore throats, and as a wash or ointment for fungal and viral skin conditions such as athlete's foot and warts. Prepared as a powder, blood root may be sniffed to treat nasal polyps.

SANGUISORBA OFFICINALIS
Rosaceae

(GREAT BURNET)

The rhizomes and sometimes the young shoots and leaves (gathered before flowering)are used medicinally. Their constituents include tannins, saponins, a glycoside (sanguisorbin), flavones and vitamin

C (in fresh material). These substances give Great Burnet astringent, haemostatic, mild antiseptic, diaphoretic and anti-inflammatory properties. In herbal medicine a decoction from the rhizomes is used for gastritis, enteritis and diarrhoea, for bleeding gums, nosebleed, strong menstrual flow and for difficulty in urinating. The young shoots and leaves are used in compresses and bath preparations to treat open wounds, rashes and ulcers and in gargles for gum infections and tonsillitis. **Great Burnet should never be taken in large doses.**
Salad Burnet has less potent medicinal actions. Its young leaves and shoots can be added to salads and soups or cooked as a vegetable. They also have several cosmetic applications.

SANICULA EUROPAEA
Apiaceae

(SANICLE)

The flowering stems, together with the basal leaves and the rhizomes are used medicinally. The constituents include tannins, saponins (primarily in the rhizomes), essential oil, mucilage and mineral salts. These substances give Sanicle astringent, haemostatic, expectorant, antispasmodic, carminative and anti-inflammatory properties. An infusion of the stems or the powdered rhizome is used for flatulence, gastritis, enteritis, internal ulcers, internal haemorrhage, urinary infections, liver disorders and for coughs and bronchitis. Externally it is used in gargles, bath preparations or com-

Sanguisorba officinalis

Sanicula europaea

presses for throat infections, inflamed gums, slow-healing wounds, skin rashes and ulcerous conditions.

SANTALUM ALBUM
(Santalaceae)

SANDALWOOD, CHANDAN

Parts Used: Wood, essential oil.

Constituents: Sandalwood contains 3-6% volatile oil (which consists predominantly of the sesquiterpenols alpha- and beta - santalol), resis and tannins.

Medicinal Actions & Uses: Sandalwood and its essential oil are used for their antiseptic properties in treating genito-urinary conditions such as cystitis and gonorrhoea. In Ayurvedic medicine, a paste of the wood is used to soothe rashes and itchy skin. In China, sandalwood is held to be useful for chest and abdominal pain.

SAPINDUS TRIFOLIATUS
(Sapindaceae)

Fruit: tonic, alexipharmac, given internally as expect., emetic, purg. and nauseant; as an errhine used in epilepsy, asthma, hysteria and hemi-crania; externally it is detergent; used as fish poison.
Saponin

SAPONARIA OFFICINALIS
Caryophyllaceae

(SOAPWORT)

The rhizomes of two to three year old plants are used medicinally. Their constituents include triterpenoid saponins (up to 5 per cent), flavonoids and sugars (up to 30 per cent). These substances give Soapwort expectorant, diuretic, diaphoretic, choleretic, cholagogic and laxative properties. In

Saponaria officinalis

herbalismm a decoction is used for coughs and bronchitis and, externally, for various skin conditions. **Soapwort is dangerous if taken in strong doses or over a long period of time** because the saponins can cause haemolysis (breakdown of the red blood cells), which results in severe irritation of the gastrointestinal tract. **It should thus be taken internally only under professional supervision**. The flowering stems have a similar but milder action.

SARACA INDICA
(Fabaceae)

Bark: astrin. used in uterine affections and in menor. in scorpion sting. Tannin and catechol.

SARCOSTEMMA ACIDUM
(Asclepiadaceae)

SOMALATA (SANSKRIT)

Dried stem: emetic
Plant: bitter, cooling, astrin. alter.

SARCOSTIGMA KLEINII
(Icacinaceae)

Oil: used in the treatment of rheumatism.

SARGASSUM FUSIFORME
(Sargassaceae)

HAI ZAO

Part Used: Whole plant.
Constituents: Hai zao contains alginic acid, mannitol, potassium and iodine.
Medicinal Actions & Uses: Hai zao is used in a similar way to its European counterpart, kelp (*Fucus vesiculosus*). In chinese medicine, it is given practically to treat thyroid problems caused by low iodine levels within the body. The herb also helps to combat other thyroid conditions that produce enlargement of the gland, for example Hashimoto's thyroditis. *hai zao* is prescribed to treat cases of scrofula (enlargement of the lymph glands in the neck due to tubercular infection) and oedema (fluid retention).

SAROTHAMNUS SCOPARIUS
(Leguminosae)

BROOM

Parts Used: Flowering tops.
Constituents: Broom contains quinolizidine alkaloids (particularly sparteine and lupanine), phenethylamines (including tyramine), isoflavones (such as genistein), flavonoids, a volatile oil, caffeic and p-coumaric acids, tannins and pigments. Sparteine reduces the heart rate and the isoflavones are oestrogenic.
Medicinal Actions & Uses: Broom is used mainly as a remedy for an irregular, fast heart beat. The plant acts on the electrical conductivity of the heart, slowing and regulating the transmission of the impulses. Broom is also strongly diuretic, stimulating urine production and thus countering fluid retention. Since broom causes the muscles of the uterus to contract, it has been used to prevent blood loss after childbirth.
Cautions: Take broom internally only under professional supervision. Do not take during pregnancy, or if suffering from high blood pressure. The plant is subject to legal restrictions in some countries.

SATUREJA HORTENSIS
Lamiaceae
(SUMMER SAVORY)

The nonwoody flowering tops are used medicinally. Their constituents include an essential oil with carvacrol and cymene as the main components, also tannins, mucilage and resin. These substances give Summer Savory astringent, antiseptic, expectorant, antispasmodic, stomachic, carminative, tonic and anthelmintic properties. It is used mainly for stomach and intestinal disorders, flatulence, to stimulate the appetite, to check diarrhoea and to destroy intestinal parasites. It also makes an effective gargle.

The distilled essential oil is used commercially as a flavouring. The fresh

Satureja hortensis

or dried can be used on their own or in herbal mixtures to season meats, fish, poultry, eggs and, especially, all kinds of beans. The herb should always be used sparingly. Winter Savory has a stronger and less-pleasant flavour and is best used only in herbal mixtures. Otherwise it has the same medicinal properties and use as Summer Savory.

SAUSSUREA LAPPA
(Compositae)

KUTH

Parts Used: Root, essential oil.

Constituents: Kuth contains a volatile oil (consisting of terpenes, sesquiterpenes and aplotaxene), an alkaloid (saussarine) and a resin. Saussarine depresses the para-sympathetic nervous system.

Medicinal Actions & Uses: Kuth is used in the Ayurvedic and Unani Tibb traditions in India for its tonic, stimulant and antiseptic properties. The root is commonly taken, in combination with other herbs, for respiratory system problems such as bronchitis, asthma and coughs. It is also used to treat cholera.

SAXIFRAGA ROTUNDIFOLIA
Saxifragaceae

(ROUND-LEAVED SAXIFRAGE)

The flowering stems are used medicinally. Their constituents include tannins, bitter compounds, resin and glycosides, which give Round-leaved Saxifrage primarily a diuretic action. It is used in herbal medicine in an infusion to help disintegrate kidney stones and uroliths. Mostly, however,

Saxifraga rotundifolia

it is a component of diuretic herbal tea mixtures with Peppermint (*Mentha × piperita*), lesser celandine (*Ranunculus ficaria*). Smooth Rupturewort (*Herniaria glabra*) and Milk Thistle (*Silybum marianum*).

SCHIZONEPETA TENUIFOLIA (Labiatae)
JING JIE

Parts Used: Aerial parts.

Constituents: *Jing jie* contains a volatile oil, the main constituents of which are menthone and limonene.

Medicinal Actions & Uses: In the Chinese tradition, *jing jie* is valued as an aromatic and warming herb. It is taken to alleviate skin conditions such as boils and itchiness. *Jing jie* also induces sweating and is used to treat fever and chills, and as a remedy for measles. It is often combined with *bo he* (*Mentha haplocalyx*).

SCHLEICHERA OLEOSA (Sapindaceae)
Bark: astrin., rubbed up with oil used as a cure for itch.

Powdered seeds: applied to ulcers of animals and for removing maggots.

Oil of the seeds: used for the cure of itch and acne; efficient and stimulating agent for the scalp, both cleansing it and promoting growth of hair. Seeds contain cyanogenetic glucd.

SCHEINFURTHIA SPHAEROCARPA (Scrophulariaceae)
Fruit and powdered leaves together with portions of the stem — prescribed in typhoid conditions.

Powdered plant: snuffed up for bleeding at the nose.

Drug contains alk. (Dymock, Warden).

SCOLOPENDRIUM VULGARE (Polypodiacea)
HARTSTONGUE

Parts Used: Fronds.

Constituents: Hartstongue contains tannins, mucilage and flavonoids (including leucodelphidin).

Medicinal Actions & Uses: Hartstongue was valued in the past for its ability to heal wounds, but today it is employed chiefly as a mild astringent. It is sometimes used in the treatment of diarrhoea and mucous colitis, and it may be of benefit to the liver and spleen. Hartstongue appears to have expectorant properties, and it is also mildly diuretic.

SCOPOLIA ANOMALA (Solanaceae)
Tincture of leaves — made in the pro-

portion of one ounce to eight ounces of alcohol, administered produces dilation of the pupil.

Hyoscyamine, hyoscine.

SCROPHULARIA NODOSA
(Scrophulariaceae)
FIGWORT

Parts Used: Aerial parts.

Constituents: Figwort contains iridoids (including aucubin, harpagoside, and acetyl harpagide), flavonoids, cardioactive glycosides and phenolic acids. Harpagoside and harpagide are though to account for its anti-arthritic activity.

Medicinal Actions & Uses: Figwort is a herb that supports detoxification of the body and may be used as a treatment for various types of skin conditions. Taken internally as an infusion or applied externally, figwort is of value in treating chronic skin diseases such as eczema and psoriasis. Applied externally, it will also help speed the healing of burns, wounds, haemorrhoids and ulcers. The traditional use of figwort as a treatment for swellings and tumours continues in Europe to this day. The herb also mildly diuretic, and it is reputed to be effective when used to expel worms.

SCUTELLARIA ALERICULATA
(Labiatae)

Decoct. of plant: used in epilepsy, intermittent fever, and agne.

Glucd. sentellarin.

SEDUM ACRE
Crassulaceae
(BITING STONECROP)

The flowering stems and leaves are used medicinally. Their constituents include several alkaloids (for example, sedamine and sedinine), the flavonoid glycoside rutin, mucilage,

Scrophularia nodosa

Sedum acre

tannins, sugar, vitamin C and calcium carbonate. These substances give Biting Stonecrop rubefacient, astringent, irritant, antisclerotic and hypotensive properties. **It should never be taken internally** because it causes headache, listlessness and vomiting. It is a component of liniments and ointments for external use. The fresh macerated aerial parts are also used in herbal medicine to treat skin disorders. They must be applied for only a short while because **they can cause blistering. The fresh plant juice may cause permanent eye damage.** In homeopathy a tincture of the fresh plant is used to treat haemorrhoids. **Biting Stonecrop should never be collected and used for self-medication.**

SELENICEREUS GRANDIFLORUS
(Cactaceae)
NIGHT-BLOOMING CEREUS

Parts Used: Flowers, young stems.
Constituents: Night-blooming cereus contains alkaloids (including cactine), flavonoids (isoharmnetin) and a pigment. Cactine's cardiotonic effect is considered similar to that of cardiac glycosides (*see* foxglove, *Digitalis* species).
Medicinal Actions & Uses: As it is in short supply, night-blooming cereus is little used at present, but it is a valuable remedy for the heart. It stimulates the action of the heart, increasing the strength of contractions while slowing heart rate. It is prescribed as a treatment for various conditions, including angina and low blood pressure, and is often given as a tonic during recovery from a heart attack.

In the Caribbean, the juice of the whole plant is used to expel worms, and the stems and flowers are used in the treatment of rheumatism.

SEMECARPUS ANACARDIUM (Anacardiaceae)

Nut: bruised and applied to osuteri to procure abortion; given as a vermifuge.
Oil from the nuts: vesic., used externally in rheumatism and leprous nodules.
Gum from bark: used in scrofulous, venereal and leprous affections and nervous debility.
Ashes of plant: in combination with other drugs used in snake-bite and scorpion-sting.
Anacardic acid, cardol, catechol, anacardol and fixed oil, semecarpol, bhilawanol.

SEMPERVIVUM TECTORUM
Crassulaceae
(HOUSELEEK)

The leaves are used medicinally. Their constituents include tannins, bitter compounds, sugars and mucilage. These substances give Houseleek astringent and vulnerary properties. A decoction has been used in the past to treat diarrhoea but the plant is nowadays rarely used internally. It has, however, several external applications. The raw macerated leaves or the pressed juice from them is applied in poultices to inflammations caused by insect bites and to itching burning skin. It is particularly recommended for softening the skin and alleviating corns. The pressed juice diluted with water makes a gargle for stomatitis.

Sempervivum tectorum

SENECIO AUREUS
(Compositae)

LIFE ROOT, SQUAW WEED

Parts Used: Aerial parts.

Constituents: Liferoot contains a volatile oil, pyrrolizidine alkaloids (includng senecine, senecionine, and otosenine), tannins and resin. In isolation, the pyrrolizidine alkaloids are highly toxic to the liver.

Medicinal Actions & Uses: Until recently, life root was employed in Anglo-American herbalism much as it was in earlier times — as a means to induce menstrual periods and to bring relief to menopausal complaints. Today, the plant is recommended only for external use, as a douche for excessive vaginal discharge.

SENECIO JACOBAEA
(Compositae)

RAGWORT

Parts Used: Aerial parts.

Constituents: Ragwort contains a volatile oil, pyrrolizidine alkaloids (including seneciphylline, senecionine and jacoline), tannins and resin. In isolation, pyrrolizidine alkaloids are highly toxic to the liver.

Medicinal Actions & Uses: While no longer taken internally, ragwort still finds use as a poultice, ointment or lotion applied to relieve pain and inflammation. Conditions treated by ragwort include rheumatism and rheumatoid arthritis, and neuralgic conditions such as sciatica.

SESAMUM INDICUM
(Pedaliaceae)

SESAME, HEI ZHI MA

Parts Used: Seeds, seed oil, root.

Constituents: The seeds are highly nutritious and contain 55% oil, comprising mainly unsaturated fats (about 43% each of oleic and linoleic acids), 26% protein, vitamins B^3 and E, folic acid and minerals (especially calcium).

Medicinal Actions & Uses: Sesame is principally used as a food and flavouring agent in China, but it is also taken to redress "states of deficiency", especially those affecting the liver and kidneys. The seeds are prescribed for problems such as dizziness, tinnitus (ringing in the ears), and blurred vision (when due to anaemia). Owing to their lubricating effect within the digestive tract, the seeds are also considered a remedy for "dry" constipation. The seeds have a marked ability to stimulate the production of breast milk. Sesame seed oil benefits the skin and is used as a base for cosmetics. A decoction of the root is used in various traditions to treat coughs and asthma.

SILYBUM MARIANUM
Asteraceae
(MILK THISTLE)

The ripe fruits are used medicinally. Their constituents include fatty oil, proteins, an essential oil, bitter compounds, and the important flavones silybin and silymarin. These sub-

Silybum marianum

stances give Milk Thistle bitter tonic, choleretic and cholagogic properties. It is processed by the pharmaceutical industry — especially on the Continent — into tinctures, drops, tablets and other preparations with specific proportions of active constituents. These preparations are used in medical practice for treating gall bladder diseases, to stimulate the flow of bile and for the regeneration of tissue in cases of liver damage. In herbal medicine the fruits are used in the form of a decoction or powder or they are chewed whole for the same complaints. A tincture of the seeds is used in homeopathy. **Treatment with Milk Thistle should be medically supervised**.

The leaves, young shoots, receptacle and root can be cooked and eaten as a vegetable.

SINAPIS ALBA
Brassicaceae
(WHITE MUSTARD)

The ripe seeds are used medicinally. They contain fatty oil (upto 30 per cent), mucilage and the glucosinolate sinalbin (up to 2 per cent), which, in the presence of moisture and the enzyme myrosinase, changes into mustard oil with a high concentration of sulphur. White Mustard has local rubefacient and irritant effects and the ground seeds are used like those of Black Mustard in compresses and plasters applied externally to ease rheumatic pain. A hot mustard plaster (up to 40 °C) is more effective but it may blister the skin. The whole seeds are laxative and antiseptic.

Sinapis alba

The seeds are preservatives and in the food industry they are used whole in pickling vegetables and ground to make various kinds of prepared mustard.

SISYMBRIUM IRIO
(Brassicaceae)

Seeds: expect., stim., restor., used in asthma; externally used as a stimulating poultice.

Infusion of leaves: given in affections of the throat and chest in Spain.

SMILAX SPP.
(Liliaceae)

SARSAPARILLA

Part Used: Root.

Constituents: Sarsaparilla contains 1-3% steroidal saponins, phytosterols (including beta and e-sitosterol), about 50% starch, resin, sarsapic and minerals.

Medicinal Actions & Uses: Sarsaparilla is anti-inflammatory and cleansing, and can bring relief to skin problems such as eczema, psoriasis and general itchiness, and help treat rheumatism, rheumatoid arthritis and gout. It has a tonic and specifically testosterogenic action on the body leading to increased muscle bulk and it has a potential use for impotence. Sarsapirilla also has a progesterogenic action, making it beneficial in premenstrual problems, and menopausal conditions such as debility and depression. In Mexico, the root is still frequently consumed for its reputed tonic and aphrodisiac properties. Native Amazonian peoples take sarsaparilla to improve virility and to treat menopausal problems.

SOLANUM DULCAMARA
Solanaceae

(BITTERSWEET)

The green shoots tips are used medicinally. Their constituents include non-glycosidic and glycosidic saponins, steroidal glycosidic alkaloids (such as soladulcine) and tannins. These substances give Bittersweet expectorant, mild diuretic, antiseptic and stimulant properties. It has been used to treat chronic bronchitis, asthma, rheumatism and prescribed by herbalists today. In homeopathy, however, preparations from the fresh plant are still used. The constituents of Bittersweet expectorant, mild diuretic, antiseptic and stimulant properties. It has been used to treat chronic

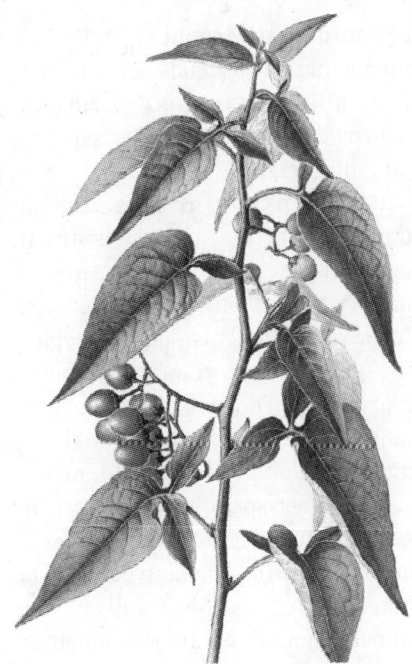

Solanum dulcamara

bronchitis, asthma, rheumatism and chronic skin disorders but is rarely prescribed by herbalists today. In homeopathy, however, preparations from the fresh plant are still used. The constituents of Bittersweet have recently been the subject of intensive study because they could be used to make steroid compounds. Bittersweet should be used only under strict medical supervision; **it should never be collected and used for self-medication**.

SOLANUM MELONGENA
(Solanaceae)
AUBERGINE

Parts Used: Fruit, fruit juice, leaves.
Constituents: Aubergine contains proteins, carbohydrates and vitamins A, B^1, B^2 and C.

Medicinal Actions & Uses: Aubergine fruit lowers blood cholesterol levels, and it is suitable as part of a diet to help regulate high blood pressure. The fruit can be applied fresh as a poultice for haemorrhoids, but it is more commonly used in the form of an oil or ointment. The fruit and its juice are effective diuretics. A soothing, emollient poultice for the treatment of burns, abscesses, cold sores and similar conditions can be made from aubergine leaves.

SOLANUM TUBEROSUM
(Solaceae)
POTATO
Part Used: Tuber.
Constituents: Potato contains starch, large amounts of vitamins A, B^1, B^2, C and K, minerals (especially potassium) and very small quantities of atropine alkaloids. One property of these alkaloids is the reduction of digestive secretions, including acids produced in the stomach.
Medicinal Actions & Uses: Taken in moderation, potato juice can be helpful in the treatment of peptic ulcers, bringing relied from pain and acidity. The juice or the mashed pulp may be used externally to soothe painful joints, headache, backache, skin rashes and haemorrhoids. Potato skins are used in India to treat swollen gums and to heal burns.

SOLANUM XANTHOCARPUM
(Solanaceae)
KANTAKARI

Parts Used: Leaves, seeds, root.
Constituents: Kantakari contains steroidal alkaloids (including solanocarpine).

Medicinal Actions & Uses: In the Ayurvedic tradition, the leaves of kantakari are taken to treat wind and constipation, and are made into a gargle for throat and gum disorders. The expectorant, anticatarrhal seeds may be taken to relieve asthma and to clear bronchial catarrh. The root is used to treat snake and scorpion bites.

Caution: Use kantakari only under professional supervision.

SOLIDAGO VIRGAUREA
(Compositae)
GOLDENROD

Parts Used: Aerial parts.
Constituents: Goldenrod contains

Solidago virgaurea

saponins, diterpenes, phenolic glucosides, acetylenes, cinnamates, flavonoids, tannins, hydroxy-benzoates and insulin. The saponins are antifungal.

Medicinal Actions & Uses: Antioxidant, diuretic and astringent, goldenrod is a valuable remedy for urinary tract disorders. It is used both for serious ailments such as nephritis, and for more common problems like cystitis. The herb also has a reputation for helping to flush out kidney and bladder stones. Goldenrod's saponins act specifically against the *Candida* fungus, the cause of vaginal and oral thrush. The herb can also be taken for conditions such as sore throats, chronic nasal catarrh and diarrhoea. Due to its mild action, goldenrod is used to treat gastroenteritis in children. Externally, it may be used as a mouthwash or douche for thrush.

SOPHORA JAPONICA
Fabaceae

(JAPANESE PAGODA TREE)

The flower buds, the leaves and the bark are used medicinally. When dry they have a bitter taste. Most important of the active constituents is the flavonoid glycoside rutin (upto 20 per cent), which decreases the permeability of the capillaries. It is extracted by the pharmaceutical industry and included in medicines prescribed for circulatory and neurological disorders. Japanese Pagoda Tree is not a suitable plant for self-medication; **it should be used only under strict medical supervision.**

Sophora japonica

Sorbus aucuparia

SORBUS AUCUPARIA
Rosaceae

(ROWAN)

The ripe fruits are used medicinally. Their constituents include a large amount of organic acids, tannins, sugars, pectin and vitamin C (particularly in fresh fruits). These substances give Rowan mild purgative, diuretic and general tonic properties. The dried fruits or the pressed juice of fresh fruits is used for constipation and kidney disorders. **Large doses of the fruits should not be taken.** The fruits are a raw material for the manufacture of sorbose, a sweetening agent for diabetics. Vitamin C (ascorbic acid) has also been extracted commercially from them.

The berries, particularly those of cultivated sweet-fruited varieties, can be used to make compotes, syrups, conserves and wines. They are also used in liqueur manufacture.

SORGHUM VULGARE
(Poaceae)

Seeds: diur., demulc., aphrodis. glucd. dhurin, leaves contain HCN.

SOYMIDA FEBRIFUGA
(Meliaceae)

Bark: astrin., bitter tonic, febge., used in general debility, intermittent fevers, diar. and dysen.
Bark contains bitter substances.

SPHAERANTHUS INDICUS
(Compositae)

Herb: tonic, deobstruent, alter., aphrodis.

Root and seed: anthelm.
Flowers: alter., cooling, tonic.
Decoct. of plant: used as a diur. in urethral discharges.
Rind of fruits: used as a fish poison.
Essen. oil in herb, alk. in leaves, stems and flowers contains alk. sphaeranthine; fresh flowering plant yields essen. oil.

SPILANTHES ACMELLA
(Compositae)
Flowers: made into a tincture used to relieve toothache; powerful mosquito larvicide.
Seeds: chewed to produce salivation when the mouth is dry.
Crushed plant: used as fish poison.
Spilanthol obtained from flowers has strong local anaesthetic action.

SPIGELIA MARILANDICA
(Loganiaceae)
PINK ROOT
Part Used: Root.
Constituents: Pink root contains alkaloids (mainly spigeleine), a volatile oil, tannin and resin. Spigeleine is emetic and irritant to the stomach.
Medicinal Actions & Uses: Pink root is used today solely to expel intestinal worms – particularly tapeworms and roundworms. It is prescribed with other herbs such as senna (*Cassia senna*), and fennel (*Foeniculum vulgare*) to ensure the elimination of both the worms and the root itself, which is potentially toxic if it is absorbed through the gut.

SPINACEA OLERACEA
(Chenopodiaceae)
Leaves: cooling, useful in febrile affections, inflam. of lungs and bowels.

Seeds: laxt., cooling, used in difficult breathing, inflam. of the liver and in jaundice.
Green plant: given for urinary calculi.
Iodine, lecithin, chlorophyll, carotin.

STACHYS OFFICINALIS
(Labiatae)
BETONY
Parts Used: Aerial parts.

Stachys officinalis

Constituents: Betony contains alkaloids (including stachydrine and betonicine), as well as betaine, choline and tannins.

Medicinal Actions & Uses: No longer regarded as a panacea, betony nevertheless has real value as a remedy for headaches and facial pain. The plant is also mildly sedative, relieving nervous stress and tension. In British herbal medicine, betony is thought to improve nervous function and to counter overactivity. It is taken to treat "frayed nerves", pre-menstrual complaints, poor memory and tension. The plant has astringent properties and in combination with other herbs, such as comfrey (*Symphytum officinale,*) and lime flowers (*Tilia species*), it is effective against sinus headaches and congestion. Betony may be taken alone or with yarrow (*Achillea millefolium*) to help staunch nosebleeds. Betony is also mildly bitter. It stimulates the digestive system and the liver, and has an overall tonic effect on the body.

STACHYTARPHETA INDICA
(Verbenaceae)
Plant: in Brazil used externally for purulent ulcers, given internally for fevers and rheum. inflam.; in Guiana used in the treatment of dysen.
Leaves: in La Reunion used as maturant.
Glucosidic substance.

STELLARIA MEDIA
(Caryophyllaceae)
Chickweed
Parts Used: Aerial parts.
Constituents: Chickweed contains triterpenoid saponins, coumarins, flavonoid carboxylic acids and vitamin C. The saponin may account for the herb's ability to help reduce itchiness.

Medicinal Actions & Uses: Chickweed is chiefly used to treat irritated skin, being applied as juice, poultice, ointment or cream. In certain cases, chickweed may soothe severe itchiness where all other remedies have failed. It is often used to relieve eczema, varicose ulcers and nettle rash (urticaria). An infusion of the fresh or dried plant may be added to bath, where the herb's emollient properties will help to reduce inflammation – in rheumatic joints, for example – and encourage tissue repair. Chickweed may also be taken internally to treat chest ailments. In small quantities, this herb also aids digestion.

STEPHANIA GLABRA
(Menispermaceae)
Root: acrid, used in Cochin-China in pulmonary tuberculosis, asthma, dysen. and fever.
Alks. gindarine, gindaricine, gindarinine.

STERCULIA FOETIDA
(Sterculiaceae)
Oil from seed: laxt. carmin.
Leaves: repellent, aper.
Decoction of fruit: mucilaginous astrin.
Seeds and fruits pulp contain fatty oil.

STEREOSPERMUM
SUAVEOLENS
(Bignoniaceae)
Root bark: considered cooling, diur., tonic; forms an ingredient in dasamula.

Flowers: rubbed up with honey given to check hiccough; taken in form of a confection as an aphrodis.
Root bark contains bitter substance.

STEREOSPERMUM TETRAGONUM
(Bignoniaceae)

Root, leaves and flowers: used in decoct. as a febge.
Juice of leaves: mixed with lime juice used in maniacal cases.
Flower and fruit: in scorpion-sting.
Bark contains crystalline bitter substance.

STREBLUS ASPER
(Moraceae)

Decoct. of bark: given in fever, dysen. and diar.
Roots: used as application to unhealthy ulcers and sinuses; antid. to snake-bite.
Milky juice: antisep., astrin., applied to chapped hands and sore heels.
Bark contains bitter substances.

STROPHANTHUS KOMBE
(Apocynaceae)

STROPHANTHUS

Parts Used: Seeds.
Constituents: Strophanthus contains up to 10% cardiac glycosides. These slow the heart rate and improve the heart's efficiency. See foxglove (*Digitalis* species).
Medicinal Actions & Uses: Strophanthus may be prescribed like foxglove in the treatment of heart disease, but strophanthus's active constituents are less well absorbed. One authority recommended it as a gentle heart tonic, of particular benefit when combined with valerian (*Valeriana*

officinalis) and deadly nightshade (*Atropa belladonna*). Like most herbs containing cardiac glycosides, strophanthus is strongly diuretic.

STILLINGIA SYLVATICA
(Euphorbiaceae)

QUEEN'S DELIGHT

Part Used: Root.
Constituents: Queen's delight contains alkaloids, diterpenes esters, fixed oil, volatile oil, resin and tannins. The fresh root is considered most active.
Medicinal Actions & Uses: Queen's delight appears to promote general detoxification. It is taken internally to help clear constipation, boils, weeping eczema and scrofula (tubercular infection of the lymph glands of the neck). The root is also taken to treat bronchitis and throat infection. Externally, it is applied as a lotion to haemorrhoids, eczema and psoriasis.
Cautions: Use only under professional supervision. Queen's delight is emetic and purgative in large doses.

STRYCHNOS NUX-VOMICA
(Loganiaceae)

NUX VOMICA

Parts Used: Seeds.
Constituents: Nux vomica contains 3% indole alkaloids (predominantly strychnine, Brucine with many others), loganin, chlorgenic acid and fixed oil. Strychnine is a lethal poison, producing intense muscle spasms.
Medicinal Actions & Uses: Through rarely used internally due to its toxicity, nux vomica can be effective nervous system stimulant, particularly in the elderly. In Chinese herbal medicine the seeds are used externally to

relieve pain, to treat various types of tumours and to relieve paralysis, including Bell's palsy (facial paralysis). Nux vomica is a common homeopathic remedy prescribed mainly for digestive problems, sensitivity to cold, irritability and melancholia.

SYMPLOCOS RACEMOSA
(Symplocaceae)
Bark: cooling, astrin., useful in menor., bowel complaints, eye diseases, ulcers; in decoct. used as a gargle for giving firmness to spongy and bleeding gums.

Bark contains two alkaloids, loturine and colloturnine which are chemically related to ahrmine found in *Peganum harmala*.

SYMPLOCARPUS FOETIDUS
(Araceae)
SKUNK CABBAGE

Parts Used: Root and rhizome.
Constituents: Skunk cabbage contains a volatile oil, serotonin, (5HT) and resins.
Medicinal Actions & Uses: Skunk cabbage continues to be used primarily as an expectorant, treating cases of asthma, bronchitis and whooping cough. It is also taken for upper respiratory problems such as nasal catarrh and hay fever. Less commonly skunk cabbage is sued as a treatment for.

SYMPHYTUM OFFICINALE
Boraginaceae
(COMMON COMFREY)
The roots and leaves are used medicinally. Their constituents include tannins, abundant mucilage, allantoin, starch, traces of puyrrolizidine alkaloids and an

Symphytum officinale

essential oil. These substances give Common Comfrey astringent, scarhealing, vulnerary, anti-inflammatory, emollient and mild sedative properties. Internally it is used in an infusion or in powder form for chronic respiratory infections, stomach and duodenal ulcers, and diarrhoea. Mostly, however, Comfrey is used externally in compresses, plasters, liniments, ointments and bath preparations. Comfrey is also used in homeopathy and it is contained in many proprietary preparations.
Note: Comfrey has been reported to cause serious liver damage if taken in large amounts over a long period of time. Although this effect is in dispute it would be best to err on the cautious side when taken Comfrey internally.

SYZYGIUM AROMATICUM AND S. CUMINI
(Myrtaceae)

Bark: astrin., used in the preparation of astrin. decoctions, gargles and washes; fresh juice given with goat's milk in the diar. of children.

Juice of leaves: used in dysen.

Juice of ripe fruit: made into a vinegar used as a stomch., carmin. and as diur.

Fruit: useful astrin. in bilious diar.

Seeds: used in diabetes.

Glucd., essen. oil in seeds, seeds contain ellagic acid, an alk. jambosine.

T

TAMARINDUS INDICA
(Leguminosae)

TAMARIND

Part Used: Fruit.

Constituents: Tamarind contains 16-18% plant acids (including nicotinic acid – vitamin B^3), a volatile oil (with geranial, geraniol and limonene), sugars, pectin, 0.8% potassium and fats. Vitamin C was formerly believed to be among the constituents of tamarind, but this is now disputed.

Medicinal Actions & Uses: Tamarind is a wholesome and cleansing fruit that improves digestion, relieves wind, soothes sore throats and acts as a mild laxative, in Ayurvedic medicine, it is given to improve the appetite and to strengthen the stom. It is also used to relieve constipation. However, mixed with cumin and sugar tamarind is also prescribed as a

treatment dysentery. In southern India, tamarind is taken to treat colds and other ailments that produce excessive catarrh. In China medicine, it is considered a cooling herb appropriate for treating "summer heat". Fruit is also given for loss of appetite, for nausea and vomiting in pregnancy and constipation.

TANACETUM VULGARE
(Compositae)

TANSY

Parts Used: Flowering tops.

Constituents: Tansy contains a volatile oil, which includes significant levels of thujone and camphor, sesquiterpene lactones, flavonoids and resin. The volatile oil strongly induces menstruation.

Medicinal Actions & Uses: Tansy is little used today because of its potential toxicity. When the plant is taken, it is chiefly is order to expel intestinal worms, and, to a lesser degree to help stimulate menstrual bleeding. Tansy may be used externally to kill scabies, fleas and lice, but even external application of tansy preparations carries the risk of toxicity.

TARAXACUM OFFICINALE
Asteraceae

(DANDELION)

The roots, flowering stems, leaves (collected before flowering) and flowerheads are used medicinally. The root is the most active part. The constituents include the terpenoid bitter compounds taraxacin and taraxacerin, a glycoside, sterols, amino acids, tannins, inulin (up to 25 per cent),

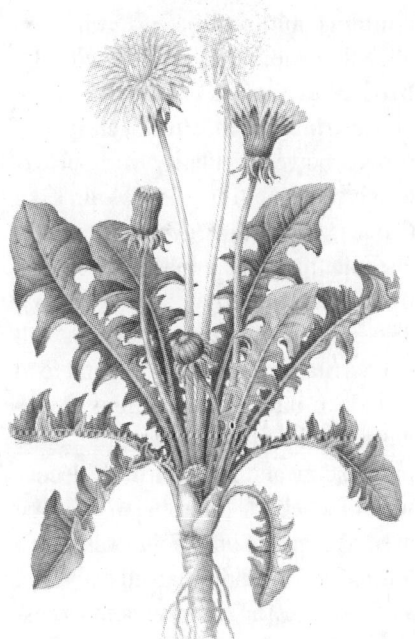

Taraxacum officinale

TAXUS BACCATA
Taxaceae

(YEW)

The needles contain the very toxic alkaloid taxine, also glycosides, bitter compounds, resin and vitamin C. In homeopathy a tincture prepared from the fresh needles is used, for example, to treat gout and rheumatism, arthritis, urinary tract infections and heart and liver conditions. The poisonous substances are absorbed within a matter of a few minutes from the gastrointestinal tract after ingestion. In small amounts they slow down the heartbeat and cause collapse and gastroenteritis; in larger doses they can cause sudden death. **Yew should never be collected and used for self medication.**

mineral substances, rubber (caoutchouc) and provitamin A, vitamins B and C (in leaves). These substances give Dandelion bitter tonic, stomachic, cholagogic, nutritive and strong diuretic properties. It is used in an infusion to stimulate the appetite, aid digestion, for biliary and liver disorders, dropsy, rheumatism and arthritis. The pressed juice from the stalks or leaves is an effective cure for warts. The fresh young leaves can be eaten raw as a spring salad. The flowers contain carotenoids and triterpenes. They are used, boiled with sugar, for coughs but honey has greater medicinal value. They can be made into an excellent wine.

The roots, dried, roasted and ground, make a caffein-free coffee substitute.

Taxus baccata

TAXUS BACCATA
(Taxaceae)

Leaves and fruits: emmen., sedative, antisp.
Leaves: used in asthma, broncht., hiccough, for indign. and epilepsy, as an aphrodis.
Plant: poisonous, used as fish poison.
Alk. taxine, the toxic principle contained in leaves, shoots and seeds, leaves contain alk. taxine, taxinine, traces of ephedrine.

TERMINALIA BELERICA
(Combretaceae)

BELERIC MYROBALAN
Part Used: Fruit.
Constituents: The fruit contains tannins and anthraquinones.
Medicinal Actions & Uses: Beleric myrobalan fruit is astringent, tonic and laxative. It is principally employed as a treatment for digestive and respiratory problems. In Indian herbal medicine, the ripe fruit is taken for diarrhoea and indigestion, and the unripe fruit is used as a laxative for chronic constipation. Beleric myrobalan is also often used to treat upper respiratory tract infections that cause symptoms of sore throats, hoarseness and coughs. Externally, the fruit is applied as a lotion for sore eyes.

TERMINALIA CHEBULA
(Combretaceae)

CHEBULIC MYROBALAN
Part Used: Fruit.
Constituents: Chebulic myrobalan contains anthraquinones, tannins, chebulic acid, resin and a fixed oil.
Medicinal Actions & Uses: Laxative and astringent, the fruit gently improves bowel regularity without excessively irritating the colon. Like Chinese rhubarb (*Rheum palmatum*), chebulic myrobalan may be used as a treatment for diarrhoea and dysentery. The fruit's tannins protect the gut wall from irritation and infection, and tend to reduce intestinal secretions. Likewise, the fruit helps to counter acidic indigestion and heartburn. A decoction of chebulic myrobalan may be used as a gargle and mouthwash, as a lotion for sore and inflamed eyes, and as a douche for vaginitis and excessive vaginal discharge.

TEUCRIUM CHAMAEDRYS
Lamiaceae

(WALL GERMANDER)
The flowering stems are used medici-

Teucrium chamaedrys

nally. Their constituents include an essential oil, flavonoids, bitter compounds and tannins. These substances give Wall Germander bitter tonic, astringent, stomachic, choleretic, antiseptic and diuretic properties. It is used to stimulate the appetite and improve digestion, and also for diarrhoea. Externally it is used to treat slow-healing wounds and haemorrhoids.

THEOBROMA CACAO
(Sterculiaceae)

Parts Used: Seeds.
Constituents: The seed pulp contains xanthines, a fixed oil and many constituents responsible for its flavours. The seeds contain very small amounts of endorphins, which are powerful painkillers that occur naturally within the body.
Medicinal Actions & Uses: Through cacao is most often used as a food, it also has therapeutic value as a nervous system stimulant. In Central America and the Caribbean, the seeds are taken as a heart and kidney tonic. The plant may be used to treat angina, and as a diuretic. Cacao butter (the fixed oil) makes a good lip salve and is often used as a base for suppositories and pessaries.

THALICTRUM FOLIOLOSUM
(Ranunculaceae)

Root: tonic, aper., purg., diur., febge., good remedy for atonic dyspep., useful in convalescence after acute diseases and as application for ophthalmia.

Berberine and thalictrine.

THUJA OCCIDENTALIS
(Cupressaceae)
ARBOR-VITAE
Parts Used: Leaves.
Constituents: Arbor-vitae contain a volatile oil (with up to 60% thujone), flavonoids, wax, mucilage and tannins.
Medicinal Actions & Uses: Arborvita has an established anti-viral activity. It is more often used to treat warts and polyps, being prescribed both internally and externally for these conditions. It is also used a part of a regime for treating cancer – especially cancer of the uterus. Arbor-vitae makes an effective expectorant and anti-catarrhal remedy, and may be used to treat acute bronchitis and other respiratory infections. It induces menstruation and can be taken to bring on delayed periods, though this use is inadvisable if menstrual pain is severe. Arbor-vitae is diuretic and is used to treat acute cystitis an bedwetting in children. Extracts may be painted on painful joints or muscles as a counter-irritant, improving local blood supply and easing pain and stiffness.

THYMUS SERPYLLUM
(Labiatae)
Herb: given in weak vision, complaints of stomach and liver, suppression of urine and menstruation; in Europe considered tonic, antisp., carmin. and in infusion used in skin diseases.
Seeds: given as a vermifuge.
Oil: applied in toothache.

Thymus serpyllum

Thymus vulgaris

Essen. oil, 0.5% containing phenols *p*-cymene, terpenes, terpene alcohols.

THYMUS VULGARIS
(Labiatae)

Volatile oil from the plant – employed in preparations for use in the treatment of broncht. and whooping cough.

Essen. oil from flowers and leaves contain about 45.0% of thymol and carvacrol, cymene *l*-pinene, horneol, linalool and bornyl acetate.

THYMUS SERPYLLUM
(Labiatae)
WILD THYME

Parts Used: Flowering tops.

Constituents: Wild thyme contains volatile oil (with thymol, carvacrol and linalool), flavonoids, caffeic acid, tannins and resin. The volatile oil's properties are similar to, but less potent than, those of thyme oil (from *Thymus vulgaris*).

Medicinal Actions & Uses: Like its close relative thyme (*Thymus vulgaris*), wild thyme is strongly antiseptic and antifungal. It may be taken as an infusion or syrup to treat flu and colds, sore throats, coughs, whooping cough, chest infections, and bronchitis. Wild thyme has anti-catarrhal properties and helps clear a stuffy nose, sinusitis, ear congestion and related complaints. It has been used to expel threadworms and round-worms in children, and is used to settle wind and colic. Wild thyme's antispasmodic action makes it useful in relieving period pain. Externally, it may be applied as a poultice to treat mastitis (inflammation of the breast), and an infusion may be used as a wash to help heal wounds and ulcers. Wild thyme is also used in herbal baths and pillows.

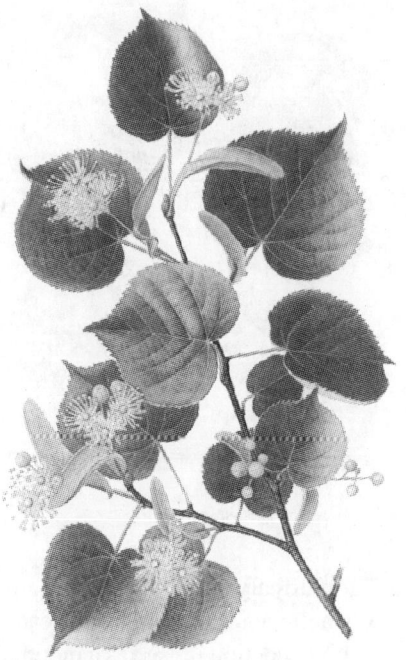

Tilia cordata

TILIA CORDATA
Tiliaceae

(SMALL-LEAVED LIME)

The flowers (linden blossoms) are used medicinally. When dry they have a pleasant and scent and a mucilaginous taste. Their constituents include an essential oil, mucilage, tannins, flavonoid glycosides, saponins and sugars. These substances give the flowers strong diaphoretic, also antispasmodic, diuretic and mild sedative properties. An infusion – linden tea – is of benefit for feverish chills and flu. The flowers are also used to stimulate the appetite, for digestive disorders, respiratory catarrh and to soothe the nerves. They are often used in diuretic tea mixtures. Charcoal made from dry lime twigs is used to expel gas from the intestines, to counteract stomach acidity, to treat gall bladder and liver disorders and in cases of poisoning. The bark is also used.

TILIA PLATYPHYLLOS
Tiliaceae

(LARGE-LEAVED LIME)

The flowers act medicinally like those of Small-leaved Lime. They contain a small amount of an essential oil, mucilage, tannins, saponins, flavonoid glycosides and sugars. The glycosides are primarily responsible for the diaphoretic action of lime flowers. The dried flowers are used on their own or in various herbal tea mixtures to stimulate the appetite, alleviate rheumatic pain, for colds and chills,

Tilia platyphyllos

and for urinary disorders. They are also beneficial for insomnia and headaches. A decoction is used as a mouthwash, as a gargle for sore throats and also in cosmetic lotions and creams.

TODDALIA ASIATICA
(Rutaceae)
Root bark: bitter, arom. tonic, stim., antiper.; given in weak infusion useful in constitutional debility and in convalescence after febrile and other exhausting diseases.
Plant: used as a febge.
Essen. oil; alk. berberine.

TRAGOPOGON PRATENSIS
(Compositae)
GOAT'S BEARD
Part Used: Root.
Constituents: The root contains inulin, inositol, mannitol and plant sterols.
Medicinal Actions & Uses: Like its cousin dandelion (*Taraxacum officinale*), goat's beard is considered

a useful remedy for the liver and gallbladder. It appears to have a detoxifying action, and may stimulate the appetite and digestion. Its high inulin content makes this herb a useful food for diabetics. Inulin is a nutrient made of fructose rather than glucose units, and therefore does arteriosclerosis and high blood pressure.

TRIBULUS TERRESTRIS
(Zygophyllaceae)
Fruits: cooling, diur., tonic, aphrodis., used in painful micturition, calculus affections, urinary discharges and impotence; in form of infusion useful as a diur. in gout, kidney diseases and gravel.
Fruits contain traces (0.001%) of an alk., a fixed oil, a small quantity of essen. oil, resins and nitrates (I.P.C.).

TRICHODESMA INDICUM
(Boraginaceae)
Plant: diur., used as an emol. poultice.
Leaves: in a cold infusion considered depurative.
Root: pounded and made into a paste applied to reduce swellings, particularly of the joints; pounded with water given as a drink to children in dysen.

TRIFOLIUM PRATENSE
(Leguminosae)
RED CLOVER
Parts Used: Flowerheads.
Constituents: Red clover contains flavonoids, phenolic acids (such as salicylic acid), volatile oil (including methyl salicylat and benzyl alcohol),

Trifolium pratense

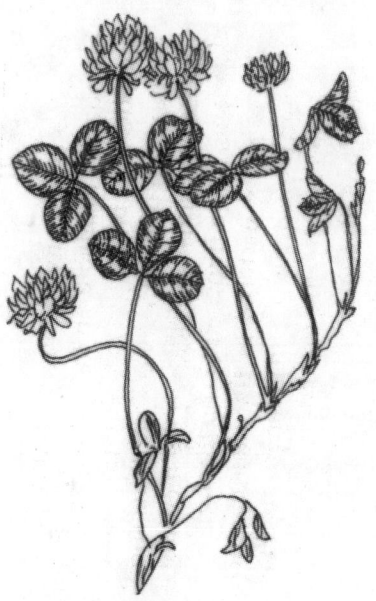

Trifolium repens

sitosterol, starch and fatty acids. Flavonoids in the flowers and leaves are oestrogenic.

Medicinal Actions & Uses: Red clover is used to treat skin conditions, normally in combination with other purifying herbs such as burdock (*Arctium lappa*), and yellow dock (*Rumex crispus*). It is also expectorant and may be used for spasmodic coughs. Red clover's oestrogenic effect may help menopausal complaints.

TRIFOLIUM REPENS
Fabaceae

(WHITE CLOVER)

The flowerheads are the medicinally active parts. When dry they have a honey-like fragrance and a slightly astringent taste. The main constituents are tannins, sugars, mucilage and organic acids. These substances give White Clover astringent, anti-inflammatory and antiseptic properties. An infusion is used to treat gastritis, enteritis, severe diarrhoea and rheumatic pains. It is also used as an inhalent for respiratory infections. The fresh flowers make a pleasant substitute for ordinary tea.

The young leaves are edible like those of Red Clover (*T. pratense*).

TRIGONELLA FOENUM-GRAECUM
(Leguminosae)

FENUGREEK

Parts Used: Seeds.

Constituents: Fenugreek contains a volatile oil, alkaloids (including trigonelline), saponins (based on diosgenin), flavonoids, mucilage

Trigonella foenum-graecum

treat abscesses, boils, ulcers and burns, or used as a douche for excessive vaginal discharge. The seeds also freshen bad breath and help restore a dulled sense of taste. In China, fenugreek is used as a pessary to treat cervical cancer.

TRIUMFETTA BARTRAMIA
(Tiliaceae)
Leaves, flowers and fruits: mucilaginous, demulc., astrin., given in gonor.
Root: bitter, diur.; a hot infusion is taken to facilitate childbirth or to hasten the inception of parturition when it is deplayed.
Bark and fresh leaves: used in diar. and dysen.

TRILLIUM ERECTUM
(Liliaceae)
BETH ROOT
Part Used: Rhizome.
Constituents: Beth root contains saponins (such as trillin), tannin, resin, fixed oil, and a trace of volatile oil.
Medicinal Actions & Uses: Beth root is a valuable remedy for heavy menstrual or intermenstrual bleeding, helping to reduce blood flow. It is also used to treat bleeding associated with uterine fibroids. Beth root may also be taken for bleeding within the urinary tubules and, less commonly, for the coughing up of blood. It remains a valuable herb in facilitating childbirth. A douche of beth root is useful for excessive, vaginal discharge and thrush.

TROPAEOLUM MAJUS
Tropaeolaceae
(NASTURTIUM)
The seeds are used medicinally. Their constituents include the glycosinolate

(about 27%), protein (about 25%), fixed oil (approximately 8%), vitamins A, B^1 and C, and minerals.
Medicinal Actions & Uses: Fenugreek is much used in herbal medicine in North Africa, the Middle East and India, being esteemed as a remedy for a wide variety of conditions. The nourishing seeds are given during convalescence and to encourage weight gain, especially in anorexia. They are also helpful in lowering fever, with some authorities comparing their ability to that of quinine. The seeds soothing effect makes them of value in treating gastritis and gastric ulcers. They are used to induce childbirth and to increase breast-milk production. Fenugreek is also thought to be antidiabetic and to lower blood cholesterol levels. Externally, the seeds may be applied as a paste to

Tropaeolum majus

glucotropaeolin (up to 1.5 per cent), which hydrolyses to yield antiseptic substances (including a mustard oil); also fatty oil (up to 20 per ent) and proteins. The seeds are effective antiseptic agents against bacteria such as *Staphylococcus*, *Proteus*, *Streptococcus* and *Salmonella*. The antiseptic substances are eliminated in the urine and partly also through the lungs and so the seeds are used for acute infections of the urinary tract and for acute bronchitis. They are usually powdered and administered in the form of pills. The fresh juice from the plant has the same antiseptic effect. The fresh leaves contain vitamin C and iron as well as antiseptic substances and they give a sharp, mustard-like flavour to salads. The open flowers, which can also be added to salads, are stimulants. The flower buds can be preserved in vinegar and used as a substitute for capers to flavour sauces and as a garnish.

TSUGA CANADENSIS
(Pinaceae)

CANADA SPRUCE

Parts Used: Bark.

Constituents: Hemlock spruce contains volatile oil (with alpha-pinene, bornyl acetate and cadinene), 10-14% tannins and resin.

Medicinal Actions & Uses: The bark of hemlock spruce is astringent and antiseptic. A decoction may be taken to treat diarrhoea, colitis, diverticulitis and cystitis. Externally, hemlock spruce can be employed as a douche to treat excessive vaginal discharge, thrush and a prolapsed uterus; as a mouthwash and gargle for gingivitis and sore throats; or as a wash to cleanse and tighten wounds.

TUSSILAGO FARGARA
Asteraceae

(COLTSFOOT)

The flowerheads (collected before they are fully open) and the young leaves are used medicinally. Their constituents include abundant mucilage, an essential oil, tannins, mineral salts, sterols, organic pigments (flowers), and a glycosidic bitter compound and inulin (leaves). These substances give Coltsfoot bitter tonic, expectorant, demulcent, astringent and anti-inflammatory properties. Because of the mucilage content it is used mainly in herbal tea mixtures for coughs, catarrh, bronchitis, laryngitis and asthma. It is included in some

Tussilago fargara

Constituents: Asmatica contains alkaloids (including tylophorine), flavonoids, sterols and tannins. Tylophorine has anti-inflammatory and anti-tumour properties.

Medicinal Actions & Uses: Considered a specific remedy for asthma, asmatica may relieve symptoms for up to 3 months. It is also beneficial in cases of hay fever, and is prescribed for acute allergic problems such as eczema and nettle rash. It holds potential as a treatment for chronic fatigue syndrome and other immune system disorders. Asmatica may relieve rheumatoid arthritis, and may also be of value in the treatment of cancer.

TYPHA ANGUSTIFOLIA
(Typhaceae)
Pu Huang, Bulrush
Part Used: Pollen.
Constituents: *Pu huang* contains isorhamnetin, pentacosane and plant sterols.
Medicinal Actions & Uses: In Chinese herbal medicine, the astringent *pu huang* pollen has been employed chiefly to stop internal or external bleeding. The pollen may be mixed with honey and applied to wounds and sores, or taken orally to reduce internal bleeding of almost any kind – for example, nosebleeds of almost any kind – for example, nosebleeds, uterine bleeding or blood in the urine. The pollen is now also used in the treatment of angina (pain in the chest or arm due to lack of oxygen to the heart muscle). *Pu huang* does not appear to have been used as a medicine in the European herbal tradition.

proprietary cough medicines too. Coltfoot is also a component of tea mixtures taken to stimulate the flow of bile, to check diarrhoea and as a general tonic. A decoction is used externally in compresses, poultices and bath preparations applied to slow-healing wounds, skin ulcers and rashes. The dried leaves are included in herbal tobaccos to relieve chest troubles. The fresh leaves, after first being washed and crushed, are applied to skin inflammations and rheumatic joints.

The flowerheads can be used to make a pleasant wine.

TYLOPHORA ASMATICA
(Asclepiadaceae)
Asmatica, Indian Lobelia
Parts Used: Leaves.

ULMUS MINOR
Ulmaceae
(SMALL-LEAVED ELM)

The yellow inner bark from young twigs of Small-leaved Elm and other elm species is used medicinally. As it dries it curls into a cylinder. It has an astringent taste. The constituents include tannins, bitter compounds and mucilage. These substances give the bark astringent, demulcent and anti-inflammatory properties and it is used in a decoction or in powder form to treat digestive disorders and severe diarrhoea. Externally a diluted decoction is used in compresses or bath preparations for inflamed wounds, haemorrhoids and as a mouthwash

Ulmus minor

and gargle. In homeopathy a tincture prepared from fresh bark is given for eczema and other skin complaints.

UNCARIA RHYNCOPHYLLA
(Rubiaceae)

GOU TENG

Parts Used: Stems, thorns.

Constituents: *Gou teng* contains alkaloids (including rhyncophylline and corynoxeine, isorhyncophyllinne, and hirsutine) and nicotinic acid.

Medicinal Actions & Uses: *Gou teng* is a sedative and antispasmodic, and is mainly used to ease symptoms such as tremors, seizure, spasms, headache and dizziness. It is also prescribed for infantile convulsions. In Chinese herbal medicine it "extinguishes [internal] wind and stops tremors". It is also used by the Chinese to reduce high blood pressure and excess liver "fire".

URGINEA MARITIMA
(Liliaceae)

SQUILL

Part Used: Bulb.

Constituents: Squill contains cardiac glycosides (0.15-2.4% bufadenolides, including scillaren A), flavonoids, anthocyanidins and mucilage. The cardiac glycosides are strongly diuretic and relatively quick-acting. They do not have the same cumulative effect as those present in foxglove (*Digitalis purpurea*).

Medicinal Actions & Uses: Squill is a diuretic, emetic , cardiotonic and expectorant plant that finds use in a wide range of conditions. It makes a good diuretic in cases of water retention. Since its active constituents do not accumulate to a great degree within the body, it is a potential substitute for foxglove in aiding a failing heart. At low dosage, squill is an effective expectorant. At higher doses, the herb acts as an emetic. Squill is also used in homeopathic preparations.

URTICA DIOICA
Urticaceae

(COMMON-NETTLE)

The flowering stems and the leaves are used medicinally. Their constituents include tannins, histamine, 5-hydroxytryptamine, organic acids, vitamin C, provitamin A and mineral salts. These substances give Common Nettle astringent, tonic, diuretic, haemostatic, antirheumatic, galactagogic and blood-purifying properties and the plant has many therapeutic applications in herbalism and homeopathy.

For example, it is used to treat urinary, liver and respiratory disorders, gastritis and enteritis, rheumatism and skin disorders.

The young shoots are rich in vitamin C and can be added to salads or cooked like spinach. They make a pleasant beverage, nettle beer. Nettle is also recommended as a hair tonic and face wash.

VACCINIUM MYRTILLUS
Ericaceae

(BILBERRY)

The leaves of non-flowering twigs and the fruits are used medicinally. The constituents of the leaves include tannins, organic acids, a glycoside (arbutin) and plant insulins. These substances give the leaves astringent, antiseptic, diuretic and weak hypoglycaemic properties and they

Urtica dioica

Vaccinium myrtillus

are used in an infusion for gastritis, enteritis and diarrhoea. They are also included in herbal tea mixtures with an antisclerotic action. **It is advisable not to take this infusion in strong doses or over a long period of time.** The ripe berries are used fresh or dried. They contain sugars, pectin, organic acids, tannins, mineral salts, vitamin B and C and organic pigments (anthocyanins). Dried berries are chewed to check diarrhoea. Wine and an alcoholic extract from the berries also have a positive action. The pressed juice from the berries and conserves are beneficial for mouth and throat infections.

The wholesome berries can be eaten raw or stewed and made into pieces.

VACCINUM MYRTILLUS
(Ericaceae)

BILBERRY

Parts Used: Fruit, leaves.

Constituents: The fruit contains about 0.5% anthocynosides, vitamins B^1 and C, pro-vitamin A, 7% tannins, and plant acids. The anthocyanosides have a tonic effect on the blood vessels.

Medicinal Actions & Uses: Ripe bilberries are mildly laxative due to their fruit sugar content. The dried fruit, however, is markedly binding and has an antibacterial action. A decoction of the dried fruit is useful for treating diarrhoea in children. Bilberry's high anthocyanin content makes it a potentially valuable treatment for varicose veins, haemorrhoids and capillary fragility. A decoction of the fruit is used as a mouthwash. The leaves

may be helpful in pre-diabetic states, but are not an alternative to conventional treatment. They may be taken for urinary tract infections.

VACCINIUM VITIS-IDAEA
Ericaceae

(COWBERRY)

The leaves and fruits are used medicinally. The constituents of the leaves include glycosides (mostly arbutin), tannins, organic acids, sugars and vitamin C. These substances give Cowberry leaves astringent, antiseptic, diuretic and hypoglycaemic properties. Their effect is much the same as Bearberry's; they are used in an infusion to treat infections of the urinary tract and bile ducts, kidney stones, rheumatism and diarrhoea. The ripe berries have a high concentration of vitamin C; they also include the glycosides arbutin and vaccinilin, sugars and organic acids. The fresh

Vaccinium vitis-idaea

or dried berries can be eaten raw or stewed as a remedy for diarrhoea, but not by individuals with inflamed kidneys or kidney stones for they contain oxalic acid. Because of the arbutin content **Cowberry should not be taken in strong doses or over a long period of time.**

VANDA ROXBURGHII
(Orchidaceae)

Root: useful in rheumatism and allied disorders; enters into the composition of various medicated oils for external application in diseases of the nervous system and rheumatism.

Leaves: made into a paste by pounding applied to the body during fever; the juice is introduced into the aural meatus as a remedy for oititis media. Whole plant contains alk.; a glucd. which stimulates all organs having autonomic cholinergic nerve supply at dose levels.

VALERIANA OFFICINALIS
Valerianaceae

(COMMON VALERIAN)

The rhizome and roots of second-year plants are used medicinally. Among the constituents are an essential oil (0.5 to 1 per cent), valepotriates (epoxy-iridoid esters, such as valtrate), traces of alkaloids, bitter compounds and tannins. In combination these substances are sedative, hypnotic and antispasmodic. They are included in pharmaceutical preparations prescribed for nervous heart disorders, convulsions, nervous heart depression, nervous exhaustion, anxiety, headaches and chronic insomnia.

Valeriana officinalis

In herbal medicine an infusion is used for the same disorders and a tincture from the fresh root is used in homeopathy. **Valerian should not be taken in strong doses for a long period of time** because it can become addictive and may also cause other side-effects.

VERATRUM ALBUM
Liliaceae

(FALSE HELLEBORINE)

The rhizome and roots are the medicinally active parts. Their constituents include several toxic alkaloids, (for example, veratrine and protoveratrine A and B), which widen the lumen of blood vessels and lower blood pressure; also bitter compounds, glycosides, resin and organic acids. The alkaloids are processed to make antihypertensive and heart preparations. The plant was once prescribed as an antispasmodic, a diaphoretic, an emetic and a purga-

Veratrum album

tive and to treat rheumatic pain and neuralgia. Herbalists cannot now prescribed it is Britain. It taken internally False Helleborine cause collapse, severe vomiting and diarrhoea and breathing difficulties. Only 1 or 2 grams are fatal. **False Helloborine should never be collected and used for self-medication.**

VERBENA OFFICINALIS
(Verbenaceae)

Fresh leaves: used as a febge., tonic and as a rubft. in rheumatism and diseases of the joints.
Plant: useful in nerve complaints and amenor.; used as a depurative and febge.

Root: considered as a remedy for scrofula and snake-bite.
Entire plant contains glucd. verbenalin.

VETIVERIA ZIZANIOIDES
(Poaceae)

Roots: in infusion considered refrig., febge., diaphor., stim., stomch. and emmen.; pulverised and made into a paste in water used as a cooling external application in fevers; their essence used as a tonic.

VERATRUM VIRIDE
(Liliaceae)

AMERICAN HELLEBORE

Parts Used: Rootstock.
Constituents: American hellebore contains steroidal and other alkaloids and chelidonic acid. Some of the alkaloids lower blood pressure and dilate the peripheral blood vessels. They have been used in conventional medicine to treat high blood pressure and rapid heart beat.
Medicinal Actions & Uses: American hellebore is a highly toxic plant that is rarely used in herbal medicine today. It is an effective insecticide, but it can cause side-effects, even when applied to unbroken skin. The plant is used in homeopathic preparations to slow heart rate.

VERBASCUM DENSIFLORUM
Scrophulariaceae

(LARGE FLOWERED MULLEIN)

The flowers: without the calyx — are the medicinally active parts. After drying they are bright-yellow with a honey-like scent and mucilaginous taste. Their constituents include saponins, mucilage, glycosides, yel-

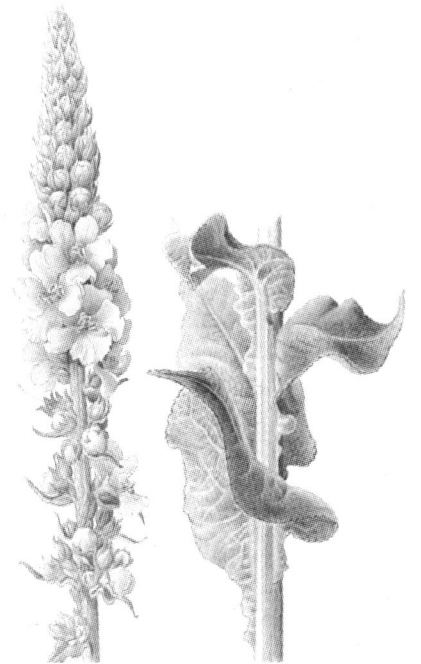

Verbascum densiflorum

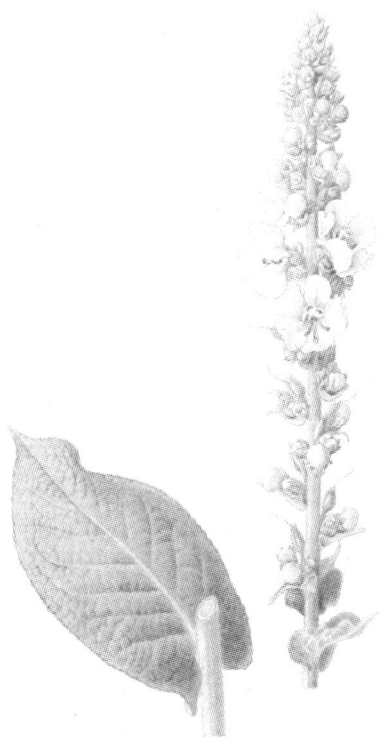

Verbascum phlomoides

low pigments and tannins. These substances give Large-flowered Mullein expectorant, emollient, demulcent, antispasmodic, diaphoretic and mild diuretic properties. It is an important ingredient of herbal tea mixtures for treating chest colds and it is also included in some proprietary cough medicines. The extract is used in making emollient ointments. Externally Large-flowered Mullein is used in compresses and bath preparations for varicose ulcers and haemorrhoids. The medicinal properties of Great Mullein and Orange Mullein (*V. phlomoides*) are similar.

VERBASCUM PHLOMOIDES
Scrophulariaceae
(ORANGE MULLEIN)

The flowers are used, like those of Large-flowered Mullein, in herbal tea mixtures for coughs and colds. The dried leaves are also used in tea mixtures. Fresh and bruised, the leaves are applied to slow-healing wounds. At one time the leaves of several mullein species were included in herbal smoking mixtures too. The root has been used to treat cramps, haemorrhoids and diarrhoea and the vinegar extract from the root was considered beneficial for toothache. In homeopathy tinctures of mulleins prepared from the fresh plants are still used to relieve earache and migraine. The flowers yield a yellow dye, once used by women of ancient Rome to dye their hair golden yellow. They are still used to give aroma to some liqueurs and are included in cosmetic preparations.

VERBASCUM THAPSUS
(Scrophulariaceae)

MULLEIN

Parts Used: Leaves, flowers.

Constituents: Mullein contains mucilage, flavonoids, triterpenoid saponins, volatile oil and tannins.

Medicinal Actions & Uses: Mullein is a valuable herb for coughs and catarrh, and is a specific treatment for tracheitis and bronchitis. The leaves and the flowers may be used as an infusion to reduce mucus formation and stimulate the coughing up of phlegm. Mullein combines well with other expectorants such as coltsfoot (*Tussilago farfara*) and thyme (*Thymus vulgaris*). Applied externally, mullein is emollient and makes a good wound healer. In Germany, the flowers are steeped in olive oil, and the resulting fixed oil is used as a remedy for ear infections and haemorrhoids.

VERBENA OFFICINALIS
Verbenaceae

(VERVAIN)

The flowering stems are used medicinally. Their constituents include the glycosides verbenalin and verbenin, tannins, an essential oil, mucilage, saponins and mineral compounds. These substances give Vervain astringent, diuretic, stomachic, tonic, diaphoretic, antispasmodic, vulnerary, mild sedative and hypnotic properties. It is used internally in an infusion for various disorders associated with the stomach, liver and kidneys. It is also excellent for stimulating the metabolism, for treating general nervous exhaustion, insomnia and migraine.

Verbena officinalis

Externally Vervain is used in gargles and in compresses and bath preparations for skin disorders. A tincture prepared from the fresh plant is used in homeopathy.

VERONICA OFFICINALIS
Scrophulariaceae

(HEATH SPEEDWELL)

The flowering stems are used medicinally. Their constituents include tannins, bitter compounds, an essential oil, organic acids, a glycoside (aucuboside) and vitamin C. These substances give Heath Speedwell astringent, expectorant, stomachic, vulnerary and diuretic properties but these actions are weak. It is still occasionally used in herbalism as a

Veronica officinalis

cough medicine and also for stomach complaints, kidney disorders and rheumatism. The decoction is sometimes used as a gargle, in hot compresses and as a bath preparation for rheumatic pain and skin disorders. It makes a pleasant tea substitute.

VIBURNUM PRUNIFOLIUM
(Caprifoliaceae)

BLACK HAW

Parts Used: Bark, root bark.
Constituents: Black haw contains coumarins (including scopoletin and aesculetin), salicin, 1-methyl-2,-3-dibutyl hemimellitate, viburnin, plant acids, a trace of volatile oil and tannin.
Medicinal Actions & Uses: Black haw is antispasmodic and astringent,

and is regarded as a specific treatment for menstrual pain. Echoing its 19th - century applications, the bark is also used to treat other gynaecological conditions, such as prolapse of the uterus, heavy menopausal bleeding, morning sickness and threatened miscarriage. Black haw's antispasmodic property makes it of values in cases where colic or other cramping pain effects the bile ducts, the digestive tract or the urinary tract.

VINCA MINOR
Apocynaceae

(LESSER PERIWINKLE)

The flowering stems are used medicinally. Their constituents include several alkaloids, tannins, saponins, pectin and organic pigments. These substances give Lesser Periwinkle tonic,

Vinca minor

astringent, hypotensive, vasodilating and diuretic properties. It is used in some proprietary preparations for cardiovascular disorders and in herbalism for treating bleeding from the nose and gums, for diarrhoea, coughing spasms and stomatitis, and in gynaecology. Greater Periwinkle has similar medicinal actions. **Both species should be used only under strict medicinal supervision: they should never be collected and used for self-medication.** Madagascan Periwinkle has been found to inhibit the growth of certain cancer-forming cells and two toxic alkaloids, vincristine and vinblastine, isolated from it are used to treat certain cancers.

Vincetoxicum hirudinaria

VINCA ROSEA
(Apocynaceae)

MADAGASCAR PERIWINKLE

Parts Used: Aerial parts, root.

Constituents: Madagascar periwinkle contains over 70 different indole alkaloids, including vinblastine, vincristine, alstonine, ajmalicine, leurocristine and reserpine.

Medicinal Actions & Uses: This plant is used in folk medicine in the Philippines as a remedy for diabetes. In the Caribbean the flowers are used as a soothing eyewash.

VINCETOXICUM HIRUDINARIA
Asclepiadaceae

(WHITE SWALLOW-WART)

The rhizomes and roots are the medically active parts but they are rarely used today. Their constituents include toxic glycosides (mostly vincetoxin), an essential oil, saponins, mucilage

and sugars. These substances give White Swallowort diaphoretic, diuretic, tonic and laxative properties. Strong doses cause retching and vomiting. **It should thus be used only under medical supervision.** A decoction is used externally in compresses for swellings and bruises.

VIOLA ODORATA
Violaceae

(SWEET VIOLET)

The leaves and flowers together or separately and the rhizome are used medicinally. The constituents of all parts include an aromatic essential oil, saponins, mucilage, a glycoside (violarutin), methyl salicylate and organic acids; the flowers also contain anthocyanin pigments. These substances give Sweet Violet expectorant and diuretic properties and the plant is used for bronchitis, whooping cough, coughs and head colds. It is

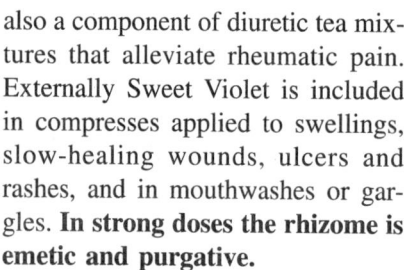

Viola odorata

Viola tricolour

also a component of diuretic tea mixtures that alleviate rheumatic pain. Externally Sweet Violet is included in compresses applied to swellings, slow-healing wounds, ulcers and rashes, and in mouthwashes or gargles. **In strong doses the rhizome is emetic and purgative.**

The fragrant essential oil, distilled from the fresh flowers, is used in perfumery and to colour medicines. The many garden varieties of violet with large, unscented flowers are not used medicinally.

VIOLA TRICOLOUR
Violaceae

(WILD PANSY)

The flowering stems are used medicinally. Their constituents include saponins, mucilage, an essential oil, a bitter compound (violin), salicylic compounds, tannins, flavonoid glycosides and organic pigments

(flowers). These substances give Wild Pansy expectorant, diuretic, diaphoretic, tonic, anti-inflammatory and blood-purifying properties. In herbal medicine an infusion is used for respiratory and urinary disorders, feverish conditions, rheumatic pain and chronic skin diseases, such as ezcema. **Strong doses may cause vomiting and allergic skin reactions.** Wild Pansy is also used as a cosmetic preparation for cleansing the skin and shampooing thinning hair, as a gargle, and it is applied in compresses or bath preparations to wounds and sores.

VISCUM ALBUM
Viscaceae

(MISTLETOE)

The young twigs, without the berries, and the leaves are used medicinally. Their constituents include a toxic sub-

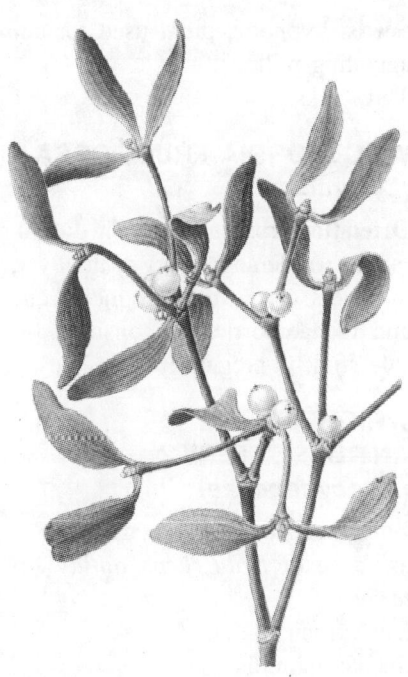

Viscum album

stance (viscotoxin), choline, acetyl-choline, alkaloids and proteins. These substances give Mistletoe hypnotic, vasodilating, cardiotonic, diuretic, sedative and anti-tumour properties. In medical practice preparations of Mistletoe are used to lower blood pressure, to stimulate heart action and to treat arteriosclerosis. Mistletoe leaves and stems (but not the berries) can be prescribed by herbalists. **In large doses Mistletoe is toxic and it should thus be used only under professional supervision.** The proteins responsible for the antitumour activities of Mistletoe have been the subject of recent research but they are unlikely to be incorporated in cancer drugs although one supposedly anticancer preparation made from

Mistletoe extracts is used by homeopathic doctors.

VITIS VINIFERA
(Vitaceae)

GRAPE VINE

Parts Used: Leaves, fruit, sap.

Constituents: Grape vine contains flavonoids, tannins, tartrates, inositol, carotenes, choline and sugars. The fruit contains tartaric and malic acids, sugars, pectin, tannin, flavone glycosides, anthocyanins (in red leaves and red grapes), vitamins A, B^1, B^2 and C and minerals. The anthocyanins reduce capillary permeability.

Medicinal Actions & Uses: Grape vine leaves, especially the red leaves, are astringent and anti-inflammatory. They are taken as an infusion to treat diarrhoea, heavy menstrual bleeding and uterine haemorrhage, as a wash for mouth ulcers and as a douche for vaginal discharge. Red leaves and grapes are helpful in the treatment of varicose veins, haemorrhoids and capillary fragility. The sap from the branches is used as an eyewash. Grapes are nourishing and mildly laxative and they support the body through illness, especially of the gastro-intestinal tract and liver. Because the nutrient content of grapes is close to that of blood plasma, grape fasts are recommended for detoxification. The dried fruit (raisins or sultanas) in mildly expectorant and emollient, with a slight effect in easing coughs. Wine vinegar is astringent, cooling and soothing to the skin.

WALSURA PISCIDIA
(Meliaceae)

Bark: stim., expect., emmen., emetic., used in skin diseases and as a fish poison.
Saponin.

WATTAKAKA VOLUBILIS
(Asclepiadaceae)

Leaves: used as an application to boils and abscesses.
Roots and tender stalks: considered emetic and expect.
Plant: used in colds and eye diseases, to cause sneezing; in snake-bite.
Glucd. dregein, alk.

WEDELIA CALENDULACEA
(Compositae)

Leaves: tonic, alter., useful in cough, cephalalgia, alopecia and in skin diseases.
Decoct. of plant: used as deobstruent and given in uterine haemor. and menor.

WITHANIA SOMNIFERA
(Solanaceae)

Root: considered alter., aphrodis., tonic, deobstruent, diur., narcotic, abortif.; used in rheumatism, consumption, debility from old age, emaciation of children, etc.
Leaves: bitter, given in infusion in fever.
Bruised leaves and ground root: used as a local application to painful swellings, carbuncles and ulcers.
Fruit: diur.

Seeds: hypnotic, diur., used for coagulating milk.
Three alks.

WOODFORDIA FRUCTICOSA
(Lythraceae)

Dried flowers: astrin., used in dysen., menor., in derangements of the liver, disorders of the mucous membrane and in haemorrhoids; considered a safe stim. in pregnancy.

WRIGHTIA ANTIDYSENTERICA
(Apocynanceae)

Bark and seeds: medicinal uses same as those of *Holarrhena antidysenterica*.
Bark: tonic.
Seeds: aphrodis.
Indican, fixed oil.

XYLIA XYLOCARPA
(Fabaceae)

Decoct. of the bark: used in worms, leprosy, vomiting, diar., gonor. and ulcers.
Oil from the seeds: given in rheumatism, piles and leprosy.

ZANTHOXYLUM ALATUM
(Rutaceae)

Seeds and bark: used as an arom. tonic, in fever, dyspep. and cholera.
Fruits, branches and thorns: used

as a fish poison; used as a remedy for toothache; considered carmin. and stomch.
Essential oil.

ZEA MAYS
Poaceae
(Maize)

The stigmas and styles of the female flowers, collected before pollination, are used medicinally. Their constituents include saponins, fatty oil, tannins, sugars, essential oil, resin, bitter compounds and mineral compounds. These substances give Maize

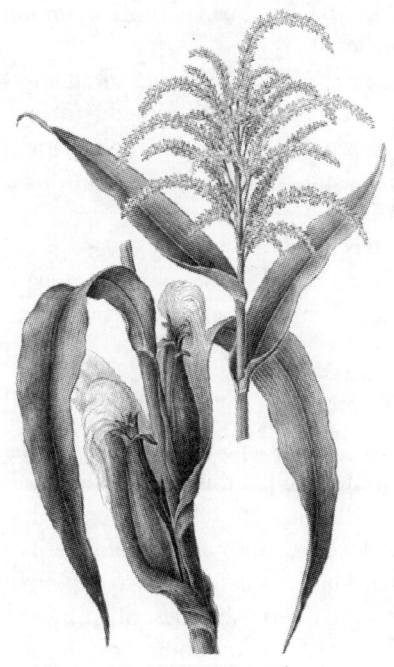
Zea mays

diuretic (in fresh state), cholagogic, cardiotonic and hypertensive properties. It is occasionally used in herbal medicine for kidney and bladder disorders, arthritis and rheumatism and for the same disorders in homeopathy, when a tincture would be prescribed. From the germinating grains maize oil is extracted for many products. When refined it can be used for cooking. It is rich in unsaturated fatty acids and is also a source of vitamin E. Defatted cornmeal is a nourishing and easily digested food recommended for convalescents and individuals allergic to wheat gluten.

ZIZIPHUS JUJUBA
(Rhamnaceae)

Jujube, Da Zao (Chinese)
Part Used: Fruit.
Constituents: Jujube contains saponins, flavonoids, sugars, mucilage, vitamins A, B^2 and C, and calcium, phosphorus and iron.
Medicinal Actions & Uses: Jujube is both a delicious fruit and an effective herbal remedy. It aids weight gain, improves muscular strength and increases stamina. In Chinese medicine, jujube is prescribed as a *qi* tonic to strengthen liver function. Mildly sedative and anti allergenic, it is given to reduce irritability and restlessness. It is also used to improve the taste of unpalatable prescriptions.

Aromatic Plants

ACHILLEA MILLEFOLIUM
(Compositae)

YARROW

Milfoil, nosebleed, herb militaris, soldier's woundwort, thousand leaf, thousand seal, field hop.

Use: folk remedy for rheumatism, toothache, haemorrhage and fever; fresh young leaves used sparingly in salads; herbal tobacco and substitute for hops in brewing.

Parts Used: top growth cut just before flowering. Dry outdoors in the shade or in gentle heat indoors.

AGRIMONIA EUPATORIA
(Rosaceae)

COMMON AGRIMONY

Church steeples, liverwort, sticklewort, cockleburr.

Use: Once regarded as a magic herb for healing jaundice; used as a spring tonic, as an astringent externally for treating wounds, and in oriental medicine to stop bleeding; a valuable yellow dye plant.

Parts Used: whole plant for dyeing; fresh leaves steeped in water to make an infusion; aerial parts except thick stems may be gathered just before flowering for drying gently in the dark; roots may be dried and grated to add to pot-pourri.

AJUGA REPTANS
(Labiatae)

COMMON BUGLE

Creeping bugle, carpet bugle, carpenter's herb (note that North American bugle, bugleweed or gipsy wort refers to Lycopus spp.)

Use: medicinally to treat bleeding from cuts and other wounds; popular ground cover among other herbs and for edging containers.

Parts Used: leaves picked as required and simmered to make an infusion; whole herb cut down to ground level in summer and dried in shade.

ALCHEMILLA VULGARIS
(Rosaceae)

COMMON LADY'S MANTLE

Lion's foot, bear's foot, common alchemil.

Use: used by alchemists in attempts to make gold; folk remedy for eye disorders and to stop bleeding; traditional treatment for menstrual disorders and in childbirth.

Parts Used: leaves and flowers shoots picked as required; aerial parts gathered in summer and dried outdoors.

ALLIUM
(Liliaceae)

The various species all possess the familiar sulphurous smell, but individually they are valued as vegetables, herbs, medicines and decorative garden bulbs. Only the most important in herbal terms are described here.

Caution: some people are slightly allergic to all forms of onions.

ALLIUM SATIVUM
(Liliaceae)

GARLIC

Description: subterranean white-skinned bulb, subdivided into numerous 'cloves'. Short, flat upright leaves, 15-30 cm/6-12 in, tall single flower stem bearing spherical head of pale pink or greenish-white blooms often mixed with tiny bulbils.

Uses: flavouring, vegetable and medicinal herb that has accumulated superstitions over the centuries. Used as an antibiotic, expectorant and digestive, and for treating high blood pressure.

Parts Used: bulbs, separated into cloves.

ALOE VERA, A. BARBADENSIS
(Liliaceae)

ALOES

Cape aloes, socotrine, medicine plant, burn plant.

Uses: according to legend the sole survivor from the Garden of Eden. Gel-like sap from leaves used externally to heal wounds, burns and mastitis.

Caution: traditional internal use as purgative bitters ill advised and may cause haemorrhage.

Parts Used: broken leaves rubbed on affected part. Sap extracted from leaves and often dried to form a resin.

ALLIUM SCHOENOPRASUM
(Liliaceae)

CHIVES

Description: small perennial bulb growing in clumps, with fine hollow dark green leaves, 20-30 cm/8-12 in., and slightly taller flower stems bearing small clusters of mauve or purple blooms. Can be found wild in moist soils, but usually cultivated.

Uses: mainly culinary for flavouring and garnishing where a mild onion flavour is required. Also a stimulant and digestive; high in vitamin C. Popular decorative herb for edging and for attracting bees. Chive tea is sometimes sprayed to prevent gooseberry mildew and apple scrab. Chive flowers can be used in decorative dried flower arrangements.

Parts Used: leaves cut as needed. Chop and freeze in bags or ice-cube trays for use out of season.

ALTHAEA OFFICINALIS
(Malvaceae)

MARSH MALLOW

Guimauve, mortification plant, schloss tea, sweet weed, wymote.

Uses: valuable and handsome herb with a long tradition of use in medicine and cosmetics, and as a vegetable and confection. Cultivated by the Romans and promoted by Charlemagne. Soothing and mucilaginous; tops of young shoots added to salads (popular with Bedu); flowers and leaves infused to treat lung and bowel disorders; roots sliced and boiled to poultice external inflammation and relieve gastric ailments: root was made into traditional 'marshmallo' (now just sugar and gelatin).

Parts Used: Flowers, leaves and young tips of shoots gathered fresh roots of least 2 years old, dug up during winter dormancy when the active constituents are most concentrated and used fresh or dried.

ALTHAEA ROSEA
(Malvaceae)

HOLLYHOCK

Uses: petals of purple-flowered variety once used to dye wine. Leaves used medicinally as a diuretic and to help some chest complaints. Provides relief for mouth ulcers; soothing to the face.

Parts Used: flowers and leaves, fresh.

ANETHUM GRAVEOLENS
(Umbellifereae)

DILL

Dillseed, dillweed

Uses: used as a condiment and flavouring and as a pickling spice. Often taken as 'dill water' to relieve digestive problems and flatulence. Occasionally used to perfume cosmetics; medicinal oil distilled from leaves, stems and seeds.

Parts Used: young leaves gathered at any time for use fresh, also flowers for adding to soups; seeds harvested when dry for use whole or crushed. Leaves may be dried, but lose much of their flavour.

ANGELICA ARCHANGELICA
(Umbelliferae)

ANGELICA

Garden angelica, Holy Ghost

Uses: all parts promote perspiration, stimulate appetite, and are used to treat ailments of the chest and digestion. Young leaves and shoots used to flavour wines and liquers, while the stout stems are candied as a cake decoration or cooked like rhubarb. Fresh or preserved roots have been added to snuff and used by Laplanders and North American Indians as tobacco.

Parts Used: young leaves can be gathered any time during the growing season, the stems in the summer of the second year. Cut seedheads and ripen until seeds are dry enough to store. Roots are dug up just before flowering and dried slowly.

ANTHRISCUS CERRFOLIUM
(Umbelliferae)

CHERVIL

Salad, chervil, garden beaked parsley

Uses: leaves used in soups, salads and vinegars, and for garnishes. Medicinally the bruised plant is applied fresh or as poultices to wounds; an infusion aids digestion and encourages perspiration.

Parts Used: leaves gathered fresh as needed, and frozen or gently dried for storing. Roots sometimes dug and used in salad preparations.

AQUILEGIA VULGARIS
(Ranunculaceae)

COLUMBINE

Granny's bonnets, European crowfoot

Uses: all parts have been used medicinally, but now usually grown only as a decorative plant for herb gardens. Roots, flowers and leaves have antiseptic properties, and roots were once used to treat ulcers.

Caution: internal use is not recommended as aquilegias contain prussic acid, and all parts are poisonous, especially the seeds.

ARCTIUM LAPPA
(Compositae)

GREATER BURDOCK

Beggar's buttons, lappa, cuckoo button, flapper-bags, bardana, clotburr, gipsy's rhizome

Uses: a plant of widespread and varied virtues, burdock is cultivated in Japan as a vegetable (*gobo*); used everywhere as a folk remedy for skin problems, especially psoriasis and eczema; grown in China for its seeds, used for throat and chest ailments. A wild vegetable used by North American Indians. The chopped root may be cooked, and the stalks treated like angelica (these contain inulin, a mildly sweet substance useful for diabetics). Roots used as flavouring.

Parts Used: roots of 1-year old plants split and dried slowly; young shoots and peeled stalks (before flowering) for salads or as a cooked vegetable; dried seeds for medicinal use.

ARMERIA MARITIMA
(Plumbaginaceae)

THRIFT

Sea pink

Uses: an infusion of fresh or dried flowers was formerly used as an antiseptic and to treat nervous disorders, but now thought to cause allergic reactions such as dermatitis. A valuable formal edging plant for herb gardens and for planting in paths; very popular with butterflies.

ARMORACIA RUSTICANA.
(Cruciferae)

HORSERADISH

Uses: used medicinally as a digestive, antiseptic and stimulant, and to make poultices for rheumatism, chest complaints and circulation problems. Young leaves may be used for flavouring in salads, or cooked: roots are often made into a sauce.

Caution: use medicinally with care, as the roots may cause internal inflammation, affect the thyroid gland or used externally, produce blisters.

Parts Used: young leaves gathered in spring: roots dug in autumn.

ARNICA MONTANA
(Compositae)

ARNICA

Mountain arnica, mountain daisy, mountain tobacco, fall dandelion, leopard's bane.

Uses: homeopathic treatment for *epilepsy* and *blood pressure*. A formerly popular Shaker salve; used in many countries to treat bruises and sprains and also for throat infections, wounds and paralysis. Included in some French herbal smoking mixtures.

Caution: poisonous and not for self medication, as the plant may be toxic and cause skin irritations.

Parts Used: whole top growth, especially flowers, either fresh or after drying slowly in shade: roots dug up in late spring or autumn and dried in artificial.

ARTEMISIA
(Compositae)

A large genus of aromatic shrubs and herbaceous perennials, some with finely cut ornamental foliage ranging in colour from grey-green to bright silver. Most shrubby kinds are adapted to hot dry places in full sun, and, apart from their medicinal uses, provide valuable colour and form in the herb garden. Only the commonest types are described here.

ARTEMISIA ABROTANUM
SOUTHERNWOOD

Lad's love, old man, Crusader herb
Uses: like all the artmisias, named after the Greek goddess Artemis, who had special care of women. Used medicinally as an infusion or tincture to regulate menstruation, but also as an antiseptic, insect repellent (in mothballs) and air freshener; used by medieval Crusaders to ward off plague. Has a reputation as a hairwash and bitter stimulant, and even as an aphrodisiac. Stems yield a yellow dye.
Parts Used: shoots and leaves, which may be dried slowly in the sun.

ARTEMISIA DRACUNCULUS
FRENCH TARRAGON

Estragon, serpentarian
Uses: popular culinary herb for stimulating the appetite and flavouring sauces, preserves and cooked dishes. Historically used for toothache and by the Romans to treat snakebite. Useful for catarrhal and digestive problems, while tarragon tea is used to cure insomnia. An ingredient of perfumes and liqueurs.

ARTEMISIA VULGARIS
MUGWORT

Felon herb, St. John's herb, moxa
Uses: one of the nine Saxon magic herbs, used to make a tea for gastritis and digestive ailments, and to treat menstrual disorders. Insect repellent, and ingredient for herbal tobaccos and Chinese treatments for rheumatism. Used in stuffings for fatty meats. Leaves have been used to make fumi-

gant candle wicks, flowers to flavour beer before the advent of hops.
Caution: may be harmful when taken internally in excessive doses.
Parts Used: all parts of the plant, either fresh or dried slowly in the shade.

ASPARAGUS OFFICINALIS
(Liliaceae)
ASPARAGUS

Sparrow grass, sperage
Uses: an ornamental plant and early summer vegetable. Medicinally used as a laxative, while a tea brewed from the mature fern has been used for rheumatic and urinary disorders and by Shakers to treat dropsy.
Parts Used: young stems ('spears or tips') and fern; roots sometimes dug up for medicinal purposes.

ATRIPLEX HORTENSIS
(Chenopodiaceae)
ORACH (E)

(Red) mountain spinach
Uses: traditional wild herb, often cultivated as a vegetable and garden ornamental: used as a spring tonic and stimulant, and in infusions to treat tiredness or exhaustion.
Parts Used: larger leaves and whole young shoots.

BORAGO OFFICINALIS
(Boraginaceae)
BORAGE

Burage, bugloss, bee bread, bee plant.
Uses: a popular reviver included in some drinks. Young leaves add the flavour of cucumber to salads, older ones can be used as a vegetable, while

in France and Italy the flowers are often cooked in batter as fritters. An infusion of leaves and seeds was a folk method of increasing the milk supply of nursing mothers, and to treat coughs, cold and depression. The roots are used to flavour wine, and the seeds are a source of gamma-linoleic acid, an essential fatty acid thought to reduce the risk of arterio-sclerosis.

Parts Used: leaves and stems (latter said to have greater stimulating properties): flowers used fresh, candied, or dried in moderate heat and kept well sealed; seeds dried for medicinal use; roots dug after flowering.

BUXUS SEMPERVIRENS
(Buxaceae)

COMMON BOX

Boxwood, box tree
Uses: leaves yield a red dye, and are used homeopathy to treat fevers and rheumatism and to promote sweating. One of the best shrubs for topiary and hedges, the compact 'Suffruticosa' being the form grown as dwarf hedging around formal flower and herb beds.
Caution: the foliage is too toxic for amateur medicinal use.
Parts Used: leaves at any time; bark and timber from larger pruned branches.

CALENDULA OFFICINALIS
(Compositae)

(POT) MARIGOLD

Common marigold, marybud, marygold, English marigold
Uses: apart from being popular old-fashioned cottage garden plants, pot marigolds have a long history of medicinal use. A tea made from the flowers is used for internal spasms and gastric disorders, but the main reputation is an *antiseptic* and *anti-inflammatory healer of wounds*; a common ingredient of many proprietary salves and ointments. Used by Shakers to treat gangrene. Petals can be used as a hair rinse, a colouring agent for butter and cheese, and a substitute for the colour of saffron. Also used in cooking and as a garnish.
Parts Used: petals fresh or dried in the shade; young fresh leaves in salads whole flowers boiled as a dye.

CAMPANULA RAPUNCULUS
(Campanulaceae)

RAMPION

Rampion bellflower
Uses: an ornamental wild flower, whose leaves are used in winter salads. Popular enough in the sixteenth and seventeenth centuries to be cultivated as a root vegetable (but note that 'German rampion' is the root of evening primrose).
Parts Used: leaves gathered any time before flowering; roots dug from autumn onwards – may be stored in sand in a cool place.Centaurium erythraeae syn.

ERYTHRAEA CENTAURIUM
(Gentianaceae)

(COMMON) CENTAURY

Bitterherb, centaury gentian, feverwort.
Uses: an ancient Greek and Celtic medicinal herb, used to make a poultice for skin disorders; taken as a bitter tonic by North American Indians. A tea made from the whole plant

treats digestive disorders including heart-burn, while homeopathic preparation is prescribed for the liver and gall bladder. One of the aromatic ingredients of vermouth.

Parts Used: whole plant gathered at flowering time and dried quickly in shade outdoors or in a warm room.

CENTAUREA NIGRA
Knapweed

Lesser knapweed, black knapweed
Uses: a medieval wound salve; used to soothe sore throats and bleeding gums. Also acts as a diuretic.

CHAMAEMELUM NOBILE
(Compositae)

Lawn Chamomile

Roman chamomile, double chamomile, common chamomile, perennial chamomile.
Uses: popular since early Egyptian times, a traditional strewing herb and often used by Arabs in the form of the essential oil. Ingredient of a famous herb tea for settling nervous disorders, stimulating the appetite and cleansing the blood. Made into herb beers and tisanes, hair rinses and eye lotions, as well as being used in the preparation of cosmetics and perfumes. Essential oil is said to revive cut flowers. Plants often grown as herbal lawns and on ornamental seats.
Parts Used: whole plant for distillation; flowers for essential oil and teas, collected as petals begin to reflex in the sun; dried rapidly in shade.

CHELIDONIUM MAJUS
(Papaveraceae)

Greater Celandine

Swallow-wort, tetterwort

Uses: once widely grown as a decorative cottage garden herb, but now regarded as a weed of waste and cultivated ground, and walls. Antispasmodic and mildly sedative, traditionally valued for its sap as a treatment for warts and source of an orange dye. Flowers are beneficial for thyroid conditions, while the roots have been used to treat liver and gall bladder disorders.
Caution: in large doses may be poisonous and an internal and external irritant.
Parts Used: top growth at flowering time, fresh or dried slowly in darkness sap at any time from top growth or dormant roots; roots dug in autumn.

CHENOPODIUM BONUS-HENRICUS
(Chenopodiaceae)

Good King Henry

All-good, mercury, poor man's asparagus
Uses: cultivated as an early green spinach-like crop, the first shoots sometimes blanched by earthing up for use like asparagus.
Parts Used: young leaves and shoots.

CHRYSANTHEMUM BALSAMINA
(Compositae)

Alecoost

Costmary, bible leaf
Uses: in medieval times used to ease childbirth; formerly used in brewing as a preservative. Medicinally, used in infusion to soothe colds and digestive disorders. Can be added in small quantities to salads, game, soups and cakes.

Parts Used: leaves and flowers; gather young leaves at any time, and flowers as they begin to open.

CICHORIUM INTYBUS
(Compositae)

CHICORY

Succory, witloof, blue sailors
Uses: a vegetable and salad ingredient with a bitter flavour; often forced and blanched (witloof or Belgian chicory) to reduce bitterness. Leaves yield a blue dye. Roots and seeds often roasted as a coffee substitute and additive. Chicory tea used to stimulate bile secretion and to treat gout, rheumatism, anaemia and liver complaints.
Caution: excessive use can lead to digestive upsets, and handling may cause dermatitis.
Parts Used: roots and flowering stems, fresh or dried slowly in the sun; woods core of dried roots discarded before rest is shredded and roasted for coffee; seeds also roasted for drinks.

CLINOPODIUM CALAMINTHA
(Labiatae)

COMMON CALAMINT

Mountain mint, mountain balm
Uses: a medicinal plant of ancient Greek and medieval physicians, used in infusions as a tonic and expectorant, and in pleasant mint-flavoured tisanes. The whole crushed plant can be used to make poultices for bruises and sprains. Flowers attract butterflies and bees.
Parts Used: leaves gathered any time, and dried slowly in warmth.

CNICUS BENEDICTUS
(Compositae)

HOLY THISTLE

Blessed thistle, spotted thistle, St. Benefict thistle
Uses: ancient remedial herb used as digestive tonic, to treat liver and gall bladder complaints, and to induce sweating. A poultice of the leaves relieves chilblains, wounds and burns. Roots used in aperitifs and may be boiled as a vegetable, flowerheads sometimes eaten like globe artichokes.
Parts Used: roots dug just after flowering: leaves picked before flowering and dried slowly; flowers used whole; seeds gathered when dry.

CONVOLVULUS ARVENSIS
(Convolvulaceae)

FIELD BINDWEED

Lesser or pink bindweed, cornbine, devil's guts
Uses: a tonic and blood cleanser, sometimes used in infusion for fevers and constipation. The flowers may be added to salads.
Parts Used: flowering top growth, fresh or dried.

CORIANDRUM SATIVUM
(Umbelliferae)

CORIANDER

cilantro
Uses: one of the oldest recorded spices, mentioned in ancient Sanskrit texts and in Exodus (coriander is one of the bitter Passover herbs). The leaves and shoots are added to salads, soups and stews, especially in India, South America and China (the

leaf is sometimes known as 'Chinese parsley'). The seeds are a stimulant and digestive, often ground and included in curries and regional meat dishes. They are used as a flavouring for bread, and yield an essential oil for soaps and perfumes. They are sometimes added to pot-pourri, and to other herbs to disguise their unpalatability. The root supplies a stronger flavouring, and is often cooked as a vegetable in South-East Asia.

Parts Used: young leaves and shoots at any time; mature seeds dried, roasted and pulverised before use; roots dug after flowering.

CYMBOPOGON CITRATUS
(Gramineae)

LEMON GRASS

Oil grass, takrai, sereh
Uses: leaf buds and chopped stems are added to oriental dishes, and made into a tea for liver complaints. The plant yields lemon-grass oil, rich in vitamin A, which is used as a tonic and stimulant, antiseptic, and oily skin cleanser, as an ingredient of cosmetics, and as an aromatherapy oil.
Parts Used: leaf buds; stems fresh and finely chopped, or dried and ground.

CYNARA CADUNCULUS
(Compositae)

CARDOON

Uses: an ornamental garden favourite, also an edible vegetable crop with a delicate flavour. Especially popular in Mediterranean countries. Best eaten in winter and spring.

Parts Used: stems, blanched, braised or fried; also added to sauces.

CYNOGLOSSUM OFFICINALE
(Boraginaceae)

HOUND'S TONGUE

Gipsy flower, rats and mice
Uses: infusion from shaved root or crushed leaves used to bathe cuts, bruises, burns and eczema, and to treat coughs and bronchitis;; leaves produce a potent poultice for external relief. A nectar-rich plant for bees and butterflies.
Caution: self-medication is not recommended, as internal use can be dangerous, and external applications may cause dermatitis.
Parts Used: flowers and leaves, fresh or dried in shade or a warm room; root fresh or dried.

DIANTHUS CARYOPHYLLUS
(Caryophyllaceae)

(CLOVE) PINK

(Clove) gillyflower, carnation, sops-in-wine, divine flower.
Uses: an outstanding ornamental plant, whose fresh flowers are used in tonics, and to flavour drinks and syrups; dried petals are often added to pot-pourri.
Parts Used: flowers, fresh or dried. Before culinary use remove the bitter petal base.

ECHIUM VULGARE
(Boraginaceae)

VIPER'S BUGLOSS

Blue weed, blue devil
Uses: an outstanding ornamental plant, originally used to treat snake-

bite (both the seeds and stem markings resemble the features of snakes), the whole plant being crushed and applied to the bite. Flowers are mildly tonic and antiseptic, and may be added to drinks or candied. Leaves may be cooked like spinach.

Parts Used: fresh young leaves and shoots before flowering; flowers when fully open.

ERUCA VESICARIA SPP. SATIVA
(Cruciferae)

(GARDEN OR SWEET) ROCKET

Rocket-salad, rocket-gentle, rocquette
Uses: salad herb with pungent flavour, especially popular in Mediterranean countries. A tonic, mild stimulant and cough remedy. Crisp seedpods are edible, seeds used to treat bruises and supply mustard oils.

Parts Used: leaves gathered in succession as soon as plants are large enough; seedpods while young and juicy; seeds from ripe seedpods.

EUPATORIUM PERFOLIATUM
(Compositae)

BONESET

Thoroughwort, agueweed
Uses: a popular tonic and stimulant, best known as a hot infusion for treating coughs and colds, or as an ointment or syrup for muscular aches.

Parts Used: whole plant as flowers open, fried or dried quickly in warmth.

FILIPENDULA ULMARIA
(Rosaceae)

MEADOWSWEET

Queen of the meadows, meadsweet, meadwort.

Uses: an early source of salicylic acid. Tea from the flowers reduces fever, and is used to treat stomach acidity, influenza and rheumatism. Flower buds produce oil for perfume, and the dried flowers are added to home-made wines. Leaves flavour drinks, especially beer and mead. Dyes made from the roots (black) and flowers (yellow).

Parts Used: Flowers fresh or dried gently is warmth; roots dried for homeopathic use; leaves for flavouring.

FOENICULUM VULGARE
(Umbelliferae)

FENNEL

Spigel
Uses: one of nine, Anglo-Saxon sacred herbs; much used by ancient Greeks and throughout the Middle Ages. Roots once boiled as a vegetable and used as an expectorant in cough mixtures. Dried stalks are an essential ingredient of Provencal cuisine. Soft growing tips are widely used to flavour and garnish fish dishes, soups and baked foods, and may be made into fennel tea to treat indigestion and colic. A popular flavouring for liqueurs and a scent for soaps and cosmetics. Oil produced from seeds is antibacterial. In warmer regions foliage attracts swallowtail butterflies.

Caution: avoid large doses.

Parts Used: growing tips, fresh or dried in early summer (leaves and stems may be cropped several times before flowering); seeds, from seedheads cut in autumn before fully mature and dried in warmth; roots dug in autumn and dried slowly.

FRAGARIA VESCA
(Rosaceae)

(WILD) STRAWBERRY

Wood strawberry

Uses: mildly laxative fruits are a nerve tonic, rich in iron. Leaves can be made into a popular tea that is a diuretic and astringent; fresh leaves sometimes added to salads.

Caution: excessive consumption sometimes leads to allergic reactions.

Parts Used: fruit when fully coloured: leaves at any time; roots dug in winter.

GALIUM ODORATUM
(Rubiaceae)

(SWEET) WOODRUFF

Kiss-me-quick, master of the woods (Waldmeister)

Uses: flavours wines and other drinks and scents linen, pot-pourri and perfumes. Medicinally used to poultice wounds and scabies, and to stimulate milk flow of nursing mothers.

Caution: excessive use may produce dizziness and respiratory allergises.

Parts Used: green plant cut at or just before flowering time, and dried slowly.

GAULTHERIA PROCUMBENS
(Ericaceae)

WINTERGREEN

Partridge berry, tea berry, chequeberry

Uses: leaves made into aromatic infusion for use as a gargle. Leaves are pain-reducing and when dried can be made into a tea. Oil distilled from the leaves rubbed in externally to treat muscular aches and pains, and also used to flavour dental preparations.

Parts Used: leaves fresh or dried in the sun, distilled for oil of wintergreen.

GENISTA TINCTORIA
(Leguminosae)

DYER'S GREENWEED

Dyer's broom

Uses: a decorative ground-cover plant to grow with heathers. Infusion made from young flowering shoots is diuretic and laxative, and used to treat dropsy and skin disorders. Young flower buds may be pickled like capers. Flowers yield a yellow dye, rich green if mixed with woad.

Parts Used: flower buds; flowers for dye; young flowering shoots fresh or dried in shade.

GERANIUM MACULATUM
(Geraniaceae)

AMERICAN CRANESBILL

Spotted cranesbill, alum root, crowfoot

Uses: a popular North American Indian herb; root is astringent, used in infusions or as a powder to treat diarrhoea, dysentery and bleeding and ulcers, both internal and external.

Parts Used: roots gathered in autumn or winter, dried and powdered.

GEUM URBANUM
(Rosaceae)

HERB BENNET

Wood avens, clove root

Uses: spicy root, a medieval pot herb and substitute for cloves, can be made into an infusion for stomach and liver disorders, to increase appetite and stop bleeding, or into a strengthening tea

to reduce fever; also used as a mouth-wash. Young leaves can be eaten in salads.
Parts Used: roots fresh or dried in shade or warmth; young leaves.

GLYCYRRHIZA GLABRA
(Leguminosae)
LIQUORICE

Licorice, sweet root
Uses: cultivated for its sweet aromatic roots since the Middle Ages; used by Arab physicians as a laxative, and for treating stomach problems such as ulcers, and bladder and kidney complaints. Infusion given for bronchial catarrh and sore throats, and to reduce fever. Popular sweetener and flavouring for confectionery, beers and tobacco. Waste fibres after processing used to make wallboard.
Caution: long-term use can raise blood pressure, lower potassium levels and lead to sodium retention.
Parts Used: roots from 3-4 year-old plants, dug in autumn or spring, peeled and dried; juice may be extracted from fresh roots.

HAMAMELIS VIRGINIANA
(Hamamelidaceae)
(COMMON) WITCH HAZEL

Winterbloom, spotted alder
Uses: traditional astringent and cooling wound herb, sacred to North American Indians. Made into an infusion for treating bruises, sprains, varicose veins and bleeding, and an ointment for relieving piles. Commercial witch hazel is made from young flowering twigs.
Caution: tinctures made from the leaves and bark may cause allergic reactions.

Parts Used: bark and leaves for tinctures; flowering twigs for distillation.

HEDERA HELIX
(Araliaceae)
IVY

English Ivy
Uses: once regarded as a protective magical plant, with numerous internal medicinal uses: berries were taken as a purgative, and leaves used to treat fevers and glandular disorders. They may safely be turned into an effective poultice for bruises and stiff joints. Plants produce a resin sometimes included in varnishes; flowers attract bees, and birds depend on the berries in winter.
Caution: whole plant is poisonous and should be restricted to external use.
Parts Used: young leaves, fresh or dried in shade.

HELICHRYSUM ITALICUM
(Compositae)
CURRY PLANT

Everlasting flower, helichrysum
Uses: intensely aromatic herb for flavouring meat dishes, drinks and jellies, and effective as an insect repellent. Essential oil used in cosmetics, and for treating respiratory problems.
Parts Used: leaves; flowers in bud dried for arrangements; flowerheads and shoots in summer for distillation.

HUMULUS LUPULUS
(Cannabinaceae)
HOP(S)

Uses: common flavouring and anti-bacterial preservative in beer; female cones used to treat insomnia and nervous tension, and included for their hormonal content in skin creams and lotions. Hop poultices applied to boils and swellings. Young shoots and leaves cooked as vegetables; fibres from the stems are used in the textile industry.

Parts Used: female cones gathered in early autumn while still green, and dried until brown, sometimes powdered; young shoots 15-20 cm/6-8 in long in spring.

HYPERICUM PERFORATUM
(Guttiferae)

(PERFORATE, OR COMMON)
ST. JOHN'S WORT

Uses: leaves mildly sedative, and stimulate gastric and bile secretions. Sometimes used to treat poor blood circulation and irregular menstruation, but best known as an antibacterial remedy for slow-healing wounds and burns. Young leaves eaten in salads; flowers yield red and yellow dyes.

Caution: avoid exposure to the sun during treatment, as the plant causes photosensitivity

Parts Used: young leaves and flowers fresh or dried in shade.

HYSSOPUS OFFICINALIS
(Labiatae)

HYSSOP

Uses: a valuable expectorant; flowering tips infused in water used to treat coughs and sore throats, also to heal bites, burns and stings. Often added for flavouring soups, stews and salads. Distilled oil used in liqueurs and perfumes.

Parts Used: leaves fresh at any time; flowering tops fresh or dried in sun.

ISATIS TINCTORIA
(Cruciferae)

WOAD

Dyer's weed

Uses: an ancient dye and healing plant, fermented leaves producing a blue dye eventually superseded by indigo. Leaves were also traditionally used to stop bleeding and heal the wounds of battle.

Caution: for external use only.

Parts Used: leaves repeatedly fermented and dried in a complicated sequence of operations.

JUNIPERUS COMMUNIS
(Cupressaceae)

JUNIPER

Uses: Oil from unripe berries used in massaging rheumatic or gouty parts of the body. Ripe berries are added as a flavouring to drinks such as gin, and meat dishes, especially game. Wood of stem and roots is burned to smoke preserved meats.

Caution: use internally only under medical supervision may also be an external irritant.

Parts Used: leafy shoots at any time; fruit gathered in late summer and dried in sun; woody stems and roots when available.

LAURUS NOBILIS
(Lauraceae)

(SWEET) BAY (TREE)

Sweet laurel, bay laurel

Uses: an ancient aromatic and antiseptic plant, used to weave a victorious Roman general's crown or 'laurels'. Leaves are a popular culinary flavouring, and stimulate the appetite;

when pulped they can be applied as an astringent to burns and bruises. Berries are pressed to make oil for liqueurs, perfume and veterinary uses. **Parts Used:** leaves at any time, dried in darkness and lightly pressed flat; ripe berries pressed for oil.

LAVANDULA ANGUSTIFOLIA
(Labiatae)
(ENGLISH) LAVENDER
True lavender
Uses: an early strewing herb, often burnt on low fires to perfume rooms. As an essential oil and in infusion used as a cough suppressant, and to treat headaches and nervous disorders. As an embrocation an external stimulant and antiseptic. Flowers attract bees, and are gathered for perfume, pot-pourri and to scent clothing.
Parts Used: flowers gathered just before fully opened and dried slowly; flowering shoots and leaves distilled for essential oil.

LEVISTICUM OFFICINALE
(Umbelliferae)
LOVAGE
Love parsley, lovage angelica, smallage
Uses: leaves can flavour soups and casseroles; stems are blanched like celery or candied like angelica; roots may be peeled and cooked as a vegetable; savoury seeds are added to bread and other baked foods. Powdered root sometimes used as a condiment. Medicinally digestive and carminative. Essence of lovage used in perfumes.
Caution: avoid taking large quantities.

Parts Used: young leaves, avoiding the central flower stem; hollow main stems before flowering; roots of 2- or 3-year old plants dug in autumn, sliced and dried; ripe seeds.

LIGUSTRUM VULGARE
(Oleaceae)
(COMMON) PRIVE
Wild privet
Uses: valuable hardwood timber for tools and charcoal; leaves once used by Shakers to make a mouthwash. Yellow dye made from the leaves.
Caution: all parts are poisonous if ingested.
Parts Used: leaves at any time during the growing season; wood of larger pruned stems.

LINARIA VULGARIS
(Scrophulariaceae)
(COMMON) TOADFLAX
Yellow toadflax, butter and eggs
Uses: a highly ornamental plant, used in the Middle Ages for laundry starch. Used homeopathically to treat diarrhoea and cystitis, and by herbalists to treat jaundice. Has a folk reputation as a fly poison when boiled in milk. Flowers yield dyes: yellow, orange, green or brown depending on the mordant used.
Parts Used: whole flowering herb, fresh or dried in shade.

LINUM USITATISSIMUM
(Linaceae)
FLAX
Flaxseed, linseed
Uses: stem varieties are soaked (retted) in water to release fibres for

making linen cloth. Linseed oil from seed varieties is one of the most important commercial drying oils, used in paints, varnishes and putty. Oil used medicinally as a laxative and vermifuge; the seeds soaked overnight and strained can be used to terat gastritis, constipation and indigestion, and in poultices for cuts, bruises and other abrasions.

Caution: large doses may be poisonous.

Parts Used: whole fresh flowering plant for medicinal purposes; stems of green plant for fibre; seeds used fresh or dried and powdered.

LIPPIA CITRIODORA
(Verbenaceae)

LEMON VERBENA

Sweet-scented verbena, herb luisa, Spanish thyme.

Uses: tea (sold in France as 'verveine') made from leaves in tonic, calming and sedative, and can be used to treat nausea, palpitations and flatulence. Essential oil ('Spanish verbena') distilled from leaves used as flavouring in cakes, drinks, stuffings and desserts in pot-pourri, and formerly in pergumes – largely replaced by lemon-grass oil. Hot leaf pulp is effective against toothache.

Caution: large doses or prolonged use can cause internal irritation.

Parts Used: leaves any time, fresh or dried in shade; shoots picked just before flowering and distilled for essential oil.

LONICERA PERICLYMENUM
(Caprifoliaceae)

(WILD) HONEYSUCKLE

Woodbine

Uses: as perfume, and a traditional remedy for a number of ailments. Bruised leaves used to treat skin disorders; flaked and infused bark for rheumatism and painful joints; flowers raw or infused in a tea for asthma and as a heart tonic; inside of the root as a veterinary vermifuge. Today only external uses are recommended.

Caution: the berries of all honeysuckles may be poisonous.

Parts Used: leaves and flowers fresh or dried in shade; fresh bark from prunings; roots dug in autumn and dried.

LYTHRUM SALICARIA
(Lythraceae)

PURPLE LOOSESTRIFE

Uses: high tannin content, so once often used in leather tanning, but now a mainly decorative and medicinal plant. Antibacterial, used as a gargle and eyewash.

Parts Used: flowering herb, fresh or dried in shade.

MALVA SYLVESTRIS
(Malvaceae)

COMMON MALLOW

Uses: whole plant is mucilaginous and therefore soothing. Leaves used to reduce inflammations and ease bee stings, and in poultices to treat ulcers and haemorrhoids. Tea made from an infusion of the flowers given for colds and bronchitis. Leaves can be added to soups or cooked like spinach, seedpods eaten raw or boiled.

Caution: large amounts may be purgative and cause indigestion.

Parts Used: flowers repeatedly over

the long season, picked when fully open and dried fast in warmth; young leaves used fresh, or the entire plant cut down at flowering time and dried; seedpods picked while still green.

MATRICARIA RECUTITA
(Compositae)
Wild Chamomile
Scented mayweed, German chamomile

Uses: popular herb with a wide range of applications. Chamomile tea is refreshing, digestive and mildly sedative. Flowers reduce inflammation; used to soothe teething pains and poultice wounds. Essential oil used to scent shampoos and soaps.
Caution: may cause a severe reaction in those with ragweed allergies.
Parts Used: flowers, fresh or dried in shade.

MELILOTUS OFFICINALIS
(Leguminosae)
(COMMON OR RIBBED) MELILOT
Yellow sweet clover

Uses: the tea is tonic and anti-colic, used to reat sleeplessness, nervous tension, thrombosis and digestive disorders. Externally, often added to relaxing baths and made into a compress for slow-healing wounds. Both flowers and seeds are used for flavouring.
Caution: excessive consumption may cause vomiting and dizziness.
Parts Used: leaves and shoots of flowering plants, fresh or dried in shade.

MELISSA OFFICINALIS
(Labiatae)
(LEMON) BALM
Bee balm, melissa, sweet balm

Uses: vigorous and indestructible herb with a strong lemon fragrance. Young leaves used to flavour teas, soups, milk, custard, sauces, and added to liqueurs; pungent oil extracted from them is used in perfumery. Medicinally, balm lowers blood pressure, and is used in infusion to treat colds and influenza, nervous tension, insomnia, indigestion and other stomach ailments. A notable bee plant. Leaves and stems are sometimes used to polish and scent wooden furniture.
Parts Used: leaves and tips picked just before or after flowering, and used fresh or dried quickly in the shade (freezing retaining more volatile oils than drying).

MENTHA
(Labiatae)
There are at least 18 species of mint and many more hybrids, most of them difficult to classify because of their variability and readiness to hybridize between each other. All are aromatic perennials, most containing menthol (essential mint oil) to some degree.
Caution: handling mints may cause skin rashes and other allergic irritations; mint teas should not be drunk in large amounts over a long period.

MENTHA × PIPERITA
PEPPERMINT

Uses: the most medicinally valuable of all mints, with great cooling properties due to its high content of menthol. Used to treat gastric and digestive disorders, and nervous complaints such as tension and insomnia. Essential oil used to flavour confectionery, liqueurs and pharmaceutical products, and to scent cosmetics.

Parts Used: leaves and stems, fresh or dried – gather before flowering for culinary use, in full flower for distillation of oil.

MENTHA PULEGIUM

PENNYROYAL

Pudding grass

Uses: a very pungent herb, once used to disguise the flavour of putrid meat and still included in a few local dishes. Valued since Roman times as a flea repellent; also popular today as an aromatic ground-cover plant. Used medicinally to treat gastric ailments, headaches, colds, bites and minor abrasions.

Parts Used: whole green plant, used fresh or dried.

MENTHA × ROTUNDIFOLIA

ROUND-LEAVED MINT

Apple mint

Uses: the main culinary mint species in continental Europe, used to flavour both savoury and sweet dishes, sauces and drinks.

Parts Used: fresh leaves as required; can also be frozen or dried.

MENTHA SPICATA

SPEARMINT

GARDEN MINT

Uses: the least pungent species, subtly fragrant; one of the main culinary mints, yielding an essential oil used for flavouring confectionery, and dental and pharmaceutical preparations.

Parts Used: fresh, frozen or dried leaves.

MONARDA DIDYMA
(Labiatae)

BERGAMOT

Bee balm, Oswego tea, horsemint

Uses: traditionally used to make a relaxing anti-depressant tea which can also be used to treat nausea and flatulence. An infusion is inhaled for colds, or used as an antiseptic for ulcers, wounds and acne. Included in potpourri mixtures and as a perfume in cosmetics.

Caution: bergamot may cause photosensitivity in some people.

Parts Used: leaves and flowers, fresh or dried.

MYRICA GALE
(Myricaceae)

BOG MYRTLE

Sweet gale, sweet willow

Uses: tonic tea made from leaves, which are sometimes used with berries for flavouring cooked dishes and treating dysentery; dried leaves used as moth and flea repellent. Roots and bark give yellow dye.

Parts Used: leaves, fresh or dried, or distilled for essential oil; berries fresh or dried; roots and bark.

MYRRHIS ODORATA
(Umbelliferae)

(EUROPEAN) SWEET CICELY

Myrrh

Uses: leaves and unripe seeds are eaten raw in salads, roots and leaves boiled as vegetables. One or two leaves at a time are included in conserves and tart fruit dishes to add sweetness and aniseed flavour; also added to brandy and liqueurs. Whole plant is tonic and gently laxative, and used in healing ointments. Dried seeds ground as a spice in Germany; leaves used to polish and scent oak furniture; horses are sometimes lured with a piece of the root.

Parts Used: leaves fresh or dried, seeds, green and immature, or dried; roots dug in autumn; whole green plant for medicinal infusions.

MYRTUS COMMUNIS
(Myrtaceae)

MYRTLE

Uses: The perfume *egu d'ange* is distilled from flowers and leaves to scent soaps and cosmetics; leaves added as flavouring to meat dishes; fruits fermented into alcoholic drinks. Medicinally, crushed leaves applied to external wounds, rashes and skin irritations; juice of berries good for stomach and digestive ailments. Dried buds and fruits used as a peppery condiment; roots and bark for tanning leather.

Parts Used: leaves fresh or dried; flower buds dried; flowers fully open; fruits fresh or dried; roots and bark.

NASTURTIUM OFFICINALE
(Cruciferae)

WATERCRESS

Uses: leaves, rich in minerals and vitamins C and A, prized since Roman times for biting, rich flavour, raw or cooked as a vegetable and in soups. Also used as a cough remedy. Crushed leaves are applied as poultice for rheumatism and gout; raw seeds used as vermifuge.

Caution: do not consume excessive quantities, and gather wild plants only from clean running water; plants in stagnant and polluted water may be host to the dangerous liver fluke.

Parts Used: older leafy stems, fresh or dried; seeds when ripe.

NEPETA CATARIA
(Labiatae)

CATMINT

Catnip, catnep

Uses: well known for its popularity with cats – volatile oil is a feline aphrodisiac and also distilled for perfumes. Leaves made into a mint-flavoured tea for colds, nervous tension, flatulence and gastric disorders, or poultices for cuts and bruises.

Parts Used: leaves and flowering stems, fresh or dried in shade.

NICOTIANA ALATA
(Solanaceae)

FLOWERING TOBACCO

Sweet-scented tobacco

Uses: occasionally used as stimulant for nervous system, but normally grown as a handsome flowering plant in herb gardens.

OCIMUM BASILICUM
(Labiatae)

(SWEET) BASIL

St. Joseph wort

Uses: popular culinary flavouring, typical of Mefiterranean cuisines and used since ancient times (remains have been found in Egyptian burial chambers). Oil of basil used in perfumery, soaps, cosmetics and liqueurs. Plant is claimed to be an insect

repellant, and can be used medicinally to soothe pain and treat vomiting, nervous stress and headaches.
Parts Used: leaves fresh or frozen, or dried in shade (flavour will change considerably).

OENOTHERA BIENNIS
(Onagraceae)

(COMMON) EVENING PRIMROSE
Evening star, king's cure-all
Uses: leaves are a winter pot herb, and may also be used to treat coughs and chest ailments. North American Indians have many uses for the plant, especially the roots, used in poultices for piles and boils; roots, sometimes known as German rampion, also eaten raw or cooked; ripe seeds used in bakery like poppy seeds. Medicinally the plant is an important source of gamma-linoleic acid, with other uses still under test.
Parts Used: leaves, shoots and flowers for fresh culinary use; roots dug in autumn or early the following spring before growth resumes.

OLEA EUROPAEA
(Oleaceae)

OLIVE

Uses: since ancient times the principal source of edible oil in the eastern Mediterranean area; biblical symbol of peace. An olive wreath was given to victors in the Olympic Games. Not only the fruits but also the leaves are edible, Oil pressed from the fruits is a major culinray and medicinal product, sometimes used as a laxative and in enemas, and in treatments for minor wounds.

Parts Used: leaves; fruits picked when green, pink or red, or fully ripe, sometimes cracked, fermented and soaked in brine, or pressed for oil.

ORIGANUM VULGARE
(Labiatae)

OREGANO
Wild marjoram, joy of the mountain, Mexican sage
Uses: tonic, digestive and expectorant herb, used to treat coughs and sore throats, indigestion and gastric upsets. Antiseptic leaves are chewed for toothache, and added to baths and poultices. Important culinary flavouring for meat dishes and salads, the leaves are also made into tea and beer, or distilled into an oil for perfumes and cosmetics.
Parts Used: sprigs of leaves and flowers, or whole clump cut almost to ground level when in flower, for use fresh or dried in shade.

OSMUNDA REGALIS
(Osmundaceae)

ROYAL FERN
Buckthorn brake
Uses: handsome ornamental fern with mucilaginous roots, often boiled in water produce royal fern jelly, once given to invalids as a nutritious, easily digested food, and also used to treat dysentery, coughs and pulmonary disorders. Dried roots (osmunda fibre) are a traditional ingredient of orchid potting composts.
Parts Used: main roots, fresh or dried, for medicinal use; thinner fibrous roots dried for compost.

PELARGONIUM GRAVEOLENS
(Geraniaceae)

ROSE GERANIUM

Uses: rose-scented leaves used to scent desserts, cakes and teas, pot-pourri, drinks and fingerbowls. Oil distilled from the leaves is an insect repellent, used in perfumes.

Parts Used: leaves, fresh or dried; all green parts, cut just before flowering for oil distillation.

PETROSELINUM CRISPUM
(Umbelliferae)

(WILD) PARSLEY

Uses: tea made from leaves or roots used to treat jaundice, coughs and menstrual problems, rheumatism, kidney stones and urinary infections; juice expressed from them soothes conjunctivitis and eye inflammations. Both seeds and dried roots are used as spices; seeds, which contain poisonous apiol, are sometimes infused to produce an external vermifuge.

Caution: avoid parsley during pregnancy or if suffering from kidney inflammation.

Parts Used: leaves, fresh, frozen or dried; roots dug in winter and dried; seeds when capsules are ripe.

PHLOMIS FRUTICOSA
(Labiatae)

Uses: mainly a handsome ornamental shrub; leaves sometimes made into an aromatic sage-flavoured tea, especially in Greece and neighbouring Mediterreanean countries.

Parts Used: leaves, fresh or dried in the sun.

PLANTAGO MAJOR
(Plantaginaceae)

GREATER PLANTAIN

Broadleaf plantain, rat's tail plantain, waybread, white man's footprint.

Uses: an old herb, one of the nine sacred Saxon species and recovered from remains of Iron Age Tollund Man. The crushed leaves are cooling and pain-relieving, used in poultices and ointments for wounds and abrasions; an infusion of leaves or boiled roots is a useful gargle and eyewash; fresh leaves may be taken for both constipation and diarrhoea, according to the dosage. Very young leaves may be cooked like spinach; seeds are popular in Chinese and Malaysian drinks, and complete seedheads are fed to caged birds.

Parts Used: fresh or dried quickly in sun or shade; roots, dug in winter and boiled; seed spikes, gathered in bags when they turn brown, and rubbed to free the seeds.

PORTULACA OLERACEA
(Portulacaceae)

(WILD) PURSLANE
Pigweed

Uses: ancient vegetable crop in India and Iran, still used as a salad herb and cooked vegetable. Infusion of the green plant is cooling and soothing, taken for fevers, headaches and chest complaints, and applied to skin rashes and abrasions. Seeds are an Australian Aboriginal wild condiment.

Parts Used: fresh leaves and stems, preferably before flowering, cut just above ground level.

POTERIUM SANGUISORBA
(Rosaceae)

SALAD BURNET

Uses: young leaves eaten as salad herb with cucumber-like flavour, and added to soups, sauces and cheeses. Leaves are digestive, and in infusions used to treat diarrhoea and haemorrhages. Decoction of root applied to cuts and burns; roots also produce a black dye and are used in tanning leather.

Parts Used: young leaves before flowering; whole green plant, fresh or dried; roots dug in spring and dried.

PRUNELLA VULGARIS
(Labiatae)

SELFHEAL

Heal-all, woundwort, carpenter's herb
Uses: used in Middle Ages according to the 'Doctrine of Signatures' to treat throat conditions and internal bleeding, and regarded as a panacea by North American Indians and Chinese physicians, the latter using the seeds for nervous complaints. Flowering plant may be eaten in salads and cooked vegetable dishes, used to prepare a styptic for wounds, and made into gargles or mouthwashes for mouth ulcers and sore throats.

Parts Used: green flowering plant, fresh or dried in shade.

PULMONARIA OFFICINALIS
(Boraginaceae)

LUNGWORT

Maple lungwort, spotted dog, Jerusalem cowslip
Uses: ancient cure for lung disorders according to 'Doctrine of Signatures'.

Young leaves used as spring pot herb in soups, stews and salads; flowering plant made into a tea for gastro-intestinal and pulmonary ailments; homeopathically used for bronchitis and colds. Powdered roots and lower leaves are wound-healing.

Parts Used: young fresh leaves gathered; green flowering plant, fresh or dried; roots dug in winter, cut, dried and powdered.

RESEDA LUTEOLA
(Resedaceae)

WELD

Dyer's rocket
Uses: traditional dye plant, once widely cultivated or gathered from the wild for brilliant yellow or reddish-yellow colour extracted from leaves, flowers and stems.

Parts Used: whole green plant gathered at flowering time.

ROSMARINUS OFFICINALIS
(LABIATAE)

ROSEMARY

Uses: ancient strewing herb and Romany charm, hung up to ward off evil: popular culinary flavouring added to meat dishes, baked foods and Mediterranean recipes. Leaves medicinally valuable and for treating depression, migraine, and disorders of the liver and digestion. Leaves also made into ointment for neuralgia, rheumatism, eczema and minor wounds, and used in hair rinses and mouthwashes.

Caution: excessive quantities or frequent use may cause poisoning.

Parts Used: leaves gathered at flowering time, used fresh or dried in shade.

RUBIA TINCTORUM
(Rubiaceae)

DYER'S MADDER

Dyer's cleavers

Uses: infusions of leaves and stems treat constipation, and liver and bladder disorders; powdered root is wound-healing, often used for skin ulcers. Homeopathically used to treat anaemia and ailments of the spleen. Most popular use of the roots is as a variable red to purple dye.

Parts Used: leaves and stems; roots peeled and dried quickly, and powdered or fermented.

RUMEX RUGOSUS
(Polygonaceae)

COMMON SORREL

Uses: popular, sharply flavoured pot herb since ancient Egyptian times, widely used by medieval apothecaries. Cooling and blood-cleansing, often taken as a spring tonic tea. Leaves made into poultices for acne and other skin complaints, and if picked very young can be eaten raw or cooked, notably in sorrel soup; juice of leaves will curdle milk and has also been used as a stain remover ('salts of sorrel'). Roots make a bitter tonic and a treatment for diarrhoea.

Caution: leaves are high in oxalic acid and should be eaten sparingly; handling them may cause skin irritations.

Parts Used: young leaves and buds, picked before flowering & used fresh or frozen; roots fresh in summer.

RUTA GRAVEOLENS
(Rutaceae)

RUE

Herb of grace

Uses: ancient medicinal herb, formerly used as an antidote to poisoning and a talisman against witchcraft. A favourite Arab herb, the only one to be blessed by Mohammed; leaves used homeopathically to treat phlebitis and varicose veins, and herbally for epilepsy, nervous complaints and uterine disorders. Essential oil used in perfumery and cosmetics; small amounts of the pungent foliage used for flavouring foods and alcoholic drinks.

Caution: to be taken internally only under medical supervision, and used externally with care as allergic skin reactions are possible.

Parts Used: leaves from flowering plant, fresh or dried in shade.

SALVIA OFFICINALIS
(Labiatae)

SAGE

Uses: an ancient herb, popular as a potent condiment for meat, fish, Mediterranean dishes. English Sage Derby cheese, and as a basis for sage tea, taken to counteract sweating. Infusion used to treat depression, nervous anxiety and liver disorders; homeopathic preparations given for circulation and menopausal problems. Leaves are also antiseptic, used in gargles for laryngitis and tonsillitis.

Parts Used: leaves fresh, or dried in shade, picked before flowering for herbal use or when in flower for oil distillation.

SAMBUCUS NIGRA
(Caprifoliaceae)

(COMMON) ELDER (BERRY)

Uses: legendary tree, long held to be guardian over all other herbs, with numerous virtues, according to the part used. Leaves are an effective insect repellent, and soothing in ointments for skin complaints. Flowers soothe the eyes, are added to cosmetics, make a calming tea, and are popular for their sweet fragrance in drinks and fruit dishes. Fruits are used in codials, syrups and preserves; medicinally to induce sweating and to treat coughs, colds, catarrh and throat infections; and as a blue-purple dye. Bark is an old treatment for epilepsy, and (together with the leaves) is laxative; root used for kidney ailments.

Parts Used: leaves, flowers and fruits, fresh or dried; root and bark gathered as needed.

SANTOLINA CHAMAECYPARISSUS
(Compositae)

SANTOLINA

Cotton lavender

Uses: grown and used since classical Greek times as a vermifuge and moth repellent. Infusion of leaves used as a rub for rheumatism and painful joints; flowers make a tonic tea. Perhaps most popular today as a decorative hedging plant for parterres and knot gardens.

Parts Used: leaves before flowering, dried and stripped from stalks; flowers.

SATUREJA HORTENSIS
(Labiatae)

SUMMER SAVORY

Bean herb

Uses: culinary herb whose use dates back to the early Romans; potent flavouring enhances all others in the same way as salt. Used sparingly in meat dishes and stuffings, with peas, beans and cabbage to improve their digestibility, and liqueurs. Infusion of leaves treats gastric upsets, indigestions and loss of appetite; tea is tonic. Spreading flowering shoots between clothing repels moths.

Parts Used: leaves gathered before flowering, fresh or dried in shade; flowering shoots fresh or dried.

SCABIOSA ARVENSIS
(Dipsaceae)

FIELD SCABIOUS

Blue buttons, pincushion flower

Uses: infusion of roots treats cuts, sores, abrasions and itching; whole herb used as a remedy for dandruff. Homeopathically used for eczema and skin disorders.

Parts Used: whole flowering herb including roots, fresh or dried; roots dug in autumn, fresh or dried.

SCUTELLARIA GALERICULATA
(Labiatae)

(COMMON) SKULLCAP

Helmet flower

Uses: bitter tonic and digestive herb; powerful nerve tonic, used in infusion to treat depression, headaches, insomnia, irritability and similar disorders. (Out of flower, skullcap is easily confused with wood sage, *Teucrium scorodonia*).

Parts Used: all green parts during flowering, fresh or dried.

SEDUM ACRE
(Crassulaceae)

BITING STONECROP

Wall pepper, golden-carpet, gold moss
Uses: plants were often grown deliberately on roofs as charms against lightning. Homeopathically used to treat piles; bruised leaves, fresh or in ointments, are soothing for wounds, abcesses, bruises and minor burns.
Caution: slightly poisonous; internal use may cause dizziness and nausea.
Parts Used: leaves, fresh or dried in warmth.

SEMPERVIVUM TECTORUM
(Crassulaceae)

HOUSELEEK

Hens-and-chickens
Uses: ancient magical herb, planted on roofs as an insurance against fire and lightning. Sliced or crushed leaves used to poultice stings, burns, rashes and itching skin, and to cure warts and corns. Sometimes used as a skin lotion. In some parts of Europe young leaves and shoots are eaten as a vegetable.
Parts Used: leaves, sliced or pulped, as needed.

SILYBUM MARIANUM
(Compositae)

MILK THISTLE

Variegated thistle, Our Lady'd milk thistle, spotted thistle
Uses: the herb has had a reputation for treating liver disorders since classical Roman times, and is included in several proprietary medicines for this purpose. Used in herbal infusions and homeopathic preparations for liver and abdominal ailments. Young shoots and leaves are edible and an Arab delicacy; roots may be cooked like parsnips, and base of flowerheads in the same way as artichokes; stems can be peeled and boiled as a vegetable; seedlings eaten raw in salads.
Parts Used: seedlings; young leaves and shoots; flower stems just before flowering; roots after flowering; flowerheads for medicinal use, fresh or dried in thin layers in warmth.

SMYRNIUM OLUSATRUM
(Umbelliferae)

ALEXANDERS

Black lovage, horse parsley
Uses: bitter herb cultivated since early Greek times; root is diuretic, seeds a condiment, crushed leaves or their juice a soothing and healing treatment for cuts and minor abrasions. Grown as a salad and pot herb before celery became popular, with the leaves, stems, shoots and flower buds all used, sometimes after blanching, in soups and fish dishes.
Parts Used: young leaves; young stems, after blanching with soil or straw if preferred, roots fresh or dried, ripe seeds.

SOLIDAGO VIRGAUREA
(Compositae)

GOLDEN ROD

Uses: a medieval Arab healing herb, also used by North American Indians; made into poultices for external wounds, and infusions for fevers and digestive upsets. Occurs in several proprietary medicines for kidney and bladder ailments, and used homeopathically to treat these, as well as arthritis and rheumatism.

Parts Used: green flowering plant before flowerheads fully opened, fresh or dried in shade.

STELLARIA MEDIA

COMMON CHICKENWEED

Uses: an ancient pot herb; seeds found in Neolithic burial sites. Traditionally fed to domesticated birds and fowls; a bitter salad and pot herb; homeopathic remedy for rheumatism. Poultice of stems and leaves used to ease arthritis and pains of the joints, cuts, skin irritations and inflammation.

Parts Used: young stesm and leaves, fresh or dried.

SYMPHYTUM OFFICINALE
(Boraginaceae)

COMFREY

Knitbone

Uses: root formerly a popular internal remedy for gastric disorders and a homeopathic treatment for ulcers. Leaves used in infusion for bronchitis, and in poultices for wounds, bruises and eczema. Young leaves and shoots were eaten as a vegetable, cooked with a change of water. Older leaves commonly used after wilting as an animal feed, for making compost and liquid fertilizer, and for mulching plants.

Caution: absolutely not to be used internally under any circumstances –

recent evidence suggests the plant is carcinogenic.

Parts Used: leaves, fresh, wilted or dried; roots fresh or split lengthways and dried in sun.

TANACETUM PARTHENIUM
(Compositae)

FEVERFEW

Uses: a bitter-flavoured herb, long used as a tonic and to treat indigestion, but currently popular as a treatment for migraine. Leaves are made into pain-soothing poultices for limb and joint aches, and whole flowering stems are an insect repellant, keeping moths away from clothing. A popular bee plant.

Caution: not to be taken during pregnancy; fresh leaves may cause mouth ulcers.

Parts Used: leaves or whole green flowering plant, fresh or dried in shade.

TARAXACUM OFFICINALE AGG.
(Compositae)

DANDELION

Blowball

Uses: leaves are a diuretic; dried to make tonic teas; added to herbal beer; blanched for salads. Flowers used in wines, schnapps, pancakes and in Arab baking; inside surface of flower stems soothes burns and stings (also stains skin). Roots roasted as a coffee substitute, cooked in Japanese cuisines, and give a magenta dye (with the leaves they produce brown dyes). Medicinally, given for gall-bladder and liver complaints.

Parts Used: whole flowering plant, fresh or dried in shade.

TEUCRIUM CHAMAEDRYS
(Labiatae)

WALL GERMANDER

Uses: leaves used in tonic teas, wines and liqueurs, and for treating digestive and gall-bladder disorders. Plants used for edging and hedges in herb and knot gardens.

Parts Used: leaves fresh or dried in warmth; flowers when fully opened in sun; roots dug in summer for medicinal use, or autumn for drying and grinding for coffee.

THLASPI ARVENSE
(Cruciferae)

FIELD PENNYCRESS

Uses: finely chopped leaves have a spicy flavour like watercress, and are used in salads and cooked dishes. Seeds were once ground and used as a mustard.

Parts Used: leaves and young shoots before flowering; seeds dried in the sun and ground.

THYMUS VULGARIS
(Labiatae)

COMMON THYME

Garden thyme

Uses: popular since classical times, thyme has a number of important uses. Leaves make a tonic and stimulating tea, used to treat digestive complaints and respiratory disorders, especially for loosening mucus. Antiseptic and vermifuge essential oil (thymol) added to disinfectants, toothpaste, perfumes, toiletries and liqueurs. A culinary herb with a powerful flavour, thyme is added sparingly to bouquets garnis, stuffings and savoury dishes.

Parts Used: leaves and flowering tips, fresh or dried in sun.

TRIFOLIUM PRATENSE
(Leguminosae)

RED CLOVER

Uses: dried flowers produce a volatile oil and a soothing tea for promoting sleep; taken medicinally as an expectorant for respiratory disorders, and to treat skin problems such as eczema and psoriasis. Externally, an infusion soothes burns and sores. Flowers also make a good wine and yield a yellow dye.

Parts Used: flowerheads, fresh or dried in shade.

TROPAEOLUM MAJUS
(Tropaeolaceae)

NASTURTIUM

Uses: antiseptic and digestive herb, also used to treat respiratory and urinary disorders; seeds are a vermifuge, and crushed for use in poultices for boils and sores. Leaves are edible and used in salads, the flowers as a garnish, and both seeds and flower buds are pickled for their pungent mustard-like flavour.

Parts Used: leaves fresh or dried; flowers; seeds while green, or when ripe for grinding as seasoning.

TUSSILAGO FARFARA
(Compositae)

COLTSFOOT

Coughwort, horse-hoof

Uses: leaves are an important cough

remedy for bronchitis and laryngitis, commonly added to herbal smoking mixtures. Both leavse (after crushing the veins) and flowers can be used in poultices for sores and ulcers; root is boiled to make coltfoot rock or candy. Flowers are made into wine, and mature leaves can be dried and burnt to an ash used as a salt substitute.

Caution: roots contain similar substances to comfrey, and may be equally dangerous if taken internally.

Parts Used: flowers before fully open, fresh or dried in shade; leaves in summer, fresh or cut up and dried in shade.

URTICA DIOICA
(Urticaceae)

STINGING NETTLE

Uses: young shoots and leaves widely used in spring soups and as a green vegetable, and added to beer. Older leaves laxative in infusion, expectorant and styptic. Made into hair restorers and used homeopathically to treat skin ailments. Stems fibres are strong enough for linen weaving, papermaking and spining into ropes. Foliage is a commercial source of chlorophyll and an effective compost activator.

Caution: handle with care, as the formic acid injected by serious stings may cause recurrent 'nettle rash'.

Parts Used: leaves gathered before flowering, fresh or dried in sun.

VALERIANA OFFICINALIS
(Valerianaceae)

VALERIAN

Garden heloptrope, cat's valerian

Uses: mildly sedative and antispasmodic, used to treat tension, anxiety, insomnia, migraine and nervous ailments, as well as colic and cramp; externally in infusion for eye problems. Attractive to cats. An occasional culinary flavouring.

Caution: large doses or extended use may lead to addiction.

Parts Used: roots at least 2 years old and gathered after leaves fall, used fresh or dried in shade.

VERATRUM VIRIDE
(Liliaceae)

GREEN FALSE HELLEBORE

American white hellebore, itchweed, Indian poke

Uses: handsome flowering plant for a prominent position, anti-parasitic; used by North American Indians as an arrow poison, and by the pharmaceutical industry in preparations for reducing blood pressure.

Caution: highly toxic and not for home preparation.

Parts Used: roots dug in autumn, cut in pieces and dried in sun or warmth.

VERBENA OFFICINALIS
(Verbenaceae)

VERVAIN

Uses: ancient herb popular with druids as a panacea, and used in the Middle Ages to ward off plague. Used homeopathically for dropsy; medicinally to treat rheumatism, and stomach and liver disorders. Tea is a stimulant, and relieves fevers and nervous tension. Externally used for sores and skin problems, and for eye complaints.

Parts Used: green flowering plant, fresh or dried in sun or warmth.

VERONICA OFFICINALIS
(Scrophulariaceae)

(HEATH) SPEEDWELL

Fluellen

Uses: a medieval healing herb, still used in a tonic tea ('Swiss tea') for liver, digestive and general intestinal complaints. Fresh juice used for skin ailments, for which there is also a homeopathic preparation. Cultivated as an ornamental ground-cover plant in the garden.

Parts Used: whole green flowering plant, fresh or dried in shade or a warm room.

VINCA MINOR
(Apocynaceae)

LESSER PERIWINKLE

Uses: used as a gargle for sore throats, as a styptic and astringent for wounds, sores and ulcers, and in a tea to reduce blood pressure and hyerptension.

Caution: large amounts may be toxic, leading to circulatory disorders.

Parts Used: green flowering plant in spring, fresh or dried.

VITEX AGNUS-CASTUS
(Verbenaceae)

CHASTE BERRY

Indian spice, monk's pepper, chaste tree.

Uses: an ancient Greek herb thought to guarantee chastity, now used to regulate female hormonal activity. Also reputed to be a male anaphrodisiac; used as symbolic stewing herb in Italian monasteries. Seeds are used as a peppery condiment, slender branches in basketwork.

Parts Used: fruits picked in autumn, fresh or dried in shade.

VIOLA TRICOLOR
(Violaceae)

WILD PANSY

Heartsease, field pansy, johnny jump up

Uses: renowned mild heart tonic, used to treat high blood pressure, indigestion and colds; cleanses blood and induces perspiration. Also used to treat dropsy and rheumatic conditions, and for skin disorder such as acne and eczema.

Parts Used: green flowering herb and root, fresh or dried in shade.